A Hitchhiker's Guide to Virtual Reality

A Hitchhiker's Guide
to Virtual Reality

Karen McMenemy
Stuart Ferguson

A K Peters, Ltd.
Wellesley, Massachusetts

Editorial, Sales, and Customer Service Office

A K Peters, Ltd.
888 Worcester Street, Suite 230
Wellesley, MA 02482
www.akpeters.com

Library of Congress Cataloging-in-Publication Data

McMenemy, Karen, 1976- –
    A hitchhiker's guide to virtual reality / by Karen McMenemy, Stuart Ferguson.
       p. cm.
    Includes bibliographical references and index.
    ISBN 13: 978-1-56881-303-5 (alk. paper)
    ISBN 10: 1-56881-303-1 (alk. paper)
     1. Human-computer interaction. 2. Virtual reality. I. Ferguson, Stuart, 1953- II. Title.

QA76.9.H85M425 2007
006.8–dc22

                                                                          2006102280

Cover image: The front cover illustrates a virtual figure exploring a virtual landscape. The
theory and practical implementation making this possible are described in Chapters 5, 7,
13 and 14. The inset images on the back cover illustrate the projection system designed in
Chapter 4 and some of the practical projects of Chapter 18.

Printed in Canada
11 10 09 08 07   10 9 8 7 6 5 4 3 2 1

To Cathy

Contents

Preface

The seeds of the ideas that have become this book were sown on the breakfast terrace of a hotel in Beverly Hills during SIGGRAPH 2005. Whilst preparing for another interesting day at the conference, we faced the immediate task of how to journey downtown, no mean task in LA. So as we turned to a small guidebook for advice, the thought struck us that it would be really useful if we had a similar informative and practical guide to virtual reality (VR). At that time, we were preparing to build the next phase of our own VR research lab, and to be able to do that, it isn't enough just to know about the background theory and technology that might be useful. We needed practical solutions— hardware and software solutions. Not finding anything suitable, we decided to compile our own—a guide that would bring together under one cover all the aspects of graphics, video, audio and haptics that have to work together to *make virtual reality a reality*.

What's in the Book?

The book has two parts, and you could read them almost independently. The CD is an integral part of the book: it contains the programs for over 30 projects in VR. These range in scope from a tool that simulates virtual sculpting, to a suite of software for the control of a four-projector immersive virtual environment.

Throughout the text we have tried to follow a logical progression: what VR fundamentally aims to achieve, to what it can be applied, what elements need to be brought together, how they work and how the theory is turned into practice. We also consider some key concepts from the allied disciplines of computer graphics and computer vision, which overlap with VR.

Part I forms the core of the book. We examine the human senses and their significance in delivering a sense of reality within the virtual world. We describe the types of interface technologies that are available and how they work. After reading the first few chapters, it should be evident that being able to see the virtual world is of prime importance. So several chapters in Part I are focused on all aspects of how we interface with the virtual world through our sense of sight.

Part II of the book is in a different format. Titled *Practical Programs for VR*, the text is tightly integrated with the CD. A wide spectrum of example programs for use in practical VR work are described and explained in Part II. The examples complement and make concrete the concepts covered in Part I. We have found them exceptionally useful, and many are in use in our own VR laboratory. The programs are written in the C or C++ languages and are targeted for the Windows PC platform.

As we stressed before, the key element of VR is the visual one, and here, perhaps more than in any other aspect of the human-computer interface, software plays the dominant role. We show you how to use the main 3D rendering libraries in interactive real-time applications. Real-time 3D is not the only source of content for VR applications; movies, DVDs and other multimedia material can also play a vital role. We show you how to use Microsoft's DirectX technology to achieve this with minimal effort.

The level and sophistication of interaction set many VR applications apart from computer games and computer-generated movies. Programming this interaction poses its own complexities. The keyboard and mouse may be very well for many things, but VR needs more: two-handed input with multiple degrees of freedom, for example. We offer some examples of how to use the USB PC interface and the input components of DirectX. Part II also discusses the challenges that a programmer faces when working with haptic devices.

You'll find more specific detail about the book's content in Chapter 1, but we hope your curiosity is sufficiently excited to read on. We found our guide very useful and enjoyed writing it; we hope you find it useful too and enjoy reading it.

Acknowledgments

Many individuals have given us much valued ideas, feedback and other assistance in preparing the book. Others have inspired a more general interest in this fascinating and engrossing subject; to all of you, many thanks.

We would like to single out the following for special mention:

Alice and Klaus Peters for encouraging us to write the book, and for publishing it. Without them, it would never have seen the light of day. Our reviewers, who took time to read at least two revisions of our manuscript and give their very helpful comments, and suggestions for topics to include. The book is much better for your thoughts; they are much appreciated. We would especially like to thank the editorial team at A K Peters, who have done a fabulous job of preparing the manuscript and making sure it conforms to their high standard.

Finally we would like to thank all our colleagues and friends at Queen's University for all your encouragement and suggestions—thank you.

Karen McMenemy and Stuart Ferguson
January 2007

Part I What, Why and How

1 Introduction

Welcome to our two-part guide to the science, technology, mathematics and practical implementation of virtual reality. Since the words *virtual reality* are going to crop up quite often, we will use the abbreviation VR from now on. As you work your way through the book, we hope you will begin to appreciate that VR is more, much more than computer graphics, more than just electronic engineering, more than just software engineering, more than mechanical engineering. In fact, it is about all these things. It is about designing and using interfaces to as many of our physical senses as is technically possible. Quite a challenge! Do it right and its potential is staggering. However, it is just as important to appreciate when it is better not to use VR. There are many instances where we don't yet have the technology to do a good job using VR, so it's better not to start.

Knowledge is supposed to be power, and we hope that our book will empower you a little so that after you finish reading it, you will feel much more comfortable in deciding for yourself whether or not to use VR. We hope that it will do this by explaining and describing the theory and technology upon which VR relies and (in the second part) by showing you *through example* how to use the best software tools and libraries to create the driving programs that power VR.

1.1 What is Virtual Reality?

Before we take a little tour of the topics and chapters in the remainder of the book it is worth pausing briefly to consider *what we mean by virtual reality.*

First, it can also go by several other names. Augmented reality, spatial reality, mixed reality, collaborative reality and even tangible reality are all used to announce some new aspect of "a computer-generated 3D environment within which users can participate in real time and experience a sensation of being there". This is what we mean by all of those realities.

The two most important things to consider when creating or using VR are the *real-time* 3D virtual environment and the *human interface devices* which connect the user to it. We are not just talking about real-time 3D graphics, important though that is; we must include the 3D audio environment and the 3D *haptic* environment.[1] So, our book fundamentally tries to delve into the fascinating subject areas of real-time 3D environments and human interface devices.

1.2   A Quick Tour of the Book

The book is in two parts, and you could read them almost independently. The CD is an integral part of the book: its contents relate primarily to the programs discussed in the second part, and you will find *readme* files with additional details about the programs that we just don't have space to put in the book.

We have tried to follow a logical progression: what VR fundamentally aims to achieve, to what it can be applied, what elements need to be brought together, how they work and how the theory is turned into the practice. We also consider some key concepts from the allied disciplines of computer graphics and computer vision which overlap with VR.

Part I is comprised of Chapters 1 through 11 and forms the core of the book. After a brief introduction in Chapter 1, we move to Chapter 2 where we examine the human senses and their significance in delivering a sense of reality within the virtual world. The chapter also explains why some senses are easier to interface with the virtual world than others.

Chapter 3 reviews a spectrum of applications where VR has proven to be an invaluable tool. It discusses in general terms the benefits to be gained in using VR, as well as the downside. Users of VR may be affected in ways that they are unaware of, and there could be wider implications which are not always obvious.

In Chapter 4, we take our first steps into the practical world of VR by describing the types of interface technologies that are available and how they work. We show how different types of human interface devices are appro-

---

[1]Haptic is the term in common use for discussing all aspects of being able to touch and feel things. We will return to it several times.

priate for different applications of VR. To really integrate the real with the virtual, we must also be able to match positions and directions in the user's reality with the computer's corresponding *virtual* reality. This brings in the human interface again, but here it is more a case of the virtual system sensing and responding to real events rather than being commanded by direct and explicit input, or a command to generate a view from such-and-such a place and in such-and-such a direction. As such, Chapter 4 also includes a section on the technology of motion tracking and capture.

After reading the first few chapters, it should be evident that being able to see the virtual world is of prime importance. It is hard to imagine VR without its 3D computer-generated graphics. Thanks to the incredible popularity of 3D computer games, the technology for visualization has achieved undreamt-of degrees of realism, and a major aspect of our book explains how this is achieved. Chapter 5 starts our discussion of visualization by explaining how the virtual world may be specified in ways the computer can understand and use to create a visual input to our senses. After a brief review of some basic mathematics for 3D graphics in Chapter 6, Chapter 7 explains the essence of the algorithms that are used to create the real-time views.

In these chapters we explore the basic principles of the two key elements to VR we identified earlier. However, before turning to consider how to program the computer to deliver a VR experience, there are a few more aspects regarding visualization that are worth exploring further. The science of computer vision is not normally associated with VR, but many of the things that computer vision makes possible (such as using a virtual environment described by pictures rather than 3D numerical data) open up some interesting applications of VR. Consequently, we feel a basic introduction to this often impenetrable mathematical subject should form part of a study of VR, and we include one in Chapter 8. We extend this discussion in Chapter 9 by introducing the concept of image-based rendering, which offers a powerful alternative to traditional geometry-based techniques for creating images.

In the penultimate chapter of Part I, we explore in detail the subject of stereopsis. If you don't already know what it is, you will learn all about it. If you do know what it is, you will already appreciate the essential enhancement it brings to any VR experience.

Of course, even generating a 3D movie in real time is not enough for VR; we must be able to generate the real-time view interactively. We must be able to move about the virtual world, navigate through it and choose the direction in which we look or the objects we look at. We examine the relevant aspects of this movement in Chapter 11.

Part I of the book is drawn to a close with a brief summary of the most significant aspects of VR introduced thus far and offers an opinion on the future for VR. It identifies the most difficult challenges in the short term and ponders just how far it may be possible to go in making a virtual reality indistinguishable from reality.

Part II of the book is in a different format. Titled *Practical Programs for VR*, the text is tightly integrated with the accompanying CD, where all the code for each application project is available. The text is concise. Most of the programs, especially the projects, have much fuller explanations embedded as comments in the source code than we have space to print. The programs are designed to cover the spectrum of applications needed to get a VR environment up and running using only a basic personal computer and the minimum of additional hardware. The examples complement and make concrete the concepts covered in Part I. We have found them exceptionally useful, and many are in use in our own VR laboratory. The programs are written in the C/C++ language and are targeted for the Windows PC platform. We hope that even if you have only a basic (no pun intended) knowledge of programming, you should still be able to modify the user-interface elements and appearance. Chapter 12 introduces the tools, libraries and programming conventions we use throughout Part II.

As we stressed before, the key element of VR is the visual one, and here, perhaps more than in any other aspect of the human-computer interface, software plays the dominant role. In Chapter 13, we show you how to use the main 3D rendering libraries in interactive real-time applications. We pursue this in Chapter 14, where we show how to take advantage of the programmability of the graphics hardware so as to significantly increase the realism of the virtual scenes we want to experience.

Real-time 3D is not the only source of content for VR applications; movies, DVDs and other multimedia material can also play a vital role. Real-time video input, sometimes mixed with 3D material, is another important source. Trying to build application programs to accommodate the huge variety of source formats is a daunting challenge. In Chapter 15, we show you how to use Microsoft's DirectX technology to achieve this with minimal effort. This chapter also provides examples of how to combine video sources with 3D shapes and content, and leads into Chapter 16, which focuses on programming for stereopsis.

The level and sophistication of interaction set many VR applications apart from computer games and computer-generated movies. Programming this interaction poses its own complexities. The keyboard and mouse may be very good for many things, but VR needs more: two-handed input with multiple degrees of freedom, for example. In Chapter 17, we offer some examples of how to use the USB PC interface and the input components of DirectX to lift restrictions on the type and formats of input. The chapter also explains the challenges that a programmer faces when working with haptic devices.

In our final chapter (Chapter 18), we bring together all the elements to form a series of projects that illustrate the concepts of our book.

1.3 Assumed Knowledge and Target Audience

This book covers a lot of ground: there is some mathematics, a little electronic engineering, a bit of science and a quite a lot of programming. We try not to assume that the reader is a specialist in any of these areas. The material in Chapters 2 through 11 could be used as the text for a first course in VR. For this we only assume that the reader is familiar with the concepts of the *vector*, has a basic knowledge of *coordinate geometry* and has studied science or engineering to "high-school" level. The chapter on computer vision may require a little more confidence in using vectors and matrices and a bit of calculus, but it should be useful to postgraduate students embarking on a research degree that needs to use computer vision in VR.

The second part of the book is intended for a mix of readers. It is basically about programming and writing computer programs to get a job done. We assume the reader is familiar with the ideas in Part I. To get the most out of this part, you should know the C/C++ programming language and preferably have a bit of experience writing applications for Windows. However, in Chapter 12 we summarize the key Windows programming concepts. We have used some of this material in our master's degree program, and it should also fit well with a final-year degree course, based on the material in Part I as a prerequisite.

In Part II, there are a lot of practically useful programs and pieces of code that research engineers or scientists might find useful. Part II could also offer a starting point for people who wish to put together their own experimental VR environment at low cost. We've used these programs to build one of our own.

1.4   Valuable Resources for VR

Throughout the book, we offer a list of specific references on particular topics. We've tried to keep these to a minimum and only refer to things that are worth looking at if you want to pursue the topic. We are not the first to write a book on this dynamic and rapidly developing and evolving topic, so we think it's worth looking at the work of some of those who have written on the topic before. Unfortunately, some of the early technology that gets pride of place can date very quickly, but that does not mean that it isn't interesting.

VR is an amalgam of different technologies so it's worth pointing out some books that cover specific technologies. The most established technology used in VR is undoubtedly graphics, and there are many excellent books that comprehensively cover this area.

We have found the following books useful in preparing our own:

- *Specifically on VR.* Vince [22], Slater et al. [19] and Bimber et al. [2].
- *On computer graphics.* Ferguson [5], Burger and Gillies [4] and Möller and Haines [10].
- *On 3D interface hardware.* Bowman et al. [3].
- *For computer vision and image processing.* Hartley and Zisserman [6], Nixon and Aguado [12] and Parker [14].
- *For general C/C++ and Windows programming.* Rector and Newcomer [16], Wright [23] and Stanfield [21].
- *For OpenGL programming.* The official ARB guide [11] and Rost [17].
- *For DirectX programming.* Comprehensive coverage of the 3D graphics aspect is given by Luna [9] and coverage of video by Pesce [15].

In addition to these books, there are a few professional journals with special-interest sections on VR. Most notable amongst these is the *IEEE Computer Graphics and Applications Journal* [7], where many a new idea in VR has appeared. It regularly publishes easily comprehensible review articles on major topics in VR.

To carry out any practical experiments in VR, one must do some computer programming. This can be done at two levels: by using a basic instruction language, or at a much higher level where one describes scenes and interactive elements, through scene graphs, for example. In either case, the same *logical concepts* apply, so it doesn't really matter whether one is creating a virtual environment by using the C programming language or a VRML Web-based description. However for greatest flexibility and highest execution speed, most VR applications are written in C or C++. Since these are general computer programming languages, one has to write an awful lot of program code in order to get even a simple VR application off the ground. This is where the VR toolkits can prove to be exceptionally helpful. One still has to use C/C++ to create the programs, but the number of lines of code we have to write is reduced from thousands to hundreds or even tens.

The open-source community provides a freely available and rich source of comprehensive VR toolkits. We will explore the use of a couple of these later in the book, but a few are worth referencing here. One of best places to get your hands on these toolkits is sourceforge.net [20]. The open-source community tends to concentrate on *non*-Windows applications. Linux is a popular development platform, but as we shall see, Windows PCs are an equally powerful base on which to build VR programs. In terms of specifically useful toolkits, OpenCV from Intel [8] provides a good library of code for carrying out image processing and computer-vision tasks. The ARToolKit [1] is a very easy-to-use source of functions for building augmented-reality applications. And Open Inventor [18] from Silicon Graphics [13] provides a collection of C++ classes that simplifies the task of writing programs to build highly detailed interactive applications that require user actions such as picking, event handling and motion control. It extends the highly successful OpenGL 3D library, developed by Silicon Graphics, in a seamless and efficient way. The only disadvantage of Open Inventor is that it has not been freely ported to the Windows environment.

1.5   The Code Listings

All the application programs that we will examine in Part II use an overall skeleton structure that is approximately the same. Each application will require some minor modifications or additions to each of the functions listed. We will not reprint duplicate code nor every line of code either. Generally, we will refer to additions and specific lines and fragments if they are significant. Using our annotated printed listings, it should be possible to gain an understanding of the program logic and follow the execution path. Using the *readme* files and comments in the code, it should be well within your grasp to adapt the complete set of application programs on the CD.

Bibliography

[1] *ARToolKit.* http://www.hitl.washington.edu/artoolkit/.

[2] O. Bimber, R. Raskar. *Spatial Augmented Reality: Merging Real and Virtual Worlds.* Wellesley, MA: A K Peters, 2005.

[3] D. A. Bowman, E. Kruijffm, J. J. LaViola and I. Poupyrev. *3D User Interfaces: Theory and Practice.* Boston, MA: Addison Wesley, 2005.

[4] P. Burger and D. Gillies. *Interactive Computer Graphics.* Reading, MA: Addison-Wesley, 1989.

[5] R. S. Ferguson. *Practical Algorithms for 3D Computer Graphics.* Natick, MA: A K Peters, 2001.

[6] R. Hartley and A. Zisserman. *Multiple View Geometry in Computer Vision.* Cambridge, UK: Cambridge University Press, 2003.

[7] *IEEE Computer Graphics and Applications Journal: IEEE Computer Society.*

[8] Intel Corp. *OpenCV: Open Source Computer Vision Library.* http://www.intel.com/technology/computing/opencv/.

[9] F. Luna. *Introduction to 3D Game Programming with DirectX 9.0.* Plano, TX: Wordware Publishing Inc., 2002.

[10] T. Möller and E. Haines. *Real-Time Rendering.* Natick, MA: A K Peters, 1999.

[11] J. Neider, T. Davis and M. Woo. *OpenGL Programming Guide*, Fifth Edition. Reading, MA: Addison-Wesley, 2006.

[12] M. Nixon and A. Aguado. *Feature Extraction and Image Processing.* Burlington, MA: Elsevier, 2005.

[13] The Open Inventor Group. *The Inventor Mentor: Programming Object-Oriented 3D Graphics with OpenInventor, Release 2 (OTL).* Reading MA: Addison-Wesley, 1994.

[14] J. R. Parker. *Algorithms for Image Processing and Computer Vision.* New York: John Wiley and Sons, 1997.

[15] M. Pesce. *Programming Microsoft DirectShow for Digital Video and Television.* Redmond, WA: Microsoft Press, 2003.

[16] B. E. Rector and J. M. Newcomer. *Win23 Programming.* Reading, MA: Addison-Wesley, 1997.

[17] R. Rost. *OpenGL Shading Language*, Fifth Edition. Reading, MA: Addison-Wesley, 2004.

[18] SGI Developer Center. *Open Inventor.* http://oss.sgi.com/projects/inventor/.

[19] M. Slater, A. Steed, Y. Chrysanthou. *Computer Graphics and Virtual Environments: From Realism to Real-Time.* Reading, MA: Addison-Wesley, 2002.

[20] SourceForge.net. http://sourceforge.net/.

[21] S. Stanfield. *Visual C++ 6 How-To.* Indianapolis, IN: Sams, 1997.

[22] J. Vince. *Virtual Reality Systems,* SIGGRAPH Series. Reading, MA: Addison-Wesley, 1995.

[23] C. Wright. *1001 Microsoft Visual C++ Programming Tips.* Roseville, CA: James Media Group, 2001.

2 The Human Senses and VR

The human body is a beautiful piece of engineering that has five major senses which allow it to gather information about the world surrounding it. These five senses were first defined by Aristotle to be sight, hearing, touch, smell and taste. Whilst these five major senses have been recognized for some time, recently other senses have been named, such as pain, balance, movement and so on. Probably the most interesting of these is the perception of body aware-ness. This essentially refers to an awareness of where one's body parts are at any one time. For example, if a person closes her eyes and moves her arm then she will be aware of where her hands are, even though they are not being detected by other senses.

The senses receive information from outside and inside the body. This information must then be interpreted by the human brain. The process of receiving information via the senses and then interpreting the information via the brain is known as perception.

When creating a virtual world, it is important to be able to emulate the human perception process; that is, the key to VR is trying to trick or deceive the human perceptual system into believing they are part of the virtual world. Thus a perfect feeling of immersion in a virtual world means that the major senses have to be stimulated, including of course an awareness of where we are within the virtual world. To do this, we need to substitute the real informa-tion received by these senses with artificially generated information. In this manner, we can replace the real world with the virtual one. The impression of being present in this virtual world is called virtual reality.

Any VR system must stimulate the key human senses in a realistic way. The most important senses by which humans receive information about the world surrounding them are vision, hearing, and touch. As such, these are the main aspects of current VR systems. In this chapter, we will concentrate on the vision and sound aspects of the VR experience, since they are the best researched fields, followed then by touch. This will be followed by a general review of the more exotic senses of smell and taste. We hope to show that the only limitation on the VR experience comes from the practical restrictions of not yet being able to artificially generate sensory stimuli with a high degree of realism.

2.1   Human Visual System

We obtain most of our knowledge of the world around us through our eyes.   Our visual system processes information in two distinct ways—conscious and preconscious processing. For example, looking at a photograph or reading a book or map requires conscious visual processing, and hence usually requires some learned skill. Preconscious visual processing, however, describes our basic ability to perceive light, color, form, depth and movement. Such processing is more autonomous, and we are less aware that it is happening. Central to our visual system is the human eye, and more specifically the retina.

2.1.1   The Retina

The retina is the light-sensitive layer at the back of the eye. It is often thought of as being analogous to the sensor in a camera that converts light energy to electrical pulses. However, its function goes beyond just creating these pulses.

Photosensitive cells called rods and cones in the retina convert incidental light energy into signals that are carried to the brain by the optic nerve. Roughly 126 million rods and cones are intermingled non-uniformly over the retina. Approximately 120 million of these are rods, which are exceedingly sensitive. They respond in light too dim for the cones to respond to, yet are unable to distinguish color, and the images they relay are not well defined. These rods are responsible for night and peripheral vision. In contrast, the ensemble of 6 million cones performs in bright conditions, giving detailed colored views. Experiments have shown that among the cones, there are three different types of color receptors. Response curves for these three cones correspond roughly to red, green and blue detectors. As such, it is convenient to use the red-green-blue (RGB) primary color model for computer graphics and therefore VR visualization displays. We explore the topic of VR visualization in more detail in Section 4.1.

Obviously, our eyes can perceive our surroundings and discriminate between the colors of objects in these surroundings. The extent of the surroundings that we can view at a single instant is governed by the field of view of our eyes. The field of view is the angle that an eye, or pair of eyes, can see either horizontally or vertically. The total horizontal field of vision of both human eyes is about 180° without eye movement, or, allowing for eye movements to the left or right, the total field of vision possible without moving the head is 270°. The vertical field of vision is typically more than 120°. While the total field is not necessary for a user to feel immersed in a virtual environment, many believe that at least 90° is necessary for the horizontal field of vision. This stipulation is important when analyzing the effectiveness of visual displays. For example, when we view a computer monitor, it usually appears as only a small feature of the real world in which it is placed. However, if we view a large projection screen that is sufficiently curved, it will dominate our field of view and thus help us to gain a feeling of immersion. In Chapter 4, we look at different display technologies and their usefulness in VR systems. In Chapter 10, we look at combining stereo vision with different display devices.

2.1.2   Depth Perception

Whilst stereo vision is usually acknowledged to be a human's main depth cue, it is not the only one. If we view the world with one eye, we can still make good assumptions about the distance of objects. This is due to monocular depth cues, which are actually the basis for the perception of depth in visual displays, and are just as important as stereo for creating images which are perceived as truly three dimensional.

Examples of monocular cues include:

- *Light and shade.* Artists can make the relationships among objects look different by adding shadows to them. There are two types of shadows, attached and cast. An attached shadow is on the object, whilst a cast shadow is projected on another surface by the object. How such shadows can change the appearance of an object is shown in Figure 2.1. The first circle does not look solid, the second looks solid and appears to be resting on a surface whilst the third also looks solid but appears to be suspended in the air.

- *Relative size.* If a familiar object casts a smaller image in our eye, we assume it has not merely shrunken, but that it is further away.

Figure 2.1. Shadows give a sense of the relative position between objects.

- *Interposition or overlap.* Sometimes this is so obvious it is taken for granted. This occurs when the image of one object partially covers the image of another. The object that is partially covered is perceived as being further away.
- *Textural gradient.* A textured material, such as the surface of an orange or a roof tile, provides a depth cue because the texture is more apparent as the object is closer to the observer.
- *Linear perspective.* When parallel lines recede in the distance, they seem to converge.

All these monocular cues can be utilized by computer graphics designers when creating virtual worlds so that they appear more realistic to the human user. However, central to many of the depth cues and three-dimensional vision in general is the phenomenon of stereoscopic vision.

Each eye captures its own view of the world, and the two separate images are sent on to the brain for processing. When the two images arrive simultaneously in the back of the brain, they are merged into one picture. The mind combines the two images by matching up the similarities and adding in the small differences. The small differences between the two images add up to a big difference in the final picture. The combined image is more than the sum of its parts; it is a three-dimensional stereoscopic picture.

A virtual environment imposes a number of requirements for visual displays, the most significant of these being the need for stereoscopic vision. There are a large number of different techniques for providing stereoscopic vision. These techniques are outlined in Chapter 10.

2.1.3 Virtual Visual System

As we have seen, VR imposes a number of requirements for visual displays. The most significant of these are stereoscopic vision and an adequate field of view. In addition, the user should be surrounded by visual stimuli of adequate resolution, in full color, with adequate brightness and high-quality motion representation. Another major challenge is display hardware that is capable of providing the necessary quality at an acceptable cost. Currently available displays for immersive VR typically require the user to wear a head-mounted display (HMD) or some form of special glasses. These introduce a range of new issues, such as ergonomics and health concerns, which are particularly critical in the case of HMDs.

Our sense of visual immersion in VR comes from several factors, which include:

- *Field of view.* Limited field of view can result in a tunnel vision feeling.

- *Frame refresh rate.* A movie or animation consists of a sequence of picture frames. A typical movie will display 25 frames per second; it is said to have a refresh rate of 25 frames per second. The refresh rate is important because it must be high enough to allow our eyes to blend together the individual frames into the illusion of motion and should limit the sense of latency between movements of the head and body and regeneration of the scene. That is, whenever we move within the virtual environment, the imagery must reflect this movement straight away.

- *Stereopsis.* Stereopsis is the ability to perceive the world around us in three dimensions. This is actually the most important mechanism by which humans can assess depth information. Essentially, each eye inputs a slightly different image of the world. These images are then blended together in the brain to produce a 3D image. In order to enhance realism, it is important that VR displays also have the capability to allow the user to perceive depth. This is achieved using the monocular cues previously described, as well as stereo-vision equipment. We approach the topic of stereo vision in much greater detail in Chapter 10.

The resolution of the image is also critical. If we only have limited resolution then images can appear grainy. If our VR system suffers from grainy images, the illusion of reality will fail. Thus when considering the visual requirements of a VR system, we must ensure that the images are of high resolution and that they can be updated very quickly or in real time.

2.2   Human Auditory System

Our ears form the most obvious part of our auditory system, guiding sound waves into the auditory canal.

- The canal itself enhances the sounds we hear and directs them to the eardrum.

- The eardrum converts the sound waves into mechanical vibrations.

- In the middle ear, three tiny bones amplify slight sounds by a factor of 30.

- The inner ear changes the mechanical vibrations into electrochemical signals which can then be processed by the brain.

- The electrochemical signals are sent to the brain via the auditory nerve.

Humans have three different types of sound cues. *Interaural time difference* is a measure of the difference in time between when a sound enters our left ear and when it enters our right ear. *Interaural intensity difference* is a measure of how a sound's intensity level drops off with distance. *Acoustic shadow* is the effect of higher frequency sounds being blocked by objects between the sound's source and us. Using these cues, we are able to sense where the sound is coming from.

2.2.1   Virtual Auditory System

Sound provides an obviously important but sometimes underutilized channel in VR systems. Speech recognition is sometimes used to provide a hands-free input alternative to the mouse and keyboard. However, we are really concerned with the output of sound. Today's multimedia-equipped computers can provide CD-quality stereo sound. Conventional stereo can easily place a sound in any position between the left and right speaker. However, with true 3D sound, the source can be placed in any location—right, left, up, down, near or far.

A virtual sound system not only requires the same sound localization cues, but must change and react in real time to move those sounds around the 3D environment. But before we consider these more complex sound systems, it would be useful to consider the more basic sound systems we are all familiar with.

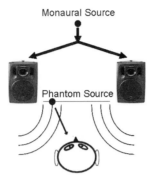

Figure 2.2. Phantom source.

The Basics

The basic type of sound system is stereo sound using two loudspeakers, where it is possible to localize a sound anywhere between the two speakers (Figure 2.2). The idea is pretty obvious:

- To place a sound on the left, send its signal to the left loudspeaker;
- to place it on the right, send its signal to the right loudspeaker;

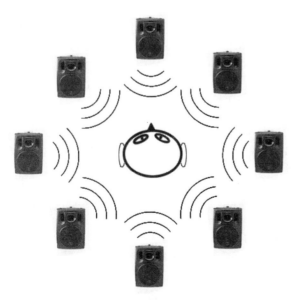

Figure 2.3. Surround sound.

- if the same signal is sent to both speakers, a "phantom source" will appear to originate from a point midway between the two loudspeakers;
- the position of the "phantom" source can also be shifted by exploiting the precedence effect. If the sound on the left is delayed relative to the sound on the right, the listener will localize the sound on the right side.

This principle can be further extended for multi-channel systems. That is, a separate channel is created for each desired location (Figure 2.3). These systems can produce impressive spatial results.

Binaural Recordings and Head-Related Transfer Functions

It has long been known that it is not necessary to have multiple channels to create 3D surround sound; two-channel binaural recordings are sufficient. All you have to do is re-create the sound pressures at the right and left eardrum that would exist if the listener were present when the sound was originally played. An easy way to do this is to put two microphones in the ear locations of a manikin and record what they pick up. These signals can then be fed to a user of a VR system via headphones. However, the major disadvantage of this type of system is that it is not interactive. That is, the sound must be prerecorded, and as such it cannot change even if the user moves. Obviously, this can detract from the user's sense of immersion in the VR world.

Binaural recording can be improved upon by using head-related transfer functions[1] or HRTFs. These HRTFs basically try to determine what a user will hear, based upon where he is in the room and on the location of the original sound. This may sound easy, but it does require some complicated mathematics! This is because a sound that originates at a location near a person is modified on its way to that person's ears by a number of factors, including objects within the room, the ceiling height and even the listener's own body! That is, the sound pressure at each person's eardrums depends on the listener's body and on the environment he is located within.

Thus, we need to relate the sound pressure of the source to the sound pressures that would be heard by a person within that environment. Initially we need to represent the sound pressure at the source and the two eardrums using sinusoidal functions, or more technically, using Fourier transforms. In Fourier analysis, one sinusoidal function can be related to another using a transfer function. Thus, we need a transfer function that relates the sound at the source to the sounds at each ear. These HRTFs are functions of frequency

---

[1]A transfer function is a mathematical expression relating one quantity, usually an input, to another quantity, usually an output.

and change depending on the location of the sound source in relation to the head. In other words, there is not a single transfer function, but a large set of transfer functions—one for each direction the sound comes from. Finally, it is important to know that there is no universal transfer function. Each person has his own unique HRTF.

To measure the HRTF of an individual:

- A loudspeaker is placed at a known location relative to the listener;
- a known signal is played through that speaker and recorded using microphones at the position of the listener.

The listener cannot tell the difference between the sound from the source and the same sound played back by a computer and filtered by the HRTF corresponding to the original source's location. Such sound technology can typically be built into head-mounted displays (HMDs) and is at present one of the best methods by which we can re-create 3D interactive sound. Interesting work in this area has been carried out by Florida International University, and further reading on the subject can be found in [6, 7].

2.3   Human Tactile System

The human sense of touch is more aptly described as our somatic sense, to better reflect all the complex behavior actually involved. We already know that touch has been classified as one of the five classic senses. However, when we touch something or someone, we actually perceive more information than can be gained from simple touch. For example, we can perceive pressure, shape, texture, temperature, movements and so on. Thus, the simple term touch involves many senses. However, we need to refine this complex topic, and so we limit what we mean by our sense of touch to our haptic system. *Haptic*, derived from the Greek word *haptesthai*, meaning to touch, emerged in the 1990s as the term used to describe the sensation of communicating with a computer through the sense of touch. Haptic sensory information incorporates both tactile and kinesthetic information.

So what happens when we touch something? Initially, the sense of contact is provided by the touch receptors in the skin, which also provide information on the shape and texture of the object. When the hand applies more force, kinesthetic information comes into play, providing details about the position and motion of the hand and arm, and the forces acting on them, to give a sense of total contact forces and even weight.

There are basically four kinds of sensory organs in the skin of the human hand that mediate the sense of touch. These are the *Meissner's corpuscles*, *Pacinian corpuscles*, *Merkel's disks* and *Ruffini endings*. The Meissner's corpuscles are found throughout the skin, with a large concentration on fingertips and palms. They are sensitive to touch, sharp edges and vibrations, but cannot detect pain or pressure. The Pacinian corpuscles are larger and fewer in number than the Meissner's corpuscles and are found in deeper layers of the skin. They are sensitive to pressure and touch. The Merkel's disks are found in clusters beneath ridges of the fingertips. They are the most sensitive receptor to vibrations. Finally, the Ruffini endings are slowly adapting receptors that can sense or measure pressure when the skin is stretched.

The sense of kinesthesis is less well understood than the sense of touch, although joint angle receptors in our fingers do include the Ruffini endings and the Pascinian corpuscles. Thus, haptic interaction between a human and a computer requires a special type of device that can convert human movement into quantities that can be processed by the computer. The result of the processing should then be converted back into meaningful forces or pressure applied to the human as a result of her initial action. This is not an easy thing to achieve.

2.3.1 Virtual Tactile System

The human haptic system has an important role to play in interaction with VR. Unlike the visual and auditory systems, the haptic sense is capable of both sensing and acting on the real and virtual environment and is an indispensable part of many human activities.

As we have said, the human haptic system incorporates both tactile and kinesthetic information, and this can be dealt with through the study of tactile feedback and force feedback. *Tactile feedback* refers to devices that can interact with the nerve endings in the skin to relate information about heat, pressure, texture etc. *Force feedback* refers to devices which interact with muscles that will then give the user the sensation of pressure being applied.

In order to provide the realism needed for effective and compelling applications, VR systems need to provide inputs to, and mirror the outputs of, the haptic system. However, this is not easy to achieve, and it still has problems that need to be resolved. The principal problem is that whilst we know the hand's sensory capabilities, they are not yet totally understood.

However, several attempts have been made at creating realistic haptic devices, and these operate under a number of different principles:

- *Piezoelectric crystals.* Crystals that expand and contract in response to changes in an electric field. Work in this area has long been conducted by the University of North Carolina [4].
- *Shape memory alloy.* Shape memory alloy wires and contract springs contract when heated and then expand again as they cool. An interesting haptic device based on this technology was developed at Harvard [10].
- *Pneumatic systems.* This type of system operates using air jets with an array of air nozzles that can be grafted to a display device. For example, the Rutgers Master II operates using pneumatic pistons which are positioned in the palm of the hand and can be activated to apply a force to both the fingers and thumb [2]. Another device is called the AirGlove; it was developed at Washington State University. The AirGlove can apply an arbitrary point force to the user's hand via controlled air jets. The device is said to give a realistic weight sensation [5].

These devices are all examples of tactile interfaces. Kinesthetic interfaces can be either static desktop devices or wearable exoskeletal devices. They are generally more developed than the tactile interfaces. The most well known is probably the Phantom device, developed by SensAble Technologies, Inc. Force feedback is achieved via a thimble where the user can place her thumb. The force is generated by brushed motors in the three main coordinate axes. We talk about this device and other kinesthetic interfaces in Section 4.2.1.

2.4   Olfaction—The Sense of Smell

In VR, the sense of smell has been largely ignored as an input for the VR participant, in spite of the fact that smell provides a rich source of information for us [1], be it pleasant or unpleasant. Indeed, Cater [3] has conducted research on how important the ambient smell of the physical environment used to house the VR system is to enhancing the sense of presence within a virtual environment. For example, a participant in the virtual environment might be entering a virtual forest but still be able to smell the laboratory in which he is sitting. Thus, we are going to take a quick look at some of the devices used to enhance the sense of smell within VR. But before doing that, it is useful to have some idea about how our olfactory system, which gives us a sense of smell, actually works.

Odors are formed by the small molecules of gaseous chemicals that come from many different things, such as flowers or food. These odors are inhaled through the nasal cavity and diffuse through it until they reach its ceiling. This is covered with a layer of microscopic hair-like nerve cells that respond to the presence of these molecules by generating electrical impulses. The impulses are transmitted into the olfactory bulb, which then forwards the electrical activity to other parts of the olfactory system and the rest of the central nervous system via the olfactory tract. This then senses the odor.

Devices have been developed to automatically detect various odors for use within virtual environments. Some of these devices consist of sensors which try to identify the smell. These sensors may be electrochemical, mechanical, thermal and/or magnetic, and can measure different characteristics of the smell. Each characteristic can then be compared against a list of characteristics for predetermined smells, until a match can be found. However, this type of sensing system becomes difficult when there are many odors to be sensed, and the electrochemical sensors required are expensive [9].

These devices have been superseded by newer technologies such as the *air cannon*, developed by the Advanced Telecommunications Research Institute in Japan [12]. This device is capable of tracking an individual and shooting an aroma directly into his nose. Again, one of the main drawbacks of the system is that it can currently only synthesize a limited number of odors.

Others types of devices range from those that simply inject odors into the physical environment to those which transfer scented air to the nose via tubes, which are usually attached to a head-mounted display. Thus, devices are now available to include the sense of smell in our VR worlds. Researchers are now concentrating their efforts into verifying whether this additional sense does play an important role in VR and what applications it has. One of the biggest areas being researched is in surgical simulation, with many people believing that smell, in particular, is very important to medical practitioners when making a diagnosis [11].

2.5   Gustation—The Sense of Taste

Interestingly, your sense of taste very much depends on your sense of smell. When eating, odors from your food stimulate your olfactory system. If you like what you are smelling then you should like what you are tasting! Taste, or gustation, is again based on chemical sensing. There are five well-known types of receptors in the tongue that can detect sweet, sour, salt and bitter. The fifth receptor, called the Umami receptor, can detect the amino acid glutamate. Each of these receptors then conveys information on the taste they have detected to different regions of the brain.

Whilst VR devices have been built that try to satisfy our senses of vision, hearing, touch and smell, it has been rather more complicated to develop devices which can mimic taste. That was until researchers at the University of Tsukuba in Japan developed the first food simulator [8]. This device tries to simulate the taste and feel of food whilst in the mouth, including the act of chewing.

Initially, to calibrate the system, different types of food are placed in a user's mouth. Biological sensors placed in the user's mouth can record the major chemical components of food. A microphone also records the audible vibrations produced in the jawbone whilst chewing. These parameters serve as inputs to the food simulator.

During the VR experience, a mechanical device is inserted into the mouth, intended to resist the user's bite. When the user chews on the device, asensor registers the force of the bite and a motor provides appropriate resistance. In addition, a thin tube squirts a mixture of flavorings onto the tongue and a tiny speaker plays back the sound of a chewing jawbone in the user's ear.

Other devices that simulate the sense of taste are still rather primitive, and to date there is little research to show or prove that developing devices to provide a virtual sense of taste enhances the virtual experience. These is no doubt, however, that this research will come!

2.6   Summary

The successful development and application of VR systems depend on an entire matrix of interrelated factors. Central to this is the need for improved hardware technologies, especially within haptic devices. A real haptic device needs to be able to stimulate the nerve endings in our hands using air jets, vibrations etc. Whilst there are a few devices in development that can do this, we are mostly restricted to using mechanical devices (joystick, Phantom, etc.) which do not address the somatic senses directly. Of course, the main problem is that relatively few individuals have the background skills to be able to develop the necessary hardware. Most individuals in the VR area are involved in software development or in the assessment of human perception and performance. One thing is clear: if we want the futuristic experiences that are experienced in movies such as *The Matrix*, we have a long way to go.

But before we exert the effort required for such compelling VR experiences, we need to be sure that the effort is justified. In the next chapter, we look at different applications of VR and ask whether the technology has scientific justification or whether it is just a play thing for the rich.

Bibliography

[1] W. Barfield and E. Danas. "Comments on the Use of Olfactory Displays for Virtual Environments". *Presence* 5:1 (1995) 109–121.

[2] M. Bouzit et al. "The Rutgers Master II—New Design Force Feedback Glove". *IEEE/ASME Transactions on Mechatronics* 7:2 (2002) 256–263.

[3] J. Cater. "The Nose Have It!" Letters to the Editor, *Presence* 1:4 (1992) 493–494.

[4] M. Finch et al. "Surface Modification Tools in a Virtual Environment Interface to a Scanning Probe Microscope". In *Proceedings of the 1995 Symposium on Interactive 3D Graphics*, pp. 13–18. New York: ACM Press, 1995.

[5] H. Gurocak et al. "Weight Sensation in Virtual Environments Using a Haptic Device with Air Jets". *Journal of Computing and Information Science in Engineering* 3:2 (2003) 130–135.

[6] N. Gupta et al. "Improved Localization of Virtual Sound by Spectral Modification of HRTFs to Simulate Protruding Pinnae". In *Proceedings of the 6th World Multiconference on Systemic, Cybernetics and Informatics*, pp. III-291–III-296. Caracas, Venezuela: International Institute of Informatics and Systemics, 2002.

[7] N. Gupta et al. "Modification of Head-Related Transfer Functions to Improve the Spatial Definition of Virtual Sounds". In *Proceedings of the 15th Florida Conference on Recent Advances in Robotics*, CD-ROM format. Miami, Florida: Florida International University, 2002.

[8] H. Iwata et al. "Food Simulator: A Haptic Interface for Biting". In *Proceedings of the IEEE Virtual Reality 2004*, pp. 51–57. Washington, D.C.: IEEE Computer Society, 2004.

[9] P. Keller, R. Kouzes and L. Kangas. "Transmission of Olfactory Information for Telemedicine". In *Interactive Technology and the New Paradigm for Healthcare*, edited by R. M. Satava et al., pp. 168–172. Amsterdam: IOS Press, 1995.

[10] P. Wellman et al. "Mechanical Design and Control of a High-Bandwidth Shape Memory Alloy Tactile Display". In *Proceedings of the Fifth International Symposium of Experimental Robotics*, pp. 56–66. London: Springer-Verlag, 1997.

[11] S. Spencer. "Incorporating the Sense of Smell into Patient and Haptic Surgical Simulators". *IEEE Transactions on Information Techology in Biomedicine* 10:1 (2006) 168–173.

[12] Y. Yanagida et al. "Projection-Based Olfactory Display with Nose Tracking". In *Proceedings of the IEEE Virtual Reality 2004*, pp. 43–50. Washington, D.C.: IEEE Computer Society, 2004.

3 Applications and Implications of VR

According to engineering legend, this guy Fubini came up with a law (excuse the lack of reference here, but it is a legend after all). It may not be a law in the true sense of the word, but it certainly is an interesting observation that goes something like this:

- People initially use technology to do what they do now—only faster;
- then they gradually begin to use technology to do new things;
- the new things change lifestyles and work styles;
- the new lifestyles and work styles change society;
- and eventually this changes technology.

Looking back on the incredible advances that have been made in technology, we can see this simple law is breathtakingly accurate. Look at the Internet—it has revolutionized the world of communications, like nothing has done before it. The Internet began in the 1960s with an arbitrary idea of connecting multiple independent networks together in order to share scientific and military research. As such, the early Internet was only used by scientists and engineers. Today, nearly every home has an Internet connection, and we use the Internet for our banking needs, shopping desires etc. In fact, we don't even need to venture outside our homes anymore! However, this increased usage of the Internet has necessitated the development of high-speed processors, larger bandwidth communications and more. Now try to tell us that the Internet has not changed lifestyles and work styles, and a result our future technology needs.

So what about VR? Does it have the same potential to impact upon our daily lives the way the Internet has? Whilst VR is still a rapidly evolving technology undergoing active research and development so that it can be made more robust, reliable and user-friendly, many potential applications have been identified for its use. Indeed, it is already used in some specialized areas. In this chapter, we would like to outline some of the current and potential uses of VR. And we ask the question that if this new technology has the potential to change lifestyles and work styles, will it do so for the betterment of society? To put it another way, the main idea of this chapter is to give a brief outline of the current and potential future uses for VR so that you can begin to think about how you could use this technology. Then, as you progress through the remainder of the book, where we show you what hardware and software you need to achieve various degrees of realism within numerous types of virtual worlds, we hope you will be enthused to implement some of your own ideas and start creating your own VR system.

3.1   Entertainment

Cinema, TV and computer games fall into this category. In fact, computer game development has somewhat driven the development of VR technology. In computer games, being *fast and furious* is the name of the game, with as many twists and turns as can be accomplished in real time as possible. The quality of images produced from games' rendering engines improves daily, and the ingenuity of game developers ensures that this is a very active area of research. And the market of course is huge. In 2000, some 60 percent of Americans had a PC connected to the Internet, and 20 percent had a gaming console. Gaming companies now make more money than the movie industry ($6.1 billion annually) [5], and of course a large portion of this profit is poured into research and development to keep pushing forward the boundaries of gaming.

This is good news, because any developments in computer gaming are almost certainly replicated in VR technology. Indeed, it is sometimes difficult to understand the difference between computer gaming and VR. In fact, many researchers are customizing existing computer games to create VR worlds. Researchers at Quebec University suggest that computer games might be a cheap and easy-to-use form of VR for the treatment for phobias. The main reason for this is that the games provide highly realistic graphics and can be easily adapted to an individual's particular fears. Graphics chips have now become so powerful that anyone can make virtual worlds [8].

Of course, computer gaming and VR technology go hand in hand with computer animation. In computer animation, we hope to produce realistic pictures or a sequence of pictures that would be very difficult, impossible or too expensive to obtain by conventional means. The movie, TV and advertising industries have adopted this use of 3D computer animation with great enthusiasm, and, therefore, perhaps it is they who have provided the main driving force behind the rapid improvements in realism that can be achieved with computer graphics. If we link this with developments in graphics processors, we now are able to produce highly realistic and complex imagery in real time, which undoubtedly benefits VR. In particular, this benefits VR where the imagery must be updated in real time to reflect any movements made by the human user.

Amusement arcades and adventure parks are also prime examples of how the entertainment industry has benefited from VR. Take the example of amusement arcades. They started out with just the notorious *fruit machine*. Then in 1976 came the first computer arcade game. Then in the mid '90s came the *simulator rides*. Before this, simulators were a curiosity of high-tech businesses. The first simulators built were used to train pilots, and were often incredibly expensive. Nowadays, anyone who had been to an arcade or an adventure park has more than likely been on a simulator ride (*Back to the Future* at Universal Studios being a prime example). This is testimony to the fact that arcade games and simulators are moving towards VR in terms of the realism of their visual display and the extent to which interactions are possible.

3.2 Visualization

It isn't just the world of entertainment that is driving VR technology; psychologists, medical researchers and manufacturing industries are all showing a keen interest. Scientific visualization is the process of using VR technology (and in particular computer graphics) to illustrate experimental or theoretical data, with the aim of bringing into focus trends, anomalies or special features that might otherwise go unnoticed if they were presented in simple tabular form or by lists of numbers. In this category, one might include medical imaging, presentations of weather or economic forecasts and interpretations of physical phenomena.

Imagine: *a blue series BMW slams against a concrete wall at 56 kilometers per hour without braking. The bonnet on the driver's side heaves upwards and is crushed almost as far as the windscreen. The car is a write-off* [9]. This is thankfully not reality, but a simulation carried out at BMW's virtual reality center. They crash cars approximately 100 times before they even reach the prototype stage! Users of this system say that this type of crash visualization is essential for understanding the results of all the calculations necessary to see what effect different types of forces and impact collisions will have on the car's structure and, of course, on the passengers.

This type of simulation represents the visualization of a real object under virtual conditions. Indeed, the virtual environment can represent *any* three-dimensional world that is either real or abstract. This includes real systems such as cars, buildings, landscapes, underwater shipwrecks, spacecraft, archae-ological excavation sites, solar systems and so on. Abstract systems might in-clude magnetic fields, molecular models, mathematical models and so on. As such, VR seems a natural progression to the use of computer graphics that en-ables engineers and designers to visualize structures before actually building them. In addition to building and visualizing these structures, the designers are also able to interact with them, which can allow designers to conceptual-ize relations that might not otherwise be apparent. Thus VR can also be an invaluable tool for:

- The radiologist trying to visualize the topology of a tumor [11];

- the chemist exploring ways that molecules can be combined [1];

- the urban planner trying to re-create simulations of a growing city [2].

One area of special interest is visualization for medical applications. Take for example keyhole surgery or laparoscopy. This surgical technique leaves minimal scars, allows the patient to get up and about sooner and reduces the chance of infections. It makes the whole process of having an operation safer. Of course, nothing comes for free; laparoscopy requires an even greater degree of skill in the surgeon, as well as extra training. Sometimes even very proficient surgeons just can't adjust to the constraints imposed on them by laparoscopy. (Imagine trying to sew up a hole in your socks; now imagine trying to do the same thing under the duvet in bed at night while wearing them, using a feeble torch for light, only able to see what you are doing by putting a video camera down the bed and connecting it to the TV. For good measure, you are also wearing thick skiing mittens!)

In laparoscopic operations, there are many standard procedures: needle insertion through a port, suturing, cutting, stapling, suction, holding and organ displacement. During each of these tasks, the surgeon must carry out close inspection, as well as using the camera to obtain a broad overview of the operation and tool delivery volumes. A broad camera overview permits the surgeon to become oriented and find complex structures that are often difficult to identify. Positioning the camera requires careful synchronization and understanding between the surgeon and the endoscopic camera assistant. Unfortunately, no matter how good the video image augmentation is, it is a fact that vital anatomical structures remain concealed beneath the immediately visible tissue. The ultimate challenge is to map the visible field of view against a 3D description of the surrounding anatomy, ideally the anatomy of the actual patient. (This is somewhat analogous to the way in which Google Maps blends street and highway detail with satellite images of the same area to give an extremely useful, informative and striking result.) The first and most obvious way to do this is to use scan data, MRI and CT.

3D reconstruction of patient anatomy from CT and MRI images can be utilized to project images of the patient's anatomy that is obstructed from the surgeon's view. Using this special technology, surgeons can interactively manipulate the view they have in real time. In addition to advances made to the actual surgery, training simulators have also been developed which allow surgeons to practice many different types of surgery prior to the operation. For example, Dr. Joseph Rosen [7] of Dartmouth University Medical Center has a VR model of a face with deformable skin which allows surgeons to practice a plastic surgical procedure and demonstrate the final outcome before making the incision on a patient. And indeed, Immersion Medical has a whole range of surgical simulators available for purchase, those being endoscopy, endovascular, hysteroscopy and laparoscopy simulators. Studies show that physicians who train on simulators which are lifelike mannequins with hearts that beat, lungs that breathe, and veins that respond to injection make fewer errors and work more quickly than those who practiced on animals or learned by observation. And since we are talking about training, we can move onto the next large application area of VR technology.

3.3   Training

Training is one of the most rapidly growing application areas of VR. We all know about the virtual training simulators for pilots. These simulators rely on visual and haptic feedback to simulate the feeling of flying whilst being seated in a closed mechanical device. These were one of the first simulators to be developed, but now we have simulators for firefighters, surgeons, space mission controls and so on. The appeal of VR for training is that it can provide training nearly equal to practicing with real-life systems, but at a much reduced cost and in a safer environment. In fact, VR is the ideal training medium for performing tasks in dangerous or hazardous environments, because the trainee can be exposed to life-threatening training scenarios under a safely controlled computer-generated environment. If the trainee makes a mistake then the simulation can be stopped and the trainer will have the chance to explain to the trainee exactly what went wrong. Then the trainee will have the opportunity to begin the task again. This allows the trainee to develop his skills before being exposed to dangerous real-life environments.

Many areas of training can take advantage of VR. For example:

- At the University of Sheffield (UK), researchers are investigating a flexible solution for VR to be used for training police officers how to deal with traffic accidents [14];
- the Simulation and Synthetics Laboratory Environment at Cranfield University (UK) has developed a number of VR training simulators [12]. These include surgical procedure training, flight deck officer training and parachute training;
- Fifth Dimension Technologies has developed a range of driving simulators for cars, trucks, forklifts, surface and underground mining vehicles. These simulators range from a single computer screen with a game-type steering console to a top-of-the-range system which consists of a vehicle with a field of view of $360°$ mounted on a motion base.

The explosion of VR for training purposes can be well justified. It is an ideal tool for training people how to use expensive equipment. This not only safeguards the equipment, but also means that it will not have to be taken out of the production line for training purposes. VR is also cost-effective when the equipment has a high running cost that could not be justified for training purposes only.

Of course the military is also a huge user of VR technology, not just for training pilots but also for training personnel in all forms of combat simulations. Despite this, it is in fact the entertainment industry who is driving the technological advances needed for military and commercial VR systems. In many ways, the entertainment industry has grown far beyond its military counterpart in influence, capabilities and investments in this area. For example, Microsoft alone expected to spend $3.8 billion on research and development in 2001, compared to the US Army's total science and technology budget of $1.4 billion [5]. Indeed, the computer-game industry has considerable expertise in games with military content (for example, war games, simulations and shooter games) [5] and is in fact driving the next level of war-game simulations. Macedonia highlights the development of these simulations [5]. He discusses the US Army's training needs and goals, and how the Army came to realize the best way to achieve these goals was to work hand-in-hand with academia and Hollywood. The use of VR for military purposes is self-explanatory. It allows soldiers to experience battlefields without endangering their own lives. Obviously, mistakes made in a virtual battlefield are less permanent and costly than they would be in reality.

And of course, let's not forget about the use of VR in medical training. We gave an overview of the use of VR for medical visualization. This overlaps somewhat with training applications, for example, keyhole surgery etc. The benefits of VR training here again are self-evident.

And we could go on and on. Essentially, anything that people require training for can be implemented in a virtual environment. It is only the imagination which limits the applications in the area of training.

3.4   Education

Before we begin to describe the latest developments of VR for education, it would perhaps be beneficial to review the differences between education and training. Education and training might be considered to be quite similar, yet there is a subtle difference. The objective of education is usually to gain knowledge about facts, concepts, principles, rules etc. This knowledge can then be used to solve problems. Training, on the other hand, usually involves gaining a particular skill to enable you to carry out a specific task. Of course, training and education are sometimes intrinsically linked. For example, you may be trained on how to operate a VR system, but then you may go on, in time, to learn other functions of the VR system you were not specifically trained in. The hypothesis is that VR can successfully be used to support complex understanding by stimulating and exploring all human senses, whereas traditional notions of learning tend to focus on purely intellectual skills [4].

So how can VR be used to aid the learning process? The most obvious is within the long distance learning arena. Obviously the Internet and video-conferencing technology have been used for many years to assist in distance learning. Whilst this has been helpful to those, for example, taking part-time study courses in a different state or country, there are certain disadvantages. The most significant is the lack of direct interaction between the educator and the student, making it difficult for the educator to gauge the reaction of his or her students. Teleconferencing could of course be used, but in a multi-user situation, it is near impossible for the educator to keep track of multiple screens. And of course, there are all the usual problems associated with streaming real-time video across the net. VR can be utilized to overcome these disadvantages. In its most simple conception, a virtual classroom can be set up on a remote server which all students must log into. Each student can be represented by an avatar (a graphical representation of themselves), and can interact with the educator and the other students. This enhances the feeling of being involved in a multi-user learning environment.

Many software packages have also been developed that provide interactive knowledge spaces that allow users to create multimedia galleries of the pictures, video, sounds and Internet links that they use. That is, users can create their own VR environment to aid their own learning. Indeed, probably one of the most difficult aspects of education is teaching difficult concepts that students are unable to visualize. All the positive aspects of VR for visualization, detailed in Section 3.2, can be utilized to enhance the learning process.

We could go on and on. There is a wealth of literature available on the uses of VR in education and its associated benefits. These are summarized very nicely in a paper by Fallman [4] which we recommend you read if you are thinking of utilizing VR for educational purposes.

3.5   Other Applications

The list is almost endless. An easier question to pose is what can't VR technology be used for? Looking at the conference paper titles of the Virtual Reality Conference 1995, we can see a list of different areas of research that VR technology was being developed for more than 10 years ago.

- Theoretical and practical issues for the use of virtual reality in the cognitive rehabilitation of persons with acquired brain injuries.
- Using virtual and augmented reality to control an assistive mobile robot.
- Remote therapy for people with aphasia.
- Ethical pathways to virtual learning.
- Virtual reality in the assessment and treatment of body image.

- Virtual presence and autonomous wheelchair control.

- A virtual science library to accommodate needs of students with cerebral palsy.

- Virtual reality and visual disability—proceed with caution.

- Virtual reality's increasing potential for meeting needs of persons with disabilities: what about cognitive impairment?

And the list goes on. Now let's have a look at what more recent published articles are doing.

- Feasibility, motivation and selective motor control: virtual reality compared to conventional exercise in children with cerebral palsy.

- Integrating haptic-tactile feedback into a video-capture-based virtual environment for rehabilitation.

- Responses to a virtual reality grocery store in persons with and without vestibular dysfunction.

- Using virtual reality to improve spatial perception by people who are blind.

- Immersive virtual reality as a rehabilitative technology for phantom limb experience.

- Three dimensional virtual environments for blind children.

- Simulating social interaction to address deficits of autistic spectrum disorder in children.

These titles were taken from the *Journal of CyberPsychology and Behavior* (9:2, 2006). Research 10 years ago and that ongoing today have a common thread. An interesting application area which has developed over the past 10 years is the utilization of VR to help people with disabilities or to aid in rehabilitation. Particularly important is this use of VR for *workers* with disabilities. Indeed, as Weiss [15] has indicated, individuals with disabilities comprise one of the most underemployed groups of workers. Reasons for their difficulty in obtaining jobs include their limited mobility, reduced manual capabilities and limited access to educational facilities and work settings. VR can help people with disabilities overcome some of these difficulties by facilitating their ability to carry out some of the tasks required in the work setting. Many people have been researching this application area for some time. And if we look at the field of rehabilitation, it is really like using VR for training. When someone is in need of rehabilitation, they are essentially in need of cognitive or physical training or both. And as we have seen already, this is a growth area for VR.

3.6   Distributed VR

One of the hottest topics in VR research is distributed VR. This means not only can we run a simulated world on one computer, but we can run that same world on several connected computers or hosts. So in theory, people all over the world can connect and communicate within the same virtual environment. In terms of communication, the crudest method available is to communicate with other users by simple text messages. At the high-tech end, people can be represented by avatars and thus people can communicate in a more meaningful way with other people's avatars.

Distributed VR embodies all the applications we have already considered, because it is able to take them one step further. Take for example VR and simulations. The biggest advantage that VR has over typical simulations is that VR allows multiple users. With distributed VR, we can have multiple users connected up at different sites globally. Thus we can have distributed VR training facilities, distributed entertainment (to a degree, we have had this for some time if we think about the online computer-gaming industry, where multiple users connect to the same game and compete against each other) and distributed visualization (again, this is really an extension of teleconferencing, but here people can enter into the meaningful simulations from multiple sites throughout the world).

Of course, before distributed VR really takes off, there are a number of problems that must be resolved, including:

- *Compatibility.* This refers to the fact that people may be using different hardware and software at their host sites.
- *Limited bandwidth.* If we require global connections then we have to rely on the Internet, and as such the realism of the VR connection will depend on the bandwidth available at the host sites.
- *Latency.* Again, with global connections between the host sites, we have to rely on Internet communication protocols. As such, it is difficult to ensure the real-time delivery of information simultaneously to each host's site.

It won't be long before these bottlenecks are resolved and then distributed VR will know no bounds.

3.7   The Implications of VR

Weiss [15] tells us that whilst there has been considerable speculation and enthusiasm among possible users of VR technology regarding its potential, much of it is still *blue sky thinking* rather than actual applications. So, as we have seen, VR has considerable potential for many applications, but it may not be appropriate or desirable in some cases. There are a number of reasons for this. Primarily, VR is currently still very expensive as a development tool and it may not be necessary. For designers, CAD (computer-aided design) may deliver everything a designer needs without the complications of VR. In addition, limited research has been done to prove that there is effective transfer of skills developed in the virtual environment to the real world. It is this issue that we would like to discuss in more detail.

Critical to using any VR simulator for training is being able to assess how the skills leant in the virtual environment will be transferred to the real world. Rose tells us that within the VR training literature, there is a wealth of anecdotal evidence that transfer does occur. However, Rose also insists that there have been relatively few attempts to investigate empirically the virtual to real world transfer process with regards to what sort of training shows transfer, in what conditions, to what extent, and how robust that transfer is [10].

Take for example the case where VR is used to desensitize people to fears and phobias. The Virtual Reality Medical Center in California currently uses VR exposure therapy in combination with physiological monitoring and feedback to treat panic and anxiety disorders. Social phobia is the most common form of anxiety disorder, and the single greatest fear that seems to exist worldwide is that of public speaking. We can overcome these fears by practicing public speaking, and by utilizing a number of calming and breathing techniques. VR is used in many instances to provide simulations for practicing public speaking, and in essence the user becomes less sensitive to his fear or phobia. This lessening of anxiety can be an important asset in enabling people to maintain their cool under duress, but it may also lead to a loss of respect for a real-life danger, particularly where the hazard is experienced in a game format [15]. So if this suggests that practicing or training for certain situations invariably leaves us feeling less anxious about the real event then let's look at the bigger picture. Some researchers have been trying to determine the effectiveness of transferring the idea of risk through VR [6]. For example, Mitchell [6] believes that if the simulation is not entirely credible then the user departs from her VR experience with a lower perception of the hazards involved. This can be detrimental to the soldier training in a war game simulation, the surgeon practicing her surgery skills and so on. That is, they have become desensitized to the real-world dangers.

So perhaps what we can say here is that VR is useful for some applications but not for all. For example, Stanney suggests that we need to determine the tasks that are most suitable for users to perform in VR. Of course, the implication here is that not *all* tasks are suitable for VR. Indeed, Stanney [13] has suggested that an understanding of the human factor issues can provide a systematic basis by which to direct future VR research. Because VR is all about trying to trick the human user into believing he is in a different environment, the best and indeed only real way to assess the efficiency of VR is to determine how much of an impact VR has on the user. Once we can determine this, we can improve the VR experience.

They suggested the following key human factors which determine the effectiveness of VR.

- Will the user get sick or be adversely affected by VR exposure?
- Which tasks are most suitable for users to perform in VR?
- Which user characteristics will influence VR performance?
- Will there be negative social impact resulting from the user's misuse of the technology?
- How should VR technology be improved to better meet the user's needs?
- How much sensory feedback from the system can the user process?
- Will the user perceive system limitations (e.g., flicker, time lags etc.)?
- What type of design metaphors will enhance the user's performance in VR?

The other interesting point that Stanney makes is to question whether there will be a negative social impact resulting from the user's misuse of the technology. What can this mean? Let's reflect for one moment on the greatest communications invention of the past few decades—the Internet. Can we say that this technology has had a negative impact upon society? Without doubt, the answer can only be yes. Think about credit-card fraud, spam, misuse of chat rooms etc. Likewise, can we predict the kind of effect VR will have on society? It's still a bit early to guess.

However, Whitby [16] quite succinctly outlines some of the ethical issues with the use of new technologies and in particular VR. He indicates that doubts have been voiced about the implications of the sort of freedom that can be provided by VR. In particular, there are worries about users having the freedom to commit rape and murder within VR. Obviously, this is an ethical rather than technical issue. It is technically possible to construct VR in such a way that almost every possibility of the user's imagination can be fulfilled. It is also possible for designers to place arbitrary limits on what is possible within a particular VR. That is, the VR system can be designed and built in such a way to prevent people from carrying out immoral acts. Whitby goes on to argue that there are four very real reasons as to why restrictions should be placed in VR. These are [16]:

1. *People acting out scenarios in VR might then do it for real.* This may not be as crazy as first seems. We all know that crime is on the rise in the western world, and for some time people have been linking this rise to the rise in crime shown on popular TV.

2. *Some things are not acceptable even in private.* This point indicates that if a person commits a morally unacceptable act in VR, because it doesn't affect anyone, can it really be seen as unethical? Of course, the answer is yes, because every person has an ethical duty to themselves.

3. *People may end up preferring the virtual world to the real.* Again we need only remember the world of the Internet and cyber dating to know that this argument could well have substance.

4. *The designers of VR can signal social approval or disapproval.* Essentially, designers of VR have the ability to reward ethical behavior and prevent unethical behavior. For example, look at some computer games on the market. They require their users to commit murder in order to advance to the next level of the game. Here they are obviously being rewarded for unethical behavior. And again, some psychologists would question the impact this has on our youth.

Well, all this looks a bit depressing for VR! We won't delve any deeper into the ethical issues of the technology—we just wanted you to know that they do exist.

But on to some more bad news for VR. Let's not forget that the use of VR technology may also have some health concerns for the individuals involved. For example, performing tasks in a virtual environment may require increased concentration compared to performing similar tasks in the natural world [3]. This may affect heart rate and blood pressure, for example, which may adversely affect some people and not others. And it is also known that certain values of lag time[1] in a visual display can cause disorientation, which is now termed *vision-induced motion sickness* (VIMS). In fact, most people cannot use VR systems for more than 15 minutes without feeling the effects of VIMS. But again, as graphic processing units become more powerful, lag time reduces, and as such VIMS will become a less problematic phenomena of using VR.

There you have it! The negative effects of VR encapsulated in a few paragraphs. Of course, we should mention that if the technology is used in a responsible manner then the technology really does have the means to improve people's lifestyles and work styles. Just think of the automobile. So it's not all bad!

3.8   Summary

As you have seen, there is an enormous potential for VR technology to impact upon numerous industries, educational establishments and various applications. Examples include the use of VR in desensitization training with patients with phobia, in treating children with autism, in training people with learning difficulties and for rehabilitation of patients. It is impossible to cover all the potential applications for VR and we have not tried. What we have done is try to give you a broad sample of applications currently being researched and developed, to give you an idea of the breadth and flexibility of VR.

However, it is also worth noting that most potential applications for VR are still very much in the development stage. A wide and varied range of tests and development scenarios must be dreamt up and implemented to verify whether VR is useful for the application, and if so to what extent and whether it justifies the cost. And then of course we need to enter into the arena of human effects and physiological impact. This requires a whole new set of skills than those possessed by the engineer or computer scientist who can design and build the VR scenarios. Thus we enter into the realms of cross-disciplinary work, and this adds a whole new set of complications to the decision of whether to use VR or not.

Added to all that, we need not forgot that the equipment required for our VR scenarios is still not as highly developed as we would like it to be. And of course VR is just like any other technology when it comes to market-

---

[1]Lag time in this context refers to the time it takes for the visual display to be updated in order to reflect the movement of the VR user.

place success. It needs to be cost effective, efficient, robust, do what it is intended to do, be user friendly and have no nasty side effects caused by it usage. It might be a few more years before we can satisfy this list of demands for all the potential applications, but certainly it is a technology in its growth phase, and one that has captured the imagination of the public and scientists alike.

So what conclusion can we make about VR? We suggest you make your own! But by reading the rest of the book, you will be able to learn how to do things in VR. You can even try them out for yourselves. By doing all this, you should be able to make a better informed judgment about the future of VR and hopefully how you too can make a contribution.

Bibliography

[1] C. Cruz-Neira et al. "VIBE: A Virtual Biomolecular Environment for Inter-active Molecular Modeling". *Journal of Computers and Chemistry* 20:4 (1996) 469–477.

[2] B. Delaney. "Visualization in Urban Planning: They Didn't Build LA in a Day". *IEEE Computer Graphics and Applications* 20:3 (2000) 10–16.

[3] R. C. Eberhart and P. N. Kzakevich. "Determining Physiological Effects of Using VR Equipment". In *Proceedings of the Conference on Virtual Reality and Persons with Disabilities*, pp. 47–49. Northridge, CA: Center on Disabilities, 1993.

[4] D. Fallman, A. Backman and K. Holmund. "VR in Education: An Introduction to Multisensory Constructivist Learning Environments". In *Proceedings of Conference on University Pedagogy*. Available online (http://www.informatik. umu.se/~dfallman/resources/Fallman_VRIE.rtf), 1999.

[5] M. Macedonia. "Entertainment Technology and Virtual Environments for Military Training and Education". In *Forum Futures 2001*. Available online (http://www.educause.edu/ir/library/pdf/ffp0107s.pdf). Cambridge, MA: Forum for the Future of Higher Education, 2001.

[6] J. T. Mitchell. "Can Hazard Risk be Communicated through a Virtual Experience?" *Disasters* 21:3 (1997) 258–266.

[7] S. Pieper et al. "Surgical Simulation: From Computer-Aided Design to Computer-Aided Surgery". In *Proceedings of Medicine Meets Virtual Reality*. Amsterdam: IOS Press, 1992.

[8]   G. Robillard et al. "Anxiety and Presence during VR Immersion: A Comparative Study of the Reactions of Phobic and Non-Phobic Participants in Therapeutic Virtual Environments Derived from Computer Games". *CyberPsychology & Behavior* 6:5 (2003) 467–476.

[9]   B. Röthlein. "Crashes Which Don't Leave Any Dents". *BMW Magazine* (2003) 84–87. Available online (http://www.sgi.com/pdfs/3374.pdf).

[10] F. D. Rose et al. "Transfer of Training from Virtual to Real Environments". In *Proceedings of the 2nd Euro. Conf. on Disability, Virtual Reality and Assoc. Technology*, pp. 69–75. Reading, UK: ECDVRAT and University of Reading, 1998.

[11] S. Russell et al. "Three-Dimensional CT Virtual Endoscopy in the Detection of Simulated Tumors in a Novel Phantom Bladder and Ureter Model". *Journal of Endourology* 19:2 (2005) 188–192.

[12] V. Sastry et al. "A Virtual Environment for Naval Flight Deck Operations Training". In *NATO Research and Technology Organization Meeting Proceedings*, MP-058, 2001.

[13] K. Stanney et al. "Human Factors Issues in Virtual Environments: A Review of the Literature". *Presence* 7:4 (1998) 327–351.

[14] B. Subaih et al. "A Collaborative Virtual Training Architecture for Investigating the Aftermath of Vehicle Accidents". In *Proceedings of Middle Eastern Symposium on Simulation and Modeling*, 2004.

[15] P. Weiss et al. "Virtual Reality Applications to Work". *WORK* 11:3 (1998) 277–293.

[16] B. R. Whitby. "The Virtual Sky Is not the Limit: Ethics in Virtual Reality". *Intelligent Tutoring Media* 4:1 (1993) 23–28.

[17] B. Ulicny. "Crowdbrush: Interactive Authoring of Real-Time Crowd Scenes". In *Proc. ACM SIGGRAPH/Eurographics Symposium on Computer Animation*, pp. 243–252. Aire-la-Ville, Switzerland: Eurographics Association, 2004.

4 Building a Practical VR System

A first-time user of any VR system may be surprised by how much more satisfying and engaging the sensation of reality can be when more than one of their senses is involved. The experience will also differ depending on the degree of engagement with that sense. Initially, VR systems typically look at stimulating the visual senses of the user. The most basic VR system is actually a desktop computer monitor (which may or may not have stereo-scopic capability). Here, the user of the virtual environment may only view the environment rather than feel immersed within it. If we need to deliver an immersive experience to the user then the hardware required to deliver the visualization would be significantly different than a simple desktop computer.

We think it is useful to classify VR systems by whether they are desktop or immersive, since the design decisions that must be made in order to build it, and therefore the visualization equipment it requires, will differ greatly. In this chapter, we will look at the current state of the art for both desktop and immersive visualization equipment.

Of course, whether using desktop or immersive visualization, the user also needs to be able to interact with his environment in order to feel some sense of control over it. By interaction, we are really referring to the ability to position ourselves within the virtual world in order to take some action, and if possible, sense the corresponding reaction. In the real world, humans principally interact with objects using their hands. Therefore, an important goal for any interactive VR environment is to provide devices which allow the user to manipulate virtual objects in the same way. In most tasks performed by a computer, the conventional keyboard and 2D mouse have proven to be an adequate form of input. However, the 3D dexterity that we normally expect when using our hands (along with their sense of touch), requires capabilities which go beyond that which can be achieved using these conventional forms of input.

Thus for interaction in our VR world, we need an input device which can sense the position of our hands at any location in the 3D real world. Sometimes, we might also like to sense the result of any action we undertake in the virtual world. In this case, we need to use a combined 3D input and output device. This is more commonly referred to as a haptic or force-feedback device. For these input and/or output devices to work, it is essential to be able to connect the real and virtual worlds. That is, not only do we need to be able to acquire input from basic devices such as mice and keyboards (that actively send 2D data to VR software), but for more advanced VR systems, we must also consider how a user's movement in the real world can be sensed and acted upon in the virtual world to allow for real-time immersion. That is, we need 3D user interfaces which can sense things in the real world and acquire 3D information on position and orientation of the user. The hardware associated with 3D user interfaces is such a vast subject that it forms a major part of the book by Bowman et al. [6], and this would make worthwhile reading if you are contemplating creating your own VR system. The book comprehensively examines the whole subject of 3D user interfaces. In particular, Chapter 4 gives a wide-ranging review of 3D input hardware including tracking devices, Chapter 6 introduces the reader to some novel solutions for interfacing to the virtual traveling experience and Chapter 3 describes the range of devices and technologies used to present the virtual world to the real visitor.

As we have seen in Chapter 2, for a VR system to have any added value, it must interface with as many of the senses as possible. Principally, vision (including stereopsis and high dynamic range lighting) and touch (including our sense of pushing and pulling, i.e., force), and in an abstract way, just our sense of *being there*. So whether you are building a system which will have your user working at a desktop, immersed in a computer-assisted virtual environment (*cave*[1]) or wearing a head-mounted display, there is no escaping the fact that

---

[1]The term cave has achieved some popular usage to describe a virtual environment where the walls of a room or structure that encloses or partially encloses the VR system user also acts as a display surface.

software, electronic hardware and often some mechanical engineering is going to be involved in your design.

In this chapter, we will start by introducing some commonly used input and output devices for both desktop and immersive VR systems and discuss the enabling technologies which can, at reasonable cost, provide the basic facilities to deliver the virtual experience. After that, we will look at the principles behind a range of devices that are used for tracking and sensing objects and actions in the real world. Many tracking systems tend to be expensive, and in some cases they are difficult to set up and to interface with. Often one uses what are called *middleware* library codes to act as a bridge between the hardware and the VR software applications. Middleware will be discussed separately in Chapter 17. We begin now by looking at the most significant part of any VR system, the visual element.

4.1   Technology of Visualization

The central component of any VR system, whether it be desktop or immersive, is the display. The display is driven by the graphics adapter. Graphics adapters are high-powered computing engines (this is discussed in detail in Chapter 13) which usually provide at least two outputs. These outputs can either be connected to a some form of monitor (or multiple monitors) for desktop VR or to a projection system (again with one or more projectors) for an immersive VR system.

4.1.1   Desktop VR

There are three types of desktop display technology in common use today: cathode ray tube (CRT), liquid crystal display (LCD) or plasma. Which one is best suited to a particular application will depend on a number of factors, e.g., brightness, size, resolution, field of view, refresh rate and stereopsis, which is discussed in detail in Chapter 10. We will briefly look at the construction of each of these display technologies and how that influences their use. In addition, an interesting experimental aspect of a desktop display involves the possible use of projected images formed on solid shapes that sit in front of the user. We will illustrate this in the section on desktop projection, but first we turn to the conventional displays.

Desktop CRT, LCD and Plasma Displays

Display output (the color or brightness of something) is analog in nature and is typically encoded using the red-green-blue (RGB) primary color model.

Since the display output is analog, a display interface is required to connect the analog display output with the digital processor. The most common interface in use is the video graphics array (VGA). Signals carrying the pixel intensity of the three color components from the display adapter's frame buffer drive the output device's electronics. The pixel values are sent in row order from top to bottom and left to right. At the end of each row, a horizontal synchronization pulse is added into the signal, and when the full frame buffer has been sent, a vertical synchronization pulse is inserted and the process is repeated. A typical video waveform is illustrated in Figure 4.1. The greater the number of *vertical (frame) sync* pulses sent per second, the higher the display refresh rate. To avoid an unpleasant flicker effect the frame refresh rate should be at least 50 Hz, typically 75 Hz. To make a higher resolution display, more rows have to be displayed during the same time interval and so the device's electronics and underlying display technology must be able to cope with higher frequency signals.

At these large resolutions and with some digital rendering devices, such as the DLP projectors, a new standard for display connections, called the Digital Visual Interface (DVI) [22], is set to take over. DVI provides for a completely digital connection between the graphics adapter and the display device. It sends each pixel color value as a binary number, and the same scanning order still applies (row order, top to bottom and left to right within a row.) Of course, if the display technology is analog, as most are, the digital signal will need to be converted in the output device. The DVI interface offers a number of variants, and often it carries an analog VGA-compatible signal so that a simple adapter can produce the traditional VGA signal.

The three types of display technology in common use for desktop work, each with their own benefits, are:

- *CRTs.* The cathode ray tube, ubiquitous as the display device in TVs for more than 70 years may be coming to the end of its useful life with the advent of the all-digital flat-panel system. CRTs do have an advantage over most LCD desktop displays because they have a much wider field of view, are normally brighter and (of particular relevance for VR applications) have frame refresh rates in excess of 100 Hz, and so will work with shutter type glasses for stereoscopic visualization (see Chapter 10). The CRT does a good job of accurate color representation and so may still be the display device of choice if accurate color interpretation is required. The big disadvantage of this display device is purely its bulk.

Looking at Figure 4.1, one sees that the CRT is based on an evacu-
ated glass container (Figure 4.1(a)) with a wire coil at one end and a
fine metal grid at the other. When the coil is heated, a gas of charged
particles (electrons) is generated. Applying a voltage of more than 25
kV between the coil (the cathode) and the metal grid (the anode) ac-
celerates a beam of electrons from cathode to anode. When they hit
the screen at the anode, a layer of phosphorescent material emits light.
Magnetic fields induced by coils around the tube bend the beam of
electrons. With appropriate driving currents in the field generating
coils, the electron beam scans across and down the screen. Changing
the voltage of a grid in the electron path allows fewer electrons in the
beam to reach the phosphor, and less light is generated. Figure 4.1(b)
depicts the color display's use of three phosphors to emit red, green
and blue (RGB) light. Three electron guns excite the color phosphors
to emit a mix of RGB light. A shadow mask makes sure that the only
electrons from an R/G/B gun excite the R/G/B phosphors. In Figure 4.1(c), the analog signal that drives the CRT has special pulses
which synchronize the line and frame scanning circuitry so that the
position of the beam on the screen matches the pixel value being read
from the graphics adapter's video memory.

- *LCDs.* Liquid crystal displays are fast becoming the norm for desktop use, as they are very compact. But, with the exception of a few
  specialized devices, they cannot exceed refresh rates of 75 Hz. This

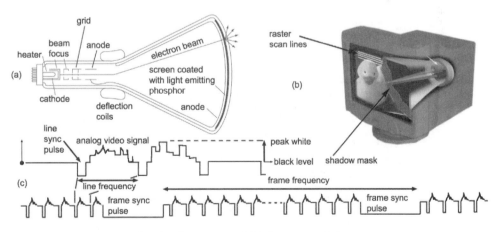

Figure 4.1. The CRT (a) layout, (b) screen and (c) driver waveform.

makes most LCDs unsuitable for stereopsis. The best technology currently available for manufacture of desktop and laptop LCD panels is the back-lit, thin film transistor, active matrix, twisted nematic liquid crystal described in Figure 4.2. The main problem with LCD film is that the color can only be accurately represented when viewed straight on. The further away you are from a perpendicular viewing angle, the more the color will appear washed out. However, now that they can be made in quite large panel sizes and their cost is dropping rapidly, several panels can be brought into very close proximity to give a good approximation of a large wide-screen desktop display.

As illustrated in Figure 4.2, the LCD screen is based on a matrix (i, j) of small cells containing a liquid crystal film. When a voltage is applied across the cell, a commensurate change in transparency lets light shine through. Color is achieved by using groups of three cells and a filter for either red, green or blue light. Each cell corresponds to image pixel (i, j) and is driven by a transistor switch to address the cells as shown in Figure 4.2(a). A voltage to give the required transparency in cell (i, j) is set on the *column data line i*. By applying another voltage to row address line j, the transistor gate opens and the column signal is applied to the LCD cell (i, j). The structure of the LCD cell is depicted in Figure 4.2(b). The back and front surfaces are coated with a 90° phased polarizing layers. This would normally block out the light. However, when no voltage is applied across the cell, the liquid crystals tend to line up with grooves in the glass that are cut parallel to the

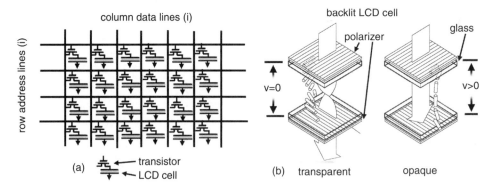

Figure 4.2. The operation of an LCD screen: (a) electrical layout, (b) physical layout.

direction of the polarizing filters, and so a twisted orientation in the liquid crystals develops across the cell. This gradual twist in crystal orientation affects the polarization of the light to such an extent that in passing from back to front, its polarization has rotated by 90° and it matches the polarizing filters, and so the light shines through. When a voltage is applied to the cell, the crystals align with the electric field, the twist effect on the polarizing light is lost, and the cell becomes opaque.

> Touch screens offer another alternative for pointer-type interaction. Their cost is not prohibitively expensive and their accuracy, whilst not down to the pixel level, can certainly locate a finger pointing to an onscreen button. The LCD or CRT surface of a touch screen is coated with either a resistive or capacitive transparent layer. In a capacitive sensor, when a finger touches the screen, the capacitance of the area around it changes, causing a small change in charge stored at that point. The (AC) electrical current flowing through the corners to accommodate this change in charge is proportional to their distance from the point of contact. Touch screens have wide use in machines that need only a small amount of user interaction, or in systems where their greater robustness compared to a keyboard or mouse is helpful. From a practical point of view, the touch screen connects to the computer via the serial or USB port, and its software driver emulates a mouse. Some models will even fit over an existing LCD panel.

- *Plasma displays.* Fluorescent display panels based on gas plasma technology have proved popular in the TV market. They have the advantage that they can deliver very large screen sizes, have a high uniform brightness from all viewing angles and allow for a higher refresh rate than most LCDs. They do not have resolution as good as a desktop LCD, they generate a lot more heat and may suffer from burn-in, so they are not normally used for desktop display. They are almost big enough to constitute a video wall with only a couple of panels. The technology underlying the plasma display panel, illustrated in Figure 4.3, is like the LCD system in that it has a matrix of cells which are addressed by applying a voltage to a row address i and column address j. Each cell contains a gas (neon or xenon), and when the voltage

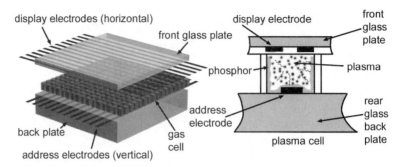

Figure 4.3. The structure of a plasma display panel and cell cross-section.

across cell (i, j) is high enough, the gas will ionize and create a plasma. Plasma is a highly energetic state of matter that gives off ultraviolet (UV) photons of light as it returns to its base state. The UV photons excite a phosphor coating on the cell walls to emit visible light. Some cells are coated with a phosphor that emits either red, green or blue light. In this way, a color image can be generated. The plasma generating reaction is initiated many times per second in each cell. Different visible light intensity in a cell is generated by changing the number of times per second that the plasma is excited.

Desktop Projection

Projection is not normally something one thinks about when considering desktop VR. But if you want something to really *stand out* then it should be. In situations where binocular stereopsis does not provide a sufficient simulation of depth perception, or you really do want to look around the back of something, projecting an image of a virtual object onto an approximately shaped surface or even onto panes of coated glass may provide an alternative. Some work on this has been done under the topic of *image-based illumination* [15]. The basic idea is illustrated in Figure 4.4, where a model (possibly clay/wood/plastic) of an approximate shape is placed on a tabletop and illuminated from three sides by projectors with appropriate images, possibly animated sequences of images. This shows surface detail on the blank shape and appears very realistic. Architectural models are a good candidate for this type of display, as are fragile or valuable museum exhibits, which can be realistically displayed on open public view in this manner. Some distortion may be introduced in the projection, but that can be corrected as illustrated in Figure 4.10.

If you are projecting onto a surface, there are a couple of issues that need to be addressed:

- *The number of projectors it takes to provide the illumination.* When projecting onto a surface that is primarily convex, good coverage will be achieved with the use of three projectors.

- *Images which are being projected may need to be distorted.* If the surface of projection is not flat and perpendicular to the direction of projection then any projected image will be distorted. The distortion may be relatively simple, such as the keystone effect, or it may be very non-linear. Whether to go to a lot of trouble to try to undistort the image is a difficult question to answer. We will examine distortion in detail shortly, but for cases where different shapes are being projected onto a generic model (for example, different textures and colors of fabric projected onto a white cloth), imperfections in alignment may not be so important.

4.1.2 Immersive VR

To get a sense of being immersed in a virtual environment, we have to leave the desktop behind, get up out of the chair and gain the freedom to move about. This poses a major set of new challenges for the designers of immersive technology. To deliver this sense of space, there are two current approaches,

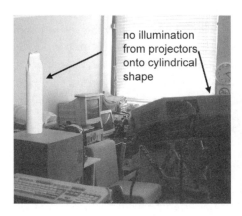

Figure 4.4. Projection of a image of a wine bottle onto a white cylinder gives the illusion of a real wine bottle being present in the room.

the cave and the head-mounted display (HMD). However, neither are perfect solutions.

Head-Mounted Displays

An HMD is essentially a device that a person can wear on her head in order to have images or video information directly displayed in front of her eyes. As such, HMDs are useful for simulating virtual environments. A typical HMD consists of two miniature display screens (positioned in front of the user's eyes) and an optical system that transmits the images from the screens to the eyes, thereby presenting a stereo view of a virtual world. A motion tracker can also continuously measure the position and orientation of the user's head in order to allow the image-generating computer to adjust the scene representation to match the user's direction of view. As a result, the viewer can look around and walk through the surrounding virtual environment. To allow this to happen, the frame refresh rate must be high enough to allow our eyes to blend together the individual frames into the illusion of motion and limit the sense of latency between movements of the head and body and regeneration of the scene.

The basic parameters that are used to describe the performance of an HMD are:

- *Field of view.* Field of view is defined as the angular size of the image as seen by the user. It is usually defined by the angular size of the diagonal of the image.

- *Resolution.* The quality of the image is determined by the quality of the optics and the optical design of the HMD, but also by the resolution of the display. The relation between the number of pixels on the display and the size of the FOV will determine how "grainy" the image appears. People with 20/20 vision are able to resolve to 1 minute of an arc. The average angle that a pixel subtends can be calculated by dividing the number of pixels by the FOV. For example, for a display with 1280 pixels in the horizontal plane, a horizontal FOV of $21.3°$ will give 1 pixel per minute of arc. If the FOV is increased beyond this, the image will appear grainy.

- *Luminance.* Luminance, or how bright the image appears, is very important for semi-transparent HMD systems which are now being used to overlay virtual data onto the user's view of the outside world. In this case, it is important that the data is bright enough to be seen over the light from the ambient scene.

- *Eye relief and exit pupil.* Eye relief is the distance of the eye from the nearest component of the HMD. This is shown in Figure 4.5. The size of the eye relief is often dependent on whether the user is required to keep his eyeglasses on, as this requires extra space between the HMD and the eye. An eye relief of 25 mm is usually accepted to be the minimum for use with eyeglasses. If the HMD is focusable such that eyeglasses do not need to be worn then the eye relief can be less. The exit pupil is the area where the eye can be placed in order to see the full display. If the eye is outside the exit pupil then the full display will not be visible. Generally, the greater the eye relief, the smaller the exit pupil will be.

An HMD can be for one or two eyes. A one-eye system is known as a monocular system, while a two-eye system is known as biocular or binocular:

- *Monocular display.* A monocular system uses one display for only one eye. This is the simplest type of system to build, as there is no requirement to match what each eye sees. This type of system generally has a limited FOV, as it is not comfortable for one eye to be scanning over a wide FOV while the other eye is not.

- *Biocular systems.* A biocular system is one in which both eyes see the same image. This can be achieved using two displays with one set of electronics to show the same image on both displays.

- *Binocular systems.* A binocular system is one in which each eye sees a different image. For this device, two displays and two sets of electronics are required. A binocular system is required for stereoscopic viewing.

Both biocular and binocular systems are more complicated than monocular systems, as they require the displays to be separated by a specific distance

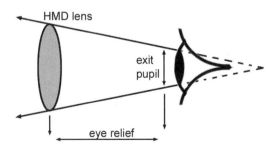

Figure 4.5. Relationship between the HMD eye relief and exit pupil size.

known as the *interpupillary distance* (IPD, the distance between the eyes). Binocular systems also require the images to be overlapped in order to produce a stereoscopic view. Today, binocular systems are more common.

At first sight, the HMD seems to offer the greatest promise, because it should be cheaper, more compact, very transportable and allows every user to be enveloped in her own little world. However, unless the HMD offers a *see-through* capability, the users can't even see their own hands. Even see-through HMDs do not provide a complete solution, because as soon as you can see through the virtual elements to see your hands, you will also be able to see other elements of the real environment such as the floor or walls. That may not be what you want—you might want the HMD wearer to believe she is standing in the middle of a football stadium. Most HMDs are also quite bulky, and because wireless links between the HMD and graphics processor are rare, the user is also constrained to within a few meters of the computer controlling the environment.

In the longer term, we have no doubt that the technical problems of miniaturization and realistic display will come to the HMD, just as accurate position sensing and photo-quality real-time rendering are with us now. This will not mean the end of the cave, for the reason alluded to above, i.e., the need to see some real-world elements (your hands, other people etc.). In the mid-term future, we have no doubt that a mix of see-through HMDs and high brightness stereoscopic enabled caves will open the door to amazing virtual/augmented/synthetic realities—but as of now, we believe the cave offers the better answer.

Computer-Assisted Virtual Environments

A computer-assisted virtual environment (a cave) is typically a surround-screen, surround-sound, projection-based VR system. The illusion of immersion is created by projecting 3D computer graphics onto the display screen, which completely surrounds the viewer.

A cave is bulky and not really transportable, but it probably gives the best approximation to the way we experience our environment. There are as many ways to design a cave as there are designers of caves. Undoubtedly the greatest constraint in the design is your budget, but of the practical constraints, the one that probably comes out on top is your space availability. Caves are big! A shower cubicle sized cave is pretty useless except for a virtual shower. Unfortunately, if you try to build a low-cost cave in your garage with back projection, the working area of your cave might indeed turn out to be no bigger than

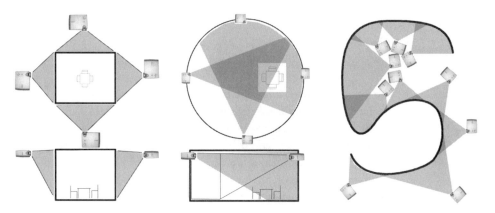

Figure 4.6. Projection geometry in a cave system.

that shower cubicle. If the cave is too small, one might as well not bother even starting to build it. Of course, one might be able to call on a little mechanical engineering and build a moving floor. Possibly something as simple as a treadmill might do the trick. Unfortunately, a treadmill is an individual thing, and making a cave for individual use somewhat negates the ability for group work, which is one of the major advantages of cave applications.

Figure 4.6 illustrates some of the key points of cave design. First amongst them is whether you use front or back projection or build the cave from video walls. An immersive cave can take on almost any shape. In Figure 4.6 (left), a back projected cubic system is depicted. Most low-cost projectors have a built-in keystone correction (as seen in the lower diagram) which allows them to be positioned near the top of the wall. In a front-projection system, such as the cylindrical configuration in Figure 4.6(center), the keystone effect helps to minimize the problem of shadow casting. A cave does not have to be either cubic or cylindrical in shape. For exhibitions or special presentations, several projectors can be linked, as in Figure 4.6 (right) to present continuous images over a complex shape. Several examples of this type of display may be found in [16].

- Front projection saves on space, and with most projectors able to throw an asymmetrical keystone projection, say from a point at roof level, anyone working in the cave will only start to cast a shadow on the display as they approach the screen.

- Back projection does not suffer from the shadow casting problem, but the construction of the walls of the cave have to be thin enough and made of appropriate materials to let the light through into the working volume. There may also be a problem of how the walls of the cave are constructed, because the supports may cast shadows.

- The video wall offers a lot of flexibility because separate tiles in the wall can carry individual displays if required. LCD and Plasma displays are getting bigger and have very small borders, so a good approximation to a whole wall in a cave could be made from as few as four displays. A system like this would have the major advantages of being very bright, occupying little space and not suffering from the shadow problem. Disadvantages include the cost and need for cooling.

The second issue involved with a cave design is the shape of the walls; there are pros and cons with this too:

- A cuboid shape is probably the most efficient use of space for a cave system, and it is relatively easy to form the necessary projections. This is especially true if the roof and floor of the room are ignored.

- A cylindrical shape for the walls of a cave has the advantage of being a seamless structure, and so any apparent corner where the walls meet is avoided. Unfortunately, a conventional projector cannot present an undistorted image over a curved screen. This is most noticeable when several projector images must be stitched together to make up the full cave. We will discuss in detail how to build and configure a semi-circular cave, which has the advantage of still giving a sense of immersion even though front projection and multiple images are stitched together.

- Spherical cave systems could have the potential of giving the most realistic projections. Most often, they are implemented for use by a single person and have an innovative projection system design that requires only one projector.

- A cave does not just have to consist of cubic or cylindrical working volume. Multiple projectors can be set up to display images on any working environment. Figure 4.6 (right) depicts an environment that might typically be used in exhibition halls and for trade shows.

Immersive Projection Technology

Unless the cave is going to be built with expensive and high resolution video walls, the only method we have to display the virtual scene to the observers is to use one or more projectors. Projection technology can vary enormously in cost and performance. For our purposes, the three most important parameters are, in order: refresh rate, brightness and resolution. Refresh rates in excess of 100 Hz are required to avoid noticeable flicker when using active stereopsis. Brightness is a fairly obvious criterion; low-cost domestic TV type projectors typically have brightness levels in the range of 500–1000 lumens. High specification projectors, typically used for public events and pop concerts, may have brightness levels of 10,000 lumens. Resolution is another fairly obvious important factor, but we rank it last because photograph/movie type output (i.e., non-text output) can still look very good even when resolutions as low as 800 × 600 are projected onto a screen that is 3 m high.

Because LCD technology does not usually offer a fast enough refresh rate, projectors built around LCDs (Figure 4.7) are not suitable for active stereopsis. However, LCD-based projectors are popular for domestic cinema use, since modest-cost LCD systems can deliver 2000 lumen brightness and subjective tests indicate that the color is more vibrant and dynamic range higher using LCDs, as opposed to the other projector technology, digital light processing (DLP).

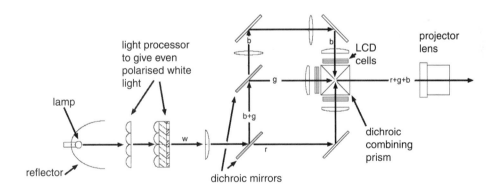

Figure 4.7. LCD projectors typically use three LCD cell matrices. Dichroic mirrors separate white light into primary colors, which pass through each of the LCDs. LCDs in projectors use active matrix cells, but they must be much smaller than the cells in a desktop LCD monitor. (*w* indicates white light, *r* red light component etc.)

DLP projectors have high refresh rates and deliver a high resolution. At the core of DLP projectors is the Texas Instruments MEMS (microelectromechanical system) chip and the digital micromirror device (DMD). Figures 4.8 and 4.9 illustrate the DMD and how DLP projectors work. The word *mirror* in DMD gives a vital clue to the idea behind the DLP. The DMD chip is an array of microscopic mirrors, one mirror for each pixel in the image. The microscopic mirrors are attached to the chip's substrate by a small hinge, and so it can be twisted into two positions. In one of these positions, it is angled in such a way that light from the projector's lamp is reflected through the projector's lens and onto the screen. In the other position, it is reflected away from the lens and no light then appears on the screen. Thus, we have a binary state for the projection of each pixel. By flicking the micromirror back and forth hundreds of times a second, intermediate shades of gray can be simulated. Color is easily added by using filters and three DMDs for the RGB primaries. For a detailed description of the structure of the DMD and a comparison of its performance against LCD and other display technology, see Hornbeck [9].

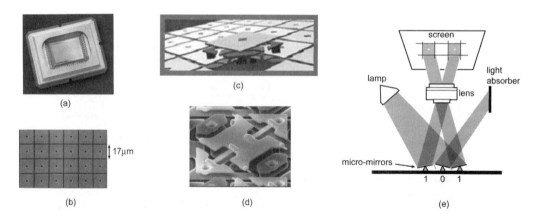

Figure 4.8. (a) The Texas Instruments DMD chip. (b) Close up of the micromirrors. (c) 3D impression of the micromirror structure. (d) Micrograph of the mechanism without mirror. (e) Principle of operation. When a pixel is on, the mirror reflects the light into the lens. When a pixel is off, the light is reflected onto a black body absorber. (Images courtesy of Texas Instruments.)

Projection Registration and Blending

If you are using a projection-based cave, it is certain that one projector will be insufficient. A cuboid environment will require at least four. It might seem that one projector per wall will do the job, but unless the aspect ratio of the wall matches that of the projector, it will be necessary to use two projectors to span the wall. Later in the book (Chapter 18), we look at practical code to allow you to pre-process images, movies and synthetic virtual 3D environments in real time using multiple projectors displaying on a half-cylindrical wall. A cylindrical display surface, such as that depicted in Figure 4.10(a), introduces an element of nonlinearity in the registration of the projected image. However, we shall see shortly that such distortions can be corrected for. This opens up the possibility of combining projected images to form a large panoramic view across non-flat display surfaces.

This brings us to the important issues of how to break up the rendered environment into pieces for projection, how to process the projection so that the pieces line up correctly without distortion and how to blend the output so that no join is visible. We will now address these one by one:

- *Slicing the display.* The easiest way to divide the output for use with multiple projectors is to use a computer video adapter with two or more outputs. The outputs can usually be configured so that one shows the

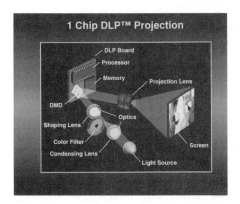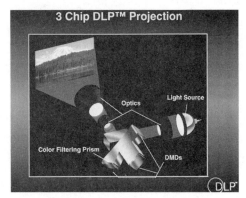

Figure 4.9. DLP projectors are built with either one or three DMD chips. In the single-chip projector, a color filter on a rotating wheel passes red, green and blue filters in the light path. In the three chip projector, a dichroic prism filters the light into primary colors, which are processed separately and reflected into the lens. (Images courtesy of Texas Instruments.)

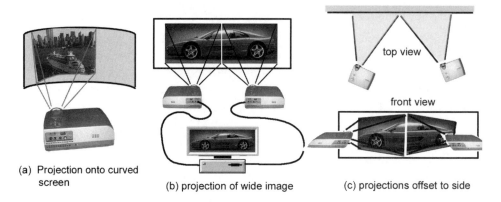

(a) Projection onto curved screen

(b) projection of wide image

(c) projections offset to side

top view

front view

Figure 4.10. (a) Projecting onto a curved surface distorts the displayed pictures. (b) To produce a wide-screen projection, a single desktop image is divided in two by a dual-output video card and presented by two projectors. (c) If the projectors are not arranged in parallel, a significant distortion will result. The application software can be changed [14] to compensate for the distortion.

left side of the display and the other the right side (Figure 4.10(b)). It is also possible to configure networked computers so that each node on the network renders one piece of projection. More programming effort is required to make sure the renderer synchronizes the views it is generating. Synchronizing the video signals sent to the projectors is vital for stereoscopic imaging, and quite a few of the stereo-ready graphics adapters offer an interface to facilitate this. In fact, the primary use of these adapters is in synchronizing rendering for ultra high resolution displays.

- *Distortion.* Unless the direction of projection is perpendicular to the screen, the display will be distorted. You are probably familiar with the idea of correcting for keystone distortion. However, if we are using two projectors as shown in Figure 4.10(c) then we will have to arrange that our software corrects for the fact that the projection is not at right angles to the screen.[2]

---

[2]Projecting two images in the way depicted in Figure 4.10(c) is often used to minimize the shadow casting effect that arises in front projection. Because the beams are directed from the side, a person standing in the middle will not cast a shadow on the screen. This is especially true if we allow for some overlap.

The problem becomes even more complex if the screen is curved [21]. In a semi-cylindrical cave, a projection will exhibit distortion, as shown in Figure 4.11(a). To overcome this, the rendering engine is programmed to negate the effect of the distortion (Figure 4.11(b)). We show how to do this in practice in Section 18.5. We can use the same idea to correct for all sorts of distortion. In a multi-projector cave, aligning adjacent images is made much easier by interactively adjusting the software anti-distortion mechanism rather than trying to mechanically point projectors in exactly the right direction (it also helps when building a portable cave).

- *Blending.* In theory, projecting an image as a set of small tiles (one tile per projector) should appear seamless, provided the distortion has been corrected. Unfortunately, in practice, there is always this annoying little gap. Often it arises because of minor imperfections in the display screen; a cylindrical shape approximated by small planar sections, for example. This annoying artefact can be removed by arranging that the tiles have small overlapped borders. Inside the border area, brightness is gradually reduced in one tile while it is increased in the other. Having to build tiles with overlapping edges complicates the projection software drivers, but as Figure 4.12 demonstrates, large panoramic images

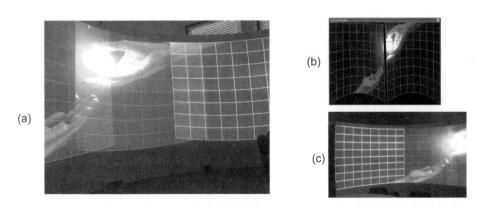

Figure 4.11. (a) A panoramic projection from three sources gives rise to nonlinear distortion on a curved screen. To correct for the distortion, the projected images are pre-processed. The image in (b) shows the changes made to the projections to correct for the distortion in (a). After pre-processing, the nonlinear distortion is greatly reduced (c). (The rectangular grid in the images is used to manually configure the pre-processor.)

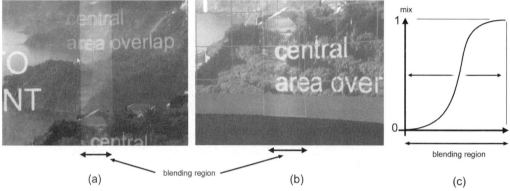

(a)   blending region   (b)   (c)

Figure 4.12. When projecting onto a non-flat surface, the edges rarely line up. Small imperfections are very noticeable to the human eye, so a better approach is to overlap the adjacent projections and gradually blend between them. In (a) we see the result of including a small overlap. By applying a gradual blend (using a sigmoid function (c)) from one projection to the other, the area of overlap becomes indistinguishable from the rest of the projection (b).

can be obtained using tiled sections that overlap and blend to make the join appear seamless.

Figure 4.13 illustrates a 180° cylindrical stereoscopic panoramic display built using low-cost components. It is driven by a pair of PCs, each sending

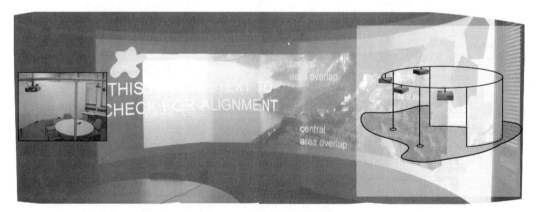

Figure 4.13. A semi-immersive wide-screen VR environment. Four low-cost DLP projectors are driven by two networked computers, each with dual video outputs. The system uses custom display software and can be interactively configured to correct for distortion.

video output to two projectors. The graphics processors are fast enough to run the distortion-correction software without the need for any special hardware.

4.2   Technology for Interaction

Once the basic decision has been made about the type of display technology that will be utilized for the virtual environment, you also have to decide how to include interaction with the other senses:

- The sound system is relatively easy to provide and control. A number of software libraries, for example Microsoft's DirectSound [13], allow application programs to easily render 3D soundscapes which match the graphics.

- We will see in Chapter 10 that the sense of stereopsis adds greatly to the perception of reality. The commonly used method of providing this effect is by using infrared (IR) controlled shutter eyewear. This technology requires the IR signal to be received by the eyewear at all times, and so the cave designer must install a sufficient number of emitters to allow all the users of the cave to pick up the signal.

- Head-mounted display VR systems *must* employ motion tracking. Without it, the correct viewpoint and orientation cannot be generated. In a cave, it might at first sight not seem so important to know where the observers are. Nevertheless, the global component (the cave) and the personal component (the head-mounted display) are likely to merge in the future and a good cave designer should recognize this and include some position-sensing technology. One way to do this is by including an ultrasonic emitter into the roof and walls of the cave, as described in Section 4.3.

- In the introduction to this chapter, we discussed the importance of user interaction through input or combined input/output devices. Within the desktop, you have the luxury of working within a relatively small working volume. In a cave, this is not the case, and as such, natural interaction is much more difficult to achieve.

We feel that touch is next most important sense (after vision) for a user to engage with whilst within the virtual environment. As such, the remaining sections will describe the current state of the art in input devices which allow us to manipulate objects within the virtual environment and output devices that allow us to sense the reaction to any changes we made. That is, they stimulate our sense of touch. These are commonly referred to as *haptic devices,* and before we describe both the input and haptic output devices, we thought we might give you a crash course in the theory of haptics.

4.2.1   Haptics

In Section 2.3, we discovered that the human haptic system has an important role to play in human interaction with VR. But we have also seen that whilst haptic technology promises much, it comes with many unresolved problems and complexities. State of the art haptic interfaces are still rather crude.

However, it is a rapidly developing topic with new ideas emerging regularly from research labs. Srinivasan and Basdagon [18] and Salisbury et al. [17] provide interesting general details of how haptic systems are classified and the challenges of trying to build haptic hardware interfaces. In designing a VR system, and especially a cave-type system, we would ideally like our haptic device to be able to work at long range, not get in the way, and offer the ability to appear to pick up a virtual object and get a sense of its weight. By extrapolation, if we could do that, most of what we would want to do with haptics would be possible.

Following [18], a useful way to describe a haptic device is in the way it is attached to a fixed location.

- *Floating* devices are things such as gloves which can perform inter-digit tasks. They can also measure finger contacts and finger-specific resistance. They cannot however measure or reproduce absolute weight or the inertial effects of a virtual object. They may only be attached to their base location via wireless link.

- *Exoskeleton* devices, typically worn on a hand, arm or leg, may have motorized devices that can resist certain motions and restrict the number of degrees of freedom. For example, such a device may prevent you from closing your hand too tightly around a virtual ball as you attempt to pick it up. This type of haptic device, like the first, does not allow you to experience a weight effect.

- *Grounded* devices behave rather like the arm of a small robot. They can be powerful and simulate multi-axial forces, including everything from a solid wall to picking up a virtual object and assessing its weight. They have the disadvantage of being bulky and very restricted in the volume in which they can move.

In the real world, we can make quite a range of movements, termed *degrees of freedom (DOF)*. In three dimensions, six DOF provides a complete specification of position and orientation. As such, a haptic device need only specify a force along three mutually perpendicular axes, but that force could depend on position and angle at the point of contact. Many haptic devices, such as the force-feedback joysticks used in computer games, can only simulate a force along two axes and at a single point of contact (since the joystick does not move).

An ideal haptic device has to have some formidable properties: low inertia and minimal friction, with device kinematics organized so that free motion can be simulated and have the ability to simulate inertia, friction and stiffness. These devices can achieve tactile stimulation in a number of different ways:

- Mechanical springs activated by solenoid, piezoelectric crystal and shape memory alloy technologies.

- Vibrations from voice coils.

- Pressure from pneumatic systems.

- Heat-pump systems.

However, these are all still technologies under active research, and most haptic feedback is offered via joystick or data gloves. In addition, a haptic device is obviously useless if it does not have haptic-enabled VR software controlling it. When developing this type of software, it is helpful to imagine haptics operating in an analogous way to the 3D graphics with which we are familiar. The term *haptic rendering* says it all. One specifies a haptic surface; it could consist of the same shapes and even be specified in the same way with planar polygons that the 3D renderer uses to visualize the shapes for us. Continuing with the analogy, we can texture the surface by describing its rigidity, friction etc. just as we do in terms of color, reflectivity etc. for the visuals.

Figure 4.14 is a block outline of the structure of the haptic-rendering algorithm. The gray box shows the main elements in the haptic-rendering loop. The device sends information about the position of its point of contact.

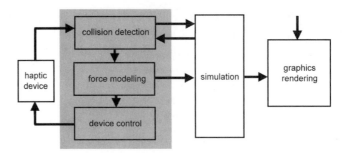

Figure 4.14. Haptic rendering.

The collision-detection block uses information from the simulation software about the shape of the haptic and visual surfaces to determine whether the virtual position of the haptic device has bumped into the felt surface. The force-response block determines an appropriate force depending on the state of the collision (hard, soft, deformable etc). It sends this signal to the block that controls the electronics of the haptic device to reproduce this force and also to the simulation engine so that changes to the shape of the felt surface can be made if necessary.

In Chapter 17, we shall look at how an application program might implement the details of these blocks. But now we will quickly summarize what each block has to do:

- The *collision-detection* block does exactly what its name implies. It detects when and where the *virtual representation of the real world point of contact* collides with the rendered haptic surface. It may also reveal other information that helps us simulate deformable objects such as the depth of penetration through the haptic surface.

- The *force-response* block calculates the force that approximates as closely as possible the real force that the avatar would experience in the collision. As stipulated in Chapter 3, an avatar is a computer generated graphical representation of the human user within the virtual environment.

- The *control* section translates the force felt by the avatar into signals to drive the haptic hardware. Because haptic devices are limited in the force they can deliver, it may not be possible to exactly simulate the force one would feel if the object had been real and we had bumped into it.

- In the *simulation* block, the effect of the haptic exchange is applied to the data that describes the haptic surface and possibly the visualization data, which might be represented by the same geometry.

Undoubtedly, haptics is going to be a major focus for R&D in the future. Most indications suggest that a lot of current research is focused on small-scale haptic topics, for example tactile perception of deformable objects and simulation of realistic haptic textures. There seems little prospect of large-scale haptic rendering such as we would like to have in a cave environment. It seems that some kind of a breakthrough is needed, because we cannot clutter up the real working space with a collection of robot arms, such as would be needed to really get to grips with objects in the parallel *virtual* universe.

After this brief aside, we now return to consider the input and haptic output devices used for both desktop and immersive VR work.

4.2.2   Desktop Interaction

In a desktop VR environment, the user may only require some basic interaction with the VR world, such as 3D navigation. Devices such as a trackball or a 3D joystick can offer this facility. Other forms of input such as light pens and drawing tablets may be only be marginally relevant for VR work. Figure 4.15 illustrates examples of some of these devices.

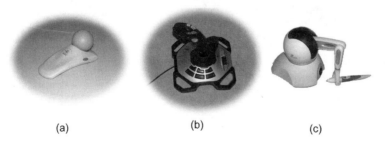

(a)   (b)   (c)

Figure 4.15. A wide range of desktop devices extend the range of input possibilities. A spaceball (a) offers the possibility to move in three dimensions. The joystick is a low-cost device with two or three degrees of freedom. Some degree of haptic feedback can be achieved with basic joysticks (b). To give a comprehensive haptic response requires a device such as SensAble's Phantom (c).

However, if the user of the desktop system also needs to experience feedback then a haptic output device is required. Most of the latest technological advances of these devices has been as a direct result of the huge commercial impact of computer games. What one would class as *cheap* force-feedback devices (joysticks, mice, steering consoles, flight simulator yokes) are in high-volume production and readily available. If building your own VR system, take advantage of these devices, even to the extent of tailoring your software development to match such devices. Microsoft's DirectInput application programming interface (API) software library was specifically designed with game programmers in mind, so it contains appropriate interfaces to the software drivers of most input devices. We provide a number of input device programming examples in Chapter 17.

There is plenty of scope for the development of custom-built haptic output devices. Flexible pressure sensors, custom motorized articulated linkages and a host of novel devices might be called on by the imaginative person to research possible new ways of interaction.

4.2.3   Immersive Interaction

Within the small working volume of the desktop, it is possible to provide a wide range of input devices and even deliver a believable sensation of force feedback. On the larger scale of a cave, where freedom of movement and

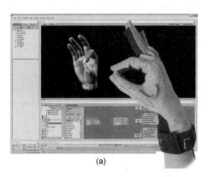

(a)   (b)

Figure 4.16. Sensing the position of individual fingers relative to one another is a complex task. (a) Immersion Corporation's Cyberglove provides joint-angle data via a wireless link. (Reproduced by permission of Immersion Corporation, Copyright © 2007, Immersion Corporation. All rights reserved.) (b) The Fakespace Pinch Glove has sensors in the fingertips to detect contact between two or more fingers. (Reproduced by permission of Fakespace Systems, Copyright © 2007, Fakespace Systems.)

lack of encumbrance with awkward prosthetics is desirable, it is much harder to acquire input, and virtually impossible to provide comprehensive force feedback.

Sometimes input may be in terms of spoken instructions, or by pointing at something or appearing to touch a particular 3D coordinate. To do this, the VR system needs to be equipped with position detection equipment, which we discuss in the next section. The cave user may be equipped with a special pointing device or a glove which can not only have its location detected, but also when it is combined with a stereoscopic display, it can give a sensation of touching something (some typical devices in this category are illustrated in Figure 4.16). These devices do allow the user to be able to gain the illusion of picking up a virtual object and detecting whether it is soft or hard. But they still do not let you sense whether it is heavy or light.

4.3   Technology for Motion Tracking and Capture

The final piece in the jigsaw of immersive VR technology concerns how position and orientation can be acquired from the real world, relative to a real world frame of reference, and then matched to the virtual world, which has its own coordinate system.

With the present state of hardware development, there is still no all-embracing method of sensing everything we would like to be able to sense within a real-world space. Some currently developed systems work better on the small scale, some work better on a large scale, some work more accurately than others and some sense things others can miss. However, there is usually a trade off between one desirable feature and another, and so how you choose to acquire your real-world data will depend on the application you are working on. One thing is certain: you can make a much better choice of which technology to use if you know something of how they work, and what advantages and disadvantages they have. We hope to give you some of this knowledge within this section.

Following on with the broad idea of real world sensing, we can identity three themes: motion capture, eye tracking and motion tracking. Although really they are just the same thing—*find the position of a point or points of interest in a given volume*—the present state of hardware development cannot provide a single solution to address them at the same time. Indeed, we have broken them down into these themes *because* a single piece of hardware tends to work well for only one of them.

Motion capture is perhaps the most ambitious technology, given that it attempts to follow the movement of multiple points across the surface of a whole object. Optical or mechanical systems can be used to gather the position of various points on a body or model as it is moved within a given space within a given timescale. Later, these template actions can be utilized in all sorts of different contexts to construct an animated movie of synthetic characters who have, hopefully, realistic behavior. This is especially important when trying to animate body movements in a natural way. Since VR systems are more concerned with real-time interactivity, the emphasis shifts slightly with the focus directed more towards *motion tracking*. It is difficult to be precise about when motion capture differs from motion tracking and vice versa. However, some examples include:

- *Continuous operation.* Often motion-capture systems need downtime for recalibration after short periods of operation.

- *Number of points sensed.* Motion tracking may need only one or two points, whereas motion capture may require tracking 20 points.

- *Accuracy.* A motion-tracking system for a haptic glove may need submillimeter accuracy; a full-body motion-capture system may only need accuracy to within a couple centimeters.

We shall consider motion capture to be the acquisition of moving points that are to be used for post-processing of an animated sequence, whilst motion tracking shall be considered the real-time tracking of moving points for real-time analysis. So, restricting the rest of our discussion to systems for motion tracking, we will look at the alternative technologies to obtain measures of position and orientation relative to a real-world coordinate system. It will be important to assess the accuracy, reliability, rapidity of acquisition and delay in acquisition. Rapidity and delay of acquisition are very important because, if the virtual elements are mixed with delayed or out-of-position real elements, the VR system user will be disturbed by it. If it does not actually make users feel sick, it will certainly increase their level of frustration.

Currently, there are five major methods of motion tracking. However, their common feature is that they all work either by triangulation or by measuring movement from an initial reference point, or perhaps best of all, by a combination of the two.

4.3.1 Inertial Tracking

Accelerometers are used to determine position and gyroscopes to give orientation. They are typically arranged in triples along orthogonal axes, as shown in Figure 4.17(a). An accelerometer actually measures force F; acceleration is determined using a known mass m and Newton's second law. The position is updated by $p_{new} = p_{old} + \int\int \frac{F}{m}dtdt$. Mechanically, an accelerometer is a spring-mass scale with the effect of gravity removed. When an acceleration takes place, it drags the mass away from its rest position so that $F = kx$, where x is the displacement and k is a constant based on the characteristic of the system. In practice, springs are far too big and subject to unwanted influences, such as shocks, and so electronic devices such as piezoelectric crystals are used to produce an electric charge that is proportional to the applied force.

A spinning gyroscope (gyro) has the interesting property that if you try to turn it by applying a force to one end, it will actually try to turn in a different direction, as illustrated in Figure 4.17(b). A simple spinning gyro does not fall under the action of gravity; it feels a force that makes it rotate about (precess) the vertical axis y. If an attempt is made to turn the gyro about the z-axis

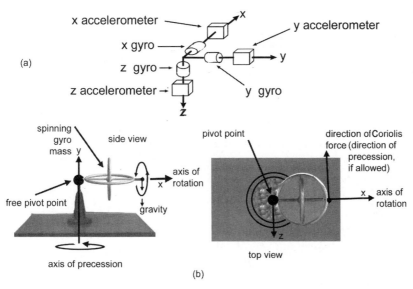

Figure 4.17. An inertial tracker uses three gyros and three accelerometers to sense position and orientation.

(coming out of the page, in the side view) by pushing the gyro down, the result is an increased Coriolis force tending to make the gyro precess faster. This extra force $\delta\mathbf{f}$ can be measured in the same way as the linear acceleration force; it is related to the angular velocity $\omega = f(\delta\mathbf{f})$ of the gyro around the z-axis. Integrating ω once gives the orientation about z relative to its starting direction.

If a gyro is mounted in a force-sensing frame and the frame is rotated in a direction that is *not* parallel with the axis of the gyro's spin, the force trying to turn the gyro will be proportional to the angular velocity. Integrating the angular velocity will give a change in orientation, and thus if we mount three gyros along mutually orthogonal axes, we can determine the angular orientation of anything carrying the gyros. In practice, traditional gyros, as depicted in Figure 4.17(b), are just too big for motion tracking.

A clever alternative is to replace the spinning disc with a vibrating device that resembles a musical tuning fork. The vibrating fork is fabricated on a microminiature scale using microelectromechanical system (MEMS) technology. It works because the in-out vibrations of the ends of the forks will be affected by the same gyroscopic Coriolis force evident in a rotating gyro whenever the fork is rotated around its base, as illustrated in Figure 4.18(a). If the fork is rotated about its axis, the prongs will experience a force pushing them to vibrate perpendicular to the plane of the fork. The amplitude of this out-of-plane vibration is proportional to the input angular rate, and it is sensed by capacitive or inductive or piezoelectric means to measure the angular rate.

The prongs of the tuning fork are driven by an electrostatic, electromagnetic or piezoelectric force to oscillate in the plane of the fork. This generates an additional force on the end of the fork $\mathbf{F} = \omega \times \mathbf{v}$, which occurs at right angles to the direction of vibration and is directly related to the angular velocity ω with which the fork is being turned and the vector $\mathbf{v}$, representing the excited oscillation. By measuring the force and then integrating it, the orientation can be obtained.

Together, three accelerometers and three gyros give the six measurements we need from the real world in order to map it to the virtual one. That is, (x, y, z) and *roll, pitch* and *yaw*. A mathematical formulation of the theory of using gyros to measure orientation is given by Foxlin [8].

The requirement to integrate the signal in order to obtain the position and orientation measures is the main source of error in inertial tracking. It tends to cause drift unless the sensors are calibrated periodically. It works very well over short periods of time and can work at high frequencies (that is, it can report position often and so can cope well in fast changing environments). Inertial tracking is best used in conjunction with one of the other methods, because of the problem of initialization and the increases in error over time without recalibration. Other problems with this method are electrical noise at low frequencies, which can give the illusion of movement where there is none, and a misalignment of the gravity correction vector. Inertial tracking has the potential to work over very large volumes, inside buildings or in open spaces.

4.3.2 Magnetic Tracking

Magnetic tracking gives the absolute position of one or more sensors. It can be used to measure range and orientation. There are two types of magnetic-tracking technologies. They use either low-frequency AC fields or pulsed DC

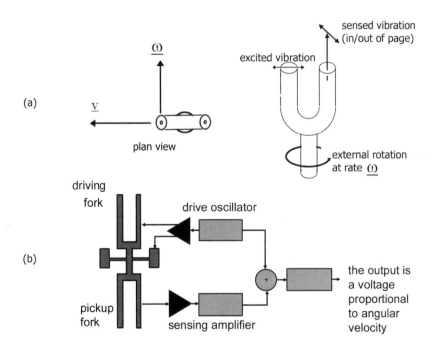

Figure 4.18. (a) A tuning-fork gyro. (b) The fork is fabricated using MEMS technology in the sensing system used in the BEI Systron Donner Inertial Divisions GyroChip technology.

fields which get around some of the difficulties that AC fields have when the environment contains conducting materials. Magnetic trackers consist of a transmitter and a receiver that both consist of three coils arranged orthogonally. The transmitter excites each of its three coils in sequence, and the induced currents in each of the receiving coils are measured continuously, so any one measurement for position and orientation consists of nine values.

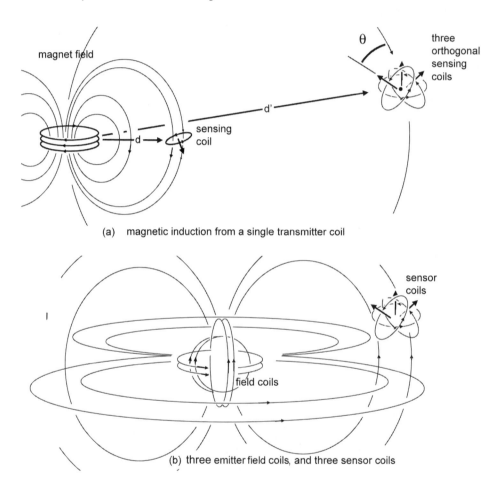

(a)   magnetic induction from a single transmitter coil

(b) three emitter field coils, and three sensor coils

Figure 4.19. (a) Magnetic induction in one sensing coil depends on the distance d or d' from the field generating coil and the angle it makes with the field ϑ. Three coils can sense fields in mutually perpendicular directions. (b) Three emitter coils and three sensing coils allow the six degrees of freedom (position (x, y, z), and orientation, pitch, roll and yaw) to be determined.

The signal strength at each receiver coil falls off with the cube of the distance from transmitter and the cosine of the angle between its axis and the direction of the local magnetic field (see Figure 4.19).

The strength of the induced signal as measured in the three receiving coils can be compared to the known strength of the transmitted signal to calculate distance. By comparing the three induced signal strengths amongst themselves, the orientation of the receiver may be determined.

Magnetic tracking sensors can feed their signals back to the data acquisition hardware and computer interface via either a wired or a wireless link. A typical configuration will use one transmitter and up to 10 sensors.

The main disadvantages of magnetic motion tracking are the problems caused by the high-strength magnetic field. Distortions can occur due to a lot of ferromagnetic material in the environment, which will alter the field and consequently result in inaccurate distances and orientations being determined. The device also tends to give less accurate readings the further the sensors are away from the transmitter, and for sub-millimeter accuracy, a range of 2–3 m may be as big as can be practically used.

4.3.3   Acoustic Tracking

Acoustic tracking uses triangulation and ultrasonic sound waves, at a frequency of about 40 kHz, to sense range. In an acoustic system, pulses of sound from at least three, and often many more, emitters placed at different locations in the tracked volume are picked up by microphones at the point being tracked. The time it takes the sound wave to travel from the source to microphone is measured and the distance calculated. By using several pulses from several emitters, the position can be determined (see Figure 4.20). If three microphones are placed at the location being tracked then the orientation can also be worked out. An alternative to this *time of flight* range determination is to measure the *phase difference* between the signals arriving from different sensors when they all emit the same ultrasonic pulse at the same time. This gives an indication of the distance moved by the target during the pulse interval. Because the speed of sound in air depends on temperature and air pressure, the equipment has to be calibrated before use or else have a built-in mechanism to sense and account for temperature and pressure.

Using the time of flight of sound waves to measure distance suffers from the problem of echoes being falsely identified as signals. It is also possible for

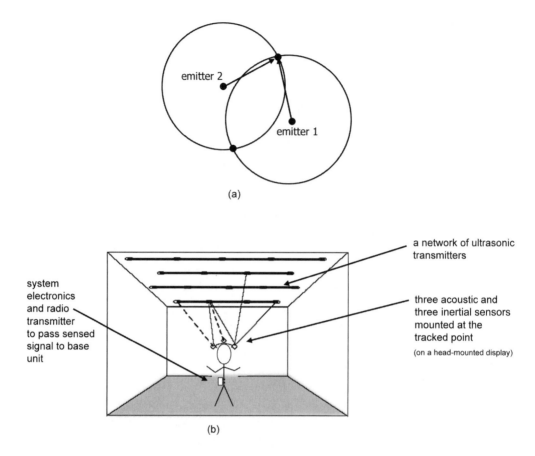

Figure 4.20. (a) Acoustic tracking takes place by triangulation. Two emitters are shown and give rise to a set of possible locations for the sensor lying on the circle of intersection between the spheres whose radius depends on time of flight and speed of sound in air. A third emitter will differentiate between these possibilities. (b) A sophisticated system has three microphone sensors mounted together, and they receive signals from an array of ultrasonic emitters or transponders.

the microphones to be obscured by objects between source and destination, so the sound pulses might not be heard at all. However, the biggest drawback with acoustic tracking is the slow speed of the sound pulse. At 0° C, the speed of sound is about 331 m/s, and thus in a typical working volume it could take 10–20 ms for a single pulse to be sent and received. With the need for multiple pulses to be checked before a valid position/orientation can be sent to the computer, it may only be possible to sample positions and orientations 10–20 times a second. This may not be fast enough to track rapid movements accurately enough.

4.3.4   Optical Tracking

One method of optical tracking works by recognizing highly visible markers placed on the objects to be tracked. If three or more cameras are used, the marker's 3D position can be obtained by simple triangulation and the known properties of the cameras. However, to achieve greater accuracy and more robust position determination, techniques such as those discussed in Chapter 8 will need to be employed. Optical tracking of point sources cannot provide orientation information, and it suffers badly from the light sources being obscured by other parts of the scene. Overcoming such difficulties might require several cameras. To obtain orientation information is more difficult. Some methods that have been tried include: putting several differently shaped markers at each tracked point, or arranging for a recognizable pattern or lights to be generated around the tracked point.

Many novel optical tracking ideas have been proposed. For example, sensors on the tracked point may be excited when a laser beam which is continually scanning the scene strikes them (like the electron beam on a TV or a bar-code reader in the supermarket). Laser ranging can give the distance to the sensor. Three scanners will determine a position either by triangulation or simply distance measures. Three detectors per sensed point will allow orientation information to be obtained. Because the speed of light is so much faster than the speed of sound in air, an optical tracking system does not suffer any of the acoustic delay.

As we shall see in Chapter 8, the science of computer vision and pattern recognition offers a reliable way of tracking visually recognizable markers in video images, which can be acquired from a network of cameras overlooking the working volume. This idea comes into its own when we wish to add virtual objects to the real-world view so that they appear to be part of the scene, as happens in augmented reality. This form of optical tracking can be used in practice by extending the open source augmented reality toolkit, the ARToolKit [5]. For example, the markers that are attached to the objects shown in the three camera views of Figure 4.21 can be tracked and located very accurately if the locations of the cameras are known and they have been calibrated. Camera calibration is discussed in Section 8.2.2, and a tracking project is described in Chapter 18.

side camera

front camera

top camera

Figure 4.21. By placing markers (the letters A, C and D) on the objects to be tracked, it is possible to determine their locations within the scene and track their movement. Using multiple camera views allows the tracking to continue so long as each marker is visible in at least one camera's field of view. (Scene courtesy of Dr. B. M. Armstrong and Mr. D. Moore.)

4.3.5   Mechanical Tracking

Mechanical tracking determines the position of a tracked point by connecting it to a reference point via an articulated linkage of some kind. Haptic devices used to give force feedback in a limited desktop working volume are going to provide, as a side effect, accurate tracking of the end effector. Trackers of this kind measure joint angles and distances between joints. Once this information is known, it is relatively simple to determine the position of the end point. Trackers of this type have been used to capture whole-body motions, and body suits which measure the positions of all major joints have been constructed. Other mechanical devices fit around fingers and hands to acquire dextrous and small-scale movement, for example in remote surgery and maintenance work in harsh environments. Figure 4.22 illustrates such mechanical tracking systems.

Figure 4.22. The Metamotion Gypsy 5 is a whole-body motion-capture system using articulated linkages to acquire orientation information. (The images are provided courtesy of Anamazoo.com).

4.3.6 Location Tracking

In Section 4.3.3, it was pointed out that the theory of acoustic tracking relied on the time of flight of a sound pulse between an emitter and sensor. The same concept applies to electromagnetic (EM) waves, but because these travel at the speed of light, it is necessary to determine the time of flight much more accurately. However, using EM waves in the microwave frequency band allows tracking devices to operate over larger distances. Because these systems are a little more inaccurate than those based on hybrid inertial/acoustic methods, we prefer to think of them as location trackers.

Perhaps one of the best-known radio-based tracking and navigation systems is the global positioning system, ubiquitously known as GPS. The GPS (and Galileo, the European equivalent) is based on a system of Earth-orbiting satellites [11] and may offer a useful alternative to the more limited range tracking systems. It has the advantage of being relatively cheap and easy to interface to handheld computers, but until the accuracy is reliably in the sub-millimeter range, its use remains a hypothetical question.

Back on the ground, Ubisense's Smart Space [19] location system uses short-pulse radio technology to locate people to an accuracy of 15 cm in three dimensions and in real time. The system does not suffer from the drawbacks of conventional radio-frequency trackers, which suffer from multipath reflections that might lead to errors of several meters. In Smart Space, the objects/people being tracked carry a UbiTag. This emits pulses and communicates with UbiSensors that detect the pulses and are placed around and within the typical coverage area (usually 400 m^2) being sensed. By using two different algorithms—one to measure the difference in time of arrival of the pulses at the sensors and the other to detect the angle of arrival of the pulses at the sensor—it is possible to detect positions with only two sensors.

The advantage of this system is that the short pulse duration makes it easier to determine which are the direct signals and which arrive as a result of echoes. The fact that the signals pass readily through walls reduces the infrastructure overhead.

Another ground-based location tracking system that works both inside buildings and in the open air is ABATEC's Local Position Measurement (LPM) system [1]. This uses microwave radio frequencies ($\sim$5–6 GHz) emitted from a group of base stations that can determine the location of up to $\sim$16,000 small transponders at a rate of more than 1000 times per second and with an accuracy of 5 cm. The base stations are connected via optical fiber links to a hub that interfaces to a standard Linux PC.

These location-tracking systems might not be the best option for tracking delicate movement or orientation, e.g., for measuring coordination between hand and eye or walking round a small room or laboratory. However, there are many VR applications such as depicting the deployment of employees in an industrial plant, health-care environment or military training exercises to which this technology is ideally suited.

4.3.7   Hybrid Tracking

Hybrid tracking systems offer one of the best options for easy-to-use, simple-to-configure, accurate and reliable motion-tracking systems. There are many possible combinations, but one that works well involves combining inertial and acoustic tracking. The inertial tracking component provides very rapid motion sensing, whilst acoustic tracking provides an accurate mechanism for initializing and calibrating the inertial system. If some of the acoustic pulses are missed, the system can still continue sending position information from the inertial sensor. In the system proposed by Foxlin et al. [7], a head-mounted tracking system (HMTS) carries an inertial sensor calibrated from an acoustic system. The acoustic system consists of a network of transponders placed around the boundaries of the tracing volume (typically a large room). The HMTS has a light emitter which sends a coded signal to trigger ultrasonic pulses from the transponders one at a time. The time of flight for the sound pulse from transponder to acoustic receptor (microphone) gives its distance. With three or four received ultrasonic pulses, the position of the sensor can be determined. Having eight or more transponders allows for missed activations, due to the light emitter not being visible to the transponder or the response not being heard by the sound sensor. The system has many other refinements, such as using multiple groups of transponders and applying sophisticated signal processing (Kalman filtering) algorithms to the ultrasound responses to eliminate noise and echoes. Figure 4.20(b) illustrates the general idea. The advantages of this system are its robustness due to transponder redundancy, noise and error elimination due to transponders being activated by the sensor itself and the electronic signal processing of the ultrasonic response.

4.3.8   Commercial Systems

Successful commercial systems developed in recent years have used optical, magnetic or combined acoustic and inertial tracking. We have seen Ascension Technology's Flock of Birds magnetic tracking system [2] used in conjunction with the human modeling and simulation package Jack [20] to make accurate measurements of torque and stress forces on people operating equipment. Its scanning range is typically up to 3 m. laserBIRD [3], also from Ascension Technology, is another successful product using an optical scanning approach. It is immune to distortion, has a wide area coverage and is typically used for head tracking up to a distance of 1.8 m. Using a combination of inertial and acoustic techniques, InterSense's IS-900 [10] and associated products can operate over very large ranges, typically up to 18 m.

For motion-capture work, Ascension Technology's MotionStar system [4] uses the same magnetic approach as the Flock of Birds and can sense from up to 108 points on 18 different performers. Meta Motion's Gypsy [12] is a mechanical tracking system that does not suffer from any occlusion problems and has a moderate cost.

There are many other commercial products for motion tracking, and an even greater number of custom systems in use in research labs. The few that we highlight here are ones that we have seen in use or used ourselves and know that they live up to the claims. For illustrative purposes, Figure 4.23 gives a sense of their physical form.

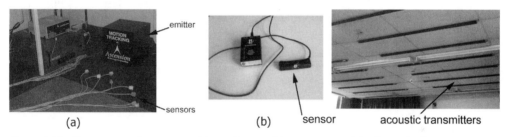

Figure 4.23. Some example commercial motion-tracking systems. (a) Ascension's Flock of Birds (magnetic tracking). (b) InterSense's IS-900 acoustic tracking system (Photographs courtesy of Dr. Cathy Craig, Queen's University Belfast).

4.4 Summary

In this chapter, we examined the current state of the technology used to build VR systems. It seems that VR on the desktop, for applications such as engineering design and the creative arts, is flourishing and providing a valuable and substantially complete service. On the other hand, any system that attempts to free the virtual visitor from her seat or desktop still has a long way to go, especially in the area of interacting with the sense of touch.

In any practical VR system, the ability to sense where things are and what they are doing in the real world is a significant problem. It requires quite complex mechanical and electronic hardware and often sophisticated software processing, too. For example, to avoid delays in obtaining position measurements, a system will often try to predict a real-world movement being tracked. The software will in many circumstances also apply signal processing techniques so that errors and missing data do not cause significant problems to VR application programs making use of real-time, real-world tracking as an input.

In the future, it is likely that display technology will continue to improve. For example, new flexible LCD-like displays may allow us to cover the surface of a cave with exceptionally bright displays that don't need space-consuming back projection or shadow-casting front projection, either. Indeed, there is no reason why immersive VR has to rely on HMDs or cave-wall projections alone. HMDs with a see-through facility cannot paint a background around the real elements in a scene, whilst in a cave, it is not possible to put 3D objects right up close to the user. However, a system involving a wireless lightweight see-through HMD worn in a stereoscopic cave is just around the corner. Most of the human interface hardware and projection equipment is already available, and any PC is powerful enough to run the software.

It is perhaps in the area of large-scale interaction with our senses of touch and inertial movement where the significant challenges lie. Thus, technically adequate hardware is still needed to realize the full potential of VR.

Bibliography

[1] ABATEC Electronic AG. *Local Position Measurement.* http://www.lpm-world.com/.

[2] Ascension Technology. *Flock of Birds.* http://www.ascension-tech.com/products/flockofbirds.php.

[3] Ascension Technology. *laserBIRD.* http://www.ascension-tech.com/products/laserbird.php.

[4] Ascension Technology. *MotionStar.* http://www.ascension-tech.com/products/motionstar.php.

[5] *ARToolKit.* http://www.hitl.washington.edu/artoolkit/.

[6] D. A. Bowman, E. Kruijffm, J. J. LaViola and I. Poupyrev. *3D User Interfaces: Theory and Practice.* Boston, MA: Addison Wesley, 2005.

[7] E. Foxlin et al. "Constellation: A Wide-Range Wireless Motion Tracking System for Augmented Reality and Virtual Reality Set Applications". In *Proceedings of SIGGRAPH 98, Computer Graphics Proceedings, Annual Conference Series*, edited by M. Cohen, pp. 371–378. Reading, MA: Addison Wesley, 1998.

[8] E. Foxlin et al. "Miniature 6-DOF Inertial System for Tracking HMDs". In *Proceedings of SPIE Vol. 3362, Helmet and Head-Mounted Displays, AeroSense 98.* Bellingham, WA: SPIE, 1998.

[9] L. J. Hornbeck. "From Cathode Rays to Digital Micromirrors: A History of Electronic Projection Display Technology". *TI Technical Journal* 15:3 (1998) 7–46.

[10] InterSense Corporation. *IS-900 System.* http://www.intersense.com/products. aspx?id=45&.

[11] E. Kaplan and C. Hegarty (editors). *Understanding GPS: Principles and Applications*, Second Edition. Norwood, MA: Artech House, 2005.

[12] Meta Motion. *Gypsy 4/5 Motion Capture System.* http://www.metamotion.com/ gypsy/gypsy-motion-capture-system.htm.

[13] Microsoft Corporation. *DirectSound.* http://www.microsoft.com/windows/ directx/.

[14] R. Raskar. "Immersive Planar Display Using Roughly Aligned Projectors". In *Proceedings of the IEEE Virtual Reality Conference*, p. 109. Washington, D.C.: IEEE Computer Society, 2000.

[15] R. Raskar. "Projectors: Advanced Geometric Issues in Applications". SIGGRAPH Course. ACM SIGGRAPH, 2003.

[16] R. Raskar et. al. "Multi-Projector Displays Using Camera-Based Registration". In *Proceedings of the IEEE Conference on Visualization*, pp. 161–168. Los Alamitos, CA: IEEE Computer Society Press, 1999.

[17] K. Salisbury et al. "Haptic Rendering: Introductory Concepts". *IEEE Computer Graphics and Applications* 24:2 (2004) 24–32.

[18] M. A. Srinivasan and C. Basdogan. "Haptics in Virtual Environments: Taxonomy, Research Status, and Challenges". *Computers and Graphics* 21:4 (1997) 393–404.

[19] Ubisense. *Smart Space.* http://www.ubisense.net.

[20] UGS *Jack.* http://www.ugs.com/products/tecnomatix/human_performance/ jack/.

[21] J. van Baar, T. Willwacher, S. Rao and R. Raskar "Seamless Multi-Projector Display on Curved Screens". In *Proceedings of the Workshop on Virtual Environments*, pp. 281–286. New York: ACM Press, 2003.

[22] Wikipedia contributors. "Digital Visual Interface." *Wikipedia, The Free Encyclopedia.* http://en.wikipedia.org/wiki/DVI.

5

Describing and Storing the VR World

There are many aspects that need to be considered when thinking about storing a practical and complete representation of a virtual world that has to include pictures, sounds, video and 3D, such as, how much detail should we include? How much storage space is it going to occupy? How easy is it going to be to read? Unlike the specific case of designing a database for computer animation programs or engineering-type computer-aided design (CAD) applications, VR involves a significant mix of data types, 3D structure, video, sounds and even specifications of touch and feel.

In this chapter, we will explore how some of the most useful storage strategies work. That is, how they can be created by applications, how they can be used in application programs and how they work together to enhance the virtual experience (for example, a video loses a lot of its impact if the soundtrack is missing).

Despite the use of the word *reality* in the title of this book, VR is not just about the real world as we perceive it. Often the use of VR is to present the world as we have never perceived it before. Visualizing and interacting with scientific data on a large projection display can and does result in surprising discoveries. The same VR data formats that we so carefully design for recording and describing a real environment can also be used to describe the most unreal synthetic environments, scenes and objects.

Since the world we live in and move around in is three dimensional, it should be no surprise to hear that the most obvious aspects of a virtual world that we have to describe and record will be a numerical description of the 3D structure one sees, touches and moves about. Thus we will begin with some suggestions for describing different shapes and geometry, using numbers.

5.1   The Geometric Description of a 3D Virtual World

In essence, there is only one way to describe the structure of a virtual world which may be stored and manipulated by computer. That way is to represent the surfaces of everything as a set of primitive shapes tied to locations in a 3D coordinate system by points along their edge. These 3D points are the vertices which give shape and body to a virtual object. The visual appearance of the structure will then depend on the properties of the primitive shapes, with color being the most important. A polygon with three sides (a triangle) is the simplest form of primitive shape. This primitive shape is also known as a *facet*. Other primitive shapes such as spheres, cubes and cylinders are potential shape builders. They too can be combined to build a numerical description of an environment or of objects decorating that environment or scene. CAD application programs tend to use more complex surface shapes, referred to as *patches*. These include Bézier patches and NURBS (nonuniform rational B-spline) patches. These are forms which have curved edges and continuously varying internal curvature. Real-world shapes can usually be accurately represented by combining a lot fewer of these more sophisticated surface patches than are necessary if using triangular polygons.

Whether one uses primitive polygons or curved patches to model the shape and form of a virtual world, one has to face up to the advantages and disadvantages of both. The triangular polygon is fast to render and easy to manipulate (they are also handled by the real-time rendering engines now implemented in the hardware of the graphics processing unit (GPU), discussed in Chapter 14). The more complex curved patches usually give a better approximation to the original shape, especially if it has many curved parts. Occasionally, a hybrid idea is useful. For example, a cube is modeled just as effectively with triangular polygons as it is with Bézier patches, but something like the famous Utah teapot, shown in Figure 5.1 with polygonal and Bézier patch representations, is much better represented by 12 patches than 960 triangular polygons.

Whatever we use to describe the surface of a 3D model, it must be *located* in space, which means attaching it to three or more *vertices* or points

Figure 5.1. Polygonal and Bézier patch surface representations.

somewhere in the 3D space. With appropriate transformations applied to the vertex coordinates, a visualization (more commonly called a rendering) can be produced. For VR work, the vitally important things are rendering the view in real-time and linking it in some way with the actions of the viewer.

The minimal description of a 3D environment model requires a list of either surface polygons or surface patches and a list of vertex coordinates. Each entry in the surface list must identify the vertices to which it is connected and have some way of allowing its surface to take on a finely detailed or richly textured look. One of the most popular ways to provide a comprehensive set of surface attributes is for each surface to store an index into a table of *textures*. A texture provides a color and many other properties, including several layers of blended image maps (including reflections, shadows, lighting and bump maps), and by using programmable GPU chips, it has become possible to create algorithmically specified (or *procedural*) textures. With careful choice of lighting models and the use of these, an extremely realistic look to the environment and object model can be achieved.

Most real-time rendering systems use a simple list of primitive planar polygons with three or four vertices to define the shape of the primitive. They do not attempt to directly render more complex curved patches. Instead, curved patches are usually triangulated, or *tessellated* before rendering as a set of triangular polygons. The data describing this geometry is most commonly organized in one of two ways:

1. This organization is driven by a list of m polygons. The n vertices of every polygon store n integers. Each integer identifies an entry in the vertex list. The vertex list stores the 3D coordinates (x, y, z) of a position vector. Most programs store (x, y, z) as single-precision floating-point numbers. To facilitate texture and image mapping, vertices also require a 2D surface coordinate (u, v); these are called the texture coordinates. If polygon edges are needed then they can be obtained by taking consecutive vertices in pairs. Each polygon also carries an index

into a list of surface material properties, which will record details such as color, shininess, reflectivity or bumpiness.

2. In this organization, m polygons are obtained from a list of $n \times m$ vertices by making an implied assumption that, in the case of triangular polygons ($m = 3$), and that three consecutive vertices in the list define a polygon. Each vertex may also hold a color value or index into a list of material properties in the same way as with organization 1.

These data-organization schemes are summarized in Figure 5.2. The first organization keeps down the number of vertices that have to be stored, but at the expense of having a second independent list of polygons. However, in cases when the mesh models of objects need to be created with different *levels of detail*,[1] the first scheme may be better because one can augment the entry for each polygon with information about its neighboring polygons and more readily subdivide the mesh to increase the detail. The second organization is the one used by hardware rendering libraries such as OpenGL or Direct3D, and some variations have been devised to account for particular mesh topologies in which the number of vertices that need to be specified can be reduced. The *triangle strip* and *triangle fan* are examples.

It is easy to obtain the data in organization 1 from the data in organization 2, but extra computation is needed to do the reverse.

5.1.1   Color, Materials and Texture Mapping

The 3D data storage schemes outlined in Figure 5.2 also include methods of assigning surface attributes to the polygons (organization 1) or the vertices (and implied polygons) in organization 2. In real-time rendering systems (which are obviously very important for VR) where we desire to keep the vertex count as low as possible, using material properties that can include color, shininess and texture mapping[2] makes a significant contribution to increasing the realism of the plain geometric data. Textures are usually regarded as a surface property of the virtual objects and scenery, and they most often follow the texture coordinates assigned to each vertex. Because textures often involve

---

[1]By storing several models of the same shape, each made to a higher degree of accuracy by using more polygonal facets, an application can speed up its rendering by using the ones with fewer polygons when the shape is further away from the camera. It can also interpolate between models made with different levels of detail as the viewpoint approaches the object so that the switch between the two models is imperceptible.

[2]Texture mapping is also known as image mapping, decalling, billboarding or just plain texturing.

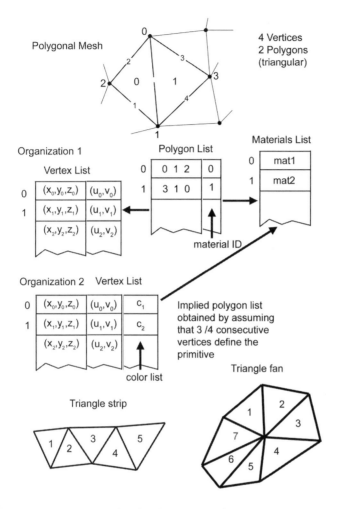

Figure 5.2. Data organization for the description of a 3D piecewise planar model. The triangle strip and triangle fan topology allow for fewer vertices to be stored in organization 1. For example, the triangle strip with 4 polygons needs only 7 vertices instead of 15, and the fan needs 8 instead of 21.

very large amounts of data, they are not stored on a per-vertex basis but in a separate list where several layers of image data and other items such as light maps, transparency maps and illumination properties can all be kept together. A virtual world may need $1,000,000$ polygons to represent its geometry, but 10–50 materials may be enough to give the feeling of reality.

5.1.2   Hierarchical Data Structures

Partitioning a polygonal model into sub-units that are linked in a parent-child-grandchild relationship is useful. For example, hierarchical linkages are very significant in character animation, robotics and chain linkages. This merges in well with simulations that involve *kinematic* and *inverse kinematic (IK)* systems where force feedback plays an important part. An easily implemented and practically useful scheme is one that is analogous to the familiar file-store structure of computer systems; i.e., a root directory with files and subdirectories which themselves contain files and subdirectories and so on, to whatever depth you like. A doubly linked list of *hierarchy* entries is the best way to organize this data. In addition to *previous* and *next* pointers, each entry will have a pointer to its *parent* (see Figure 5.3). Other data can easily be appended to the structure as necessary. When using the structure for character animation or IK robotic models, the entries can be given symbolic names. An example of this is shown in Figure 5.3.

To complete a hierarchical data structure, every polygon or vertex is identified as belonging to one of the hierarchical names. This is a similar concept to how files on a computer disk are identified by the folders in which they are stored. In a connected network of vertices and facets, it is probably more useful to associate vertices, rather than polygons, with one of the hierarchical names.

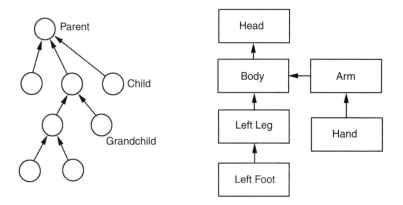

Figure 5.3.   A hierarchical data structure for character animation and articulated figures.

5.1.3 The Natural World

A VR application needs to be able to generate almost any level of detail for a scene at any given instant. This is particularly difficult when modeling the appearance of natural scenes; that is, ones with trees, clouds, landscapes etc., which have a high degree of realism. For example, the landscape depicted in Figure 5.4 can require a very large polygon count when stored as polygonal data.

Thus, while a lot of mesh models and clever rendering tricks such as preemptive culling and active viewports can speed up rendering (the topic of rendering is covered in much greater detail in Chapter 7), if you still go too close to a polygonal mesh, you will see the true nature of your data. An alternative description of the natural world has been pioneered in computer graphics over the last 25 years: *fractals*[3] [16, 17, 25]. Trees, plants, smoke, clouds and terrain can all be simulated with fractals. It may not be possible to model an actual tree, plant or mountain range, but the surface detail can be generated to almost any resolution. It takes very little data to describe a fractal scene (see Figure 5.4), and no matter how close to the surface the viewpoint is placed, one never sees anything that resembles a polygonal model. The small amounts of data needed to describe a scene take little time to load and

Figure 5.4. Modeling the natural environment using a polygonal model and a fractal description. The image on the left was generated using 120,000 polygons, while the one on the right required storing only a few fractal parameters. (Rendering still requires the scene to be polygonized, but there are no boundaries to the scene or level of detail issues.)

---

[3]A fractal is a form of computer-generated art that creates complex, repetitive, mathematically based geometric shapes and patterns that resemble those found in nature.

require little memory. Using the latest generation of programmable graphics hardware, there is the potential to generate natural scenes to any level of detail in real time.

5.2   Scene Graphs

To describe a virtual environment, especially one containing many different kinds of active elements and complex objects, it is often useful to use some form of hierarchy (not unlike that described in Section 5.1.2). This hierarchy can be used to identify the contents of the scene and how they are related to, or dependent on, each other. In the early days of 3D graphics when rendering real-time scenes was a complex task, the ability to organize the order in which the polygons used to construct the scene (see Section 5.1) are processed was of high importance. For one thing, it would allow the hardware to easily determine whether a whole object within the scene falls outside the field of view. This was achieved by creating a node in the hierarchy[4] to represent the object. This node would store a list of the polygons for the object, as well as an overall band for its size and its location within the scene. Even in the most advanced rendering hardware today, it is still often useful to rearrange the order in which the polygons are sent through the rendering pipeline (see Chapter 7). For example, it may be useful to group all polygons that are colored red, even if they come from different objects, so that they all pass through the pipeline without each one having to be accompanied by a color attribute. It was in this context that the *scene graph* came into being. The idea is illustrated in Figure 5.5, which shows the nodes of two scene graphs representing the same scene. The scene contains two cubes, two spheres and a triangle. The rendering pipeline processes the objects in the order indicated starting at the root node. The objects are given color attributes of either red or blue. In Figure 5.5(a), the ordering of the graph indicates that the color attribute has to be applied to each object. In Figure 5.5(b), the graph indicates a rendering order in which the color attribute only has to be applied twice. In a large scene, this can significantly increase rendering speed, especially when the idea is applied to the full range of attributes, e.g. materials, textures etc.

When describing a virtual world, the scene graph offers many advantages. Using scene graphs, it is possible to make, store and reuse part of a graph (termed a *sub-graph*) that not only holds structural detail but also movement

---

[4]Because the structure of this hierarchy resembles a graph (a tree graph, a cyclic graph or a directed graph), it has become known as a *scene graph*.

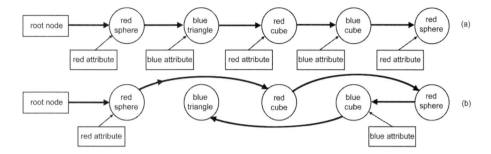

Figure 5.5. Two scene graphs representing a scene with two cubes, two spheres and a triangle.

and other attributes. For example, a network of windmills in a wind farm can be described by loading, at different locations, multiple copies of a scene graph describing the structure of an individual windmill which includes instructions as to how to animate the rotation of its blades.

If you look at Figure 5.6(a), you will see one of the classic types of graph, the *tree*. A tree graph has one node called the *root*, and all the other nodes in the graph are descended from it in a parent-child relationship. A child is only related to one parent. A second type of graph is the *cyclic* graph (Figure 5.6(b)), in which each node can have more than one parent and it is possible have to circular paths in the graph. A cyclic graph is not a good way to describe a 3D scene, because we could end up rendering the same thing over and over again. However, the *acyclic* graph (Figure 5.6(c)), which allows for a node to have more than one parent but does not have any circular paths, has merit because it allows us to apply a link *to* a node from more than one starting point. For example, we can think of applying hierarchical translational transformations to instances of the same polygon mesh object so as to move it to different locations. Or, the graph might indicate applying some attribute, such as a color or texture, to different objects, or indeed, a different color to the same object in different places, as was illustrated in Figure 5.5.

Another advantage that the scene graph offers is that a non-specialist computer programmer can build a virtual world without having to get involved in the details of coding the rendering process. VRML (Section 5.3.1) is a simple form of a scene-graph specification language. Even the experienced programmer who wants to build a VR application program can have his work greatly simplified by using a scene-graph development environment.

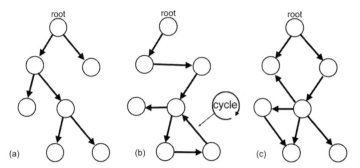

Figure 5.6. Three graph topologies. (a) The tree has a definite parent-child relation-ship. (b) The cyclic graph. (c) The acyclic graph, which has no paths leading back to a node that has been visited before.

The scene-graph developer tools offer platform independence, hierar-chical data storage, a selection of viewers and viewpoints (stereoscopic, full screen, multiple cameras etc.), texturing, billboarding, instancing, selection, picking, navigation by taking control input from a wide range of input devices etc. They are usually constructed using object-oriented programming tech-niques on top of a hardware rendering library such as OpenGL or Direct3D. There are fully supported commercial scene-graph tools, e.g., Mercury's In-ventor [12] and enthusiastically supported *open-source* packages, e.g., Open Inventor [22], OpenSG [19] and OpenSceneGraph [15]. Note that OpenSG should not be confused with OpenSceneGraph. They are entirely different scene-graph packages, even though they share similar names.

The open-source packages have the advantage that we can build them from source for use on different platforms. However, the idea of being able to extend them or add to them is rarely practical because they tend to very large. And sadly, open-source packages often have limited documentation which can be very hard to follow. In these cases, one has to make do with mi-nor modifications to the distributed examples, which is actually often sufficient.

We will briefly explore two open-source packages that encapsulate the OpenGL (see Chapter 13) 3D graphics rendering library: Open Inventor and OpenSceneGraph. Since scene graphs are *object oriented*, it is natural that both packages have been written in the C++ language. They provide a number of C++ classes with which to construct the scene graph, arrange its visualization and allow for the user to interact with it—everything we need for VR. It is interesting to compare these tools, because they are essentially doing the same thing, only using different terminology. Often some of the

```c
#include <stdlib.h>
// Header file for the X windows platform and viewer
#include <Inventor/Xt/SoXt.h>
#include <Inventor/Xt/viewers/SoXtExaminerViewer.h>
// Header file for the nodes in the scene graph. Separator
// nodes divide the graph up into its components.
#include <Inventor/nodes/SoSeparator.h>
// Header for the lighting classes
#include <Inventor/nodes/SoDirectionalLight.h>
// Header for the materials classes
#include <Inventor/nodes/SoMaterial.h>
// Header for a cube object class
#include <Inventor/nodes/SoCube.h>


int main(int , char **argv){
   // Get an X window handle
   Widget theWindow = SoXt::init(argv[0]);
   // Create the root node of the graph OI uses the
   // Separator node to divide up the graph.
   SoSeparator *rootNode = new SoSeparator;
   rootNode->ref();
   // set the materials
   SoMaterial *theMaterial = new SoMaterial;
   myMaterial->diffuseColor.setValue(1.0, 1.0, 0.0);
   rootNode->addChild(theMaterial);
   // Add a cube object
   rootNode->addChild(new SoCube);
   // Add a light
   SoDirectionalLight *theLight = new SoDirectionalLight;
   theLight->direction,setValue(0,1,1);
   rootNode->addChild(theLight);
   // Set up viewer - use default behaviour with toolbar
   // and NON full screen:
   SoXtExaminerViewer *theViewer = new SoXtExaminerViewer(theWindow);
   theViewer->setSceneGraph(rootNode);
   theViewer->setTitle("An Examiner Viewer");
   theViewer->show();
   // Show the window and wait in the loop until the user quits.
   SoXt::show(theWindow);
   SoXt::mainLoop();
   return 0;
}
```

Listing 5.1. The structure of an Open Inventor program for drawing a primitive shape and viewing it from any direction in a window on the desktop.

hardest things to come to terms with when you are confronted with a software development environment are the terms used to describe the actions of the tools. In the context of the scene graph, these are just what constitutes a node of the graph, how the scene graph is partitioned and in what order its nodes are processed.

Open Inventor (OI) [22] from Silicon Graphics (the inventors of OpenGL) has the advantage of being a well-documented scene-graph system; see the book by Wernecke [26]. The toolkit contains four principal tools:

1. A 3D scene database that allows mesh models, surface properties and elements for user interaction to be arranged in a hierarchical structure.
2. *Node kits* that are used for bringing in pre-built node components.
3. Manipulator objects that are placed in the scene to allow the user to interact with its objects. For example, a manipulator object will allow the user to select and move other objects in the scene.
4. A component library for the X Window display environment that delivers the usual range of window behavior, including a display area, an event handler (for mouse movement etc.), an execution loop that repeats continuously while the program is running and different appearances for the window container, e.g., a simple viewer or an editor window decorated with command buttons and menus.

A simple Open Inventor scene graph is illustrated in Figure 5.7, and the key instructions for a template program can be found in Listing 5.1. OI also offers a set of wrappers to allow equivalent coding in C. Nodes in the graph may carry geometry information or material properties, or be a transformation, a light or camera etc. Alternatively, nodes could be *group nodes*, which bring together other nodes or groups of nodes to form the hierarchy. *Separator group* nodes isolate the effect of transformations and material properties into a part of the graph.

5.2.1 Open Inventor

The main disadvantage of Open Inventor is that is it is almost exclusively UNIX-based and requires X Windows or another compatible graphics shell in which to execute, so it is not practical for use on a Microsoft Windows PC.

5.2.2 OpenSceneGraph

OpenSceneGraph (OSG) [15] is a vast and flexible package that covers many of the same ideas as Open Inventor. It too is composed of a number of tools that manifest themselves as the C++ classes:

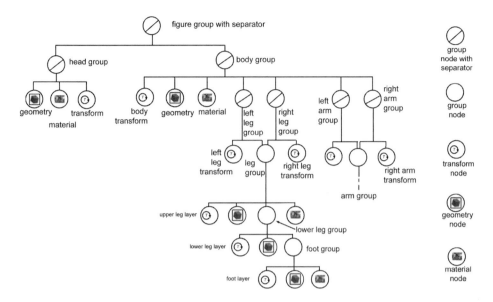

Figure 5.7. Open Inventor scene graphs allow *groups* to be either separated from or included in the inheritance of transformations and attributes from their parents. It is also possible for groups to share instances of subgroups and geometry nodes. This graph represents part of a human figure and shows that only one copy of the geometry of the leg and foot have to be kept in memory.

- *Core classes.* `osg::` These represent the nodes in the graph, a grouping of nodes, the objects in the scene (the drawables), the transformations that change the appearance and behavior of the objects and many other things such as the definition of light sources and levels of detail.

- *Data classes.* `osgDB::` These provide support for managing the *plugin* modules, which are known as *NodeKits*, and also for managing *loaders* which read 3D scene data.

- *Text classes.* `osgText::` These are NodeKits for rendering TrueType fonts.

- *Particle classes.* `osgParticle::` This is a NodeKit for particle systems. This is a particularly useful feature for VR environments requiring the simulation of smoke, cloud and flowing water effects.

- *Plugin classes.* `osgPlugins::` A large collection for reading many kinds of image file formats and 3D database formats.

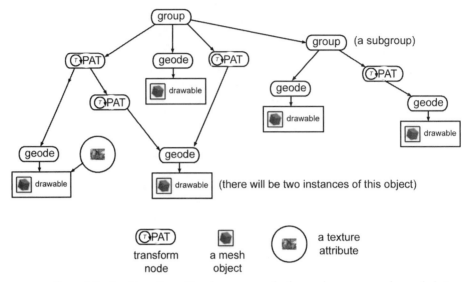

Figure 5.8. An OpenSceneGraph scene graph shows the group nodes and their attached drawable objects connected through the *Geode* class. Multiple copies of the same object at different locations are obtained by applying different transformations (called a position and attitude transformation (PAT)).

The typical structure of an OSG graph is shown in Figure 5.8, along with the C++ class names associated with the various types of nodes in a graph. Nodes can carry geometric information (called drawables), material properties (called states) or transformations. Like OI, nodes in OSG can act as grouping nodes that serve to divide the graph up. However unlike OI, the drawables are connected into the graph through another type of node, called a geode. Transformation and state nodes that change the appearance and position of the drawable elements in the scene must be applied through the drawable's geode node.

The toolbox that provides the viewing element of OpenSceneGraph is called OpenProducer. OpenSceneGraph is distributed in a form suitable for use on Windows, OS X and Linux platforms, and in this regard it is a very useful scene-graph toolbox library for VR application development. You can see how all these components are brought together in the short program given in Listing 5.2, which visualizes a simple cube.

Because both Open Inventor and OpenSceneGraph use OpenGL as their rendering engine, applications can also use the full range of OpenGL functions.

```
// Include the header files for the OSG classes
// Linking nodes for making geometry
   ...

osg::Node* createCube(){
   // Declare pointer and create Geometry object to store all the vetices and
   // lines primtive.
   osg::Geometry* cube = new osg::Geometry();

   // Declare pointer and make the cube -
   // (no attributes or surface normals are shown here. )
   osg::Vec3 cubeCoords[] = { /* array of vertices round the 6 sides of the
   // cube */};
   // set the vertices of the cube
   cube->setVertexArray(new osg::Vec3Array(24,cubeCoords));

   // Create the Geode (Geometry Node) to contain all the osg::Geometry objects.
   osg::Geode* geode = new osg::Geode();
   // add the geometry to the geode.
   geode->addDrawable(cube);

   // Make a Transformation node to place the cube in the scene
   osg::PositionAttitudeTransform* cubeXform;
   osg::Vec3 cubePosition;
   cubeXform = new osg::PositionAttitudeTransform();
   cubeXform->addChild(geode);    // add the node to the transformation
   cubePosition.set(5,0,0);
   cubeXform->setPosition( cubePosition );
   return cubeXform; // return this group to be included in the graph
}

int main( int argc, char **argv ){
   osg::Group* root = new osg::Group();// Create the root node
   osgProducer::Viewer viewer;          // Make the viewer object
   root->addChild( createCube() );     // Add the cube object graph
   // Set up the viewer
   viewer.setUpViewer(osgProducer::Viewer::STANDARD_SETTINGS);
   viewer.setSceneData( root );
   viewer.realize();
   while( !viewer.done() ){    // Message loop until user quits.
      viewer.sync();
      viewer.update();
      viewer.frame();
   }
   return 0;
}
```

Listing 5.2. The structure of an OpenSceneGraph program to display a cubic shape and view it from any direction. The osgProducer class provides a rich choice of navigation methods for traveling around the scene and options for windowed or full-screen viewing.

5.3   Putting VR Content on the WEB

In Section 5.1, we described a raw data format for 3D scenes and objects. Variations on this format are used by all the well-known computer animation and modeling packages. Even with such similar data organization, it can still be a major undertaking to take the mesh models of an object constructed in one package and use them in another package. Often quite separate conversion software has to be written or purchased. However, what happens if you want to make your 3D content available on the Web? Indeed, why not make it interactive—a virtual world on the Web. This is happening already: realtors, art galleries, museums, even cities are offering Web browsers the ability to tour their homes for sale and look at their antiquities, online, interactively and in detailed 3D. All this is possible because a standard has evolved with textural descriptions of a virtual scene and a scripting language to facilitate user interaction, such as appearing to touch objects, triggering sound effects and moving about. The ideas from the Virtual Reality Modeling Language (VRML) has evolved into an Extensible Markup Language (XML) standard called X3D.

5.3.1   X3D/VRML/XML

In the mid-1990s, Silicon Graphics Inc. (SGI), the pioneers of 3D computer graphics, began a project to bring interactive 3D graphics to the World Wide Web. The project took form as a scripting language that could be mixed in with traditional HTML webpages. A browser *plugin* would detect the codes (loaded from a `.wrl` file) written in VRML and use the computer's 3D graphics capability to obey the commands of the language, render scenes and objects and interact with them. The project reached a relatively stable state with the release of VRML 97. A full language speciation and tutorial introduction is given in [6]. VRML was not just about 3D rendering over the Web. It involved animation, interactivity, sound navigation and collision detection.

However, a decade later, the term VRML is hardly mentioned, the browser plugins have virtually disappeared and little development seems to be taking place. But VRML has not faded away—it has been subsumed into the XML-enabled 3D file format (X3D) [27], and both have also been integrated into the MPEG-4 [13] video movie standard.

MPEG-4 comes out of the ISO/IEC SC29/WG11 *MPEG* standardization activity. (We will discuss the video/audio aspects of MPEG in Section 5.5.3.) In addition to providing a standard for high-quality video compression and streaming, it also allows a time-varying 2D or 3D world to be represented, using a hierarchical scene description format based on X3D/VRML. Objects within the scene may contain 2D still and moving images, 2D and 3D graphics and sound. It supports user interaction, synchronization between objects and efficient compression of both objects and scene description. A typical MPEG-4 data stream may resemble that shown in Figure 5.9, where two video streams are mixed in with a 3D scene description. A user can interact with the objects in the scene by moving them around with a pointing device.

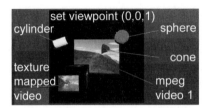

Figure 5.9. An MPEG-4 data stream can contain 3D structural elements as well as video (and audio). The video streams can be overlaid with the 3D objects moved around the output window or even texture-mapped onto 3D objects. The MPEG-4 data format is flexible enough to act as a complete interactive multimedia container for both streamed or downloaded source material.

VRML is a simple text language for describing 3D shapes and interactive environments. To describe a cylindrical shape, the commands in a .wrl file take the form:

```
#VRML V2.0 utf8
# This line tells decoders VRML code follows
# A Cylinder - lines beginning with a '#' indicate a comment
Shape {                    # A basic shape follows
    geometry Cylinder {  # The shape is a cylinder
        height 2.0
        radius 1.5
    }                      # curly brackets { } are the language delimiters
}
```

X3D and its incarnation as XML format look a little different, but they are essentially the same. An X3D encoding of the previous code fragment would look like:

```
#X3D V3.0 utf8
PROFILE IMMERSIVE
Shape {
    geometry Cylinder {
        height 2.0
        radius 1.5
    }
}
```

And in an XML format, the code would be:

```
<?xml version="1.0" encoding="utf-8"?>
<!DOCTYPE X3D PUBLIC "ISO//Web3D//DTD X3D 3.0//EN"
  "http://www.web3d.org/specifications/x3d-3.0.dtd">
<X3D profile="IMMERSIVE" >
 <Scene>
 <Shape>
 <Cylinder height=2.0 radius=1.5>
 </Shape>
 </Scene>
</X3D>
```

Note the replacement of the C-style curly-bracket syntax with pairs of XML tags. Nevertheless, the code semantics are very similar and software tools are available to convert between them.

Text file formats such as these are very inefficient for storing 3D data, and to insert the description into an MPEG-4 stream, the scene is compiled into a binary format and assembled into what is know as a *BIFS* (binary format for scenes) stream. The BIFS stream is subsequently multiplexed with the audio and video streams to form the MPEG-4 file or streaming data. BIFS data streams in MPEG-4 provide a suitable standard way to describe collaborative virtual environments [8]. One of their significant features is the synchronization layer, which offers the facility to pass timing information between collaborating participants. It is the job of an MPEG-4 player to decode the file data streams and provide the viewer with the interactivity intended by the 3D scene description. Again, see Figure 5.9 for an example.

In terms of describing virtual worlds and virtual objects, interacting with them and giving a sense of reality, the significant aspects that X3D/VRML offer are:

- An OpenGL-like 3D coordinate system maps out the scene. Polygonal mesh representations of objects with color and image mapping (called billboarding) populate the scene.

- Transformations allow objects to be positioned, rotated and scaled relative to a parent coordinate system.

- *Behaviors* are used to make shapes move, rotate, scale and more.

- There is a means to trigger events when the viewer appears to touch something.

- Lights can be positioned, oriented and colored.

- Sounds, for example horn honks, can be triggered by viewer actions. Sounds can also be continuous in the background, e.g., wind and crowd noises. Sounds can be emitted from a given location, in a given direction and within a given area.

- Collision detection is used when the viewer's avatar gets close to a shape. VRML can detect when the viewer collides with any shape and can automatically stop the viewer from going through the shape.

- To control the level of detail seen in the scene, *level-of-detail (LOD)* techniques can be deployed. Two or three models of the same shape are usually provided. The best one is automatically selected for rendering, depending on how close the objects are away from the viewpoint.

Listing 5.3 shows the description in VRML syntax of a simple plane consisting of two image mapped triangular polygons. It was exported from OpenFX [3], a modeling and computer animation software package. It is shown as opposed to XML syntax for ease of reading. Note the 3D vertex list, the index list of triangular polygons and the 2D texture coordinate list with its own polygon index list.

5.4   Storing Images and Textures

So far in this chapter, we have focused on methods to numerically describe and store the shapes used in VR. We pointed out in Section 5.1.1 that shape is not, on its own, enough to give a realistic look to a virtual environment, or for any virtual models either. The additional concept of surface texturing has to be employed, and a key element of this is image mapping. Images can be used not only to provide surface detail, but also for lighting effects, reflection

```
#VRML V2.0 utf8
# Produced by OpenFX 2006    '#' indicates a comment
DEF OFX-MODEL-ROOT Transform {# define the model with a name and expect a
                                # transformation
 translation 0 0 0                  # no translation
 Shape {                            # the model has a shape
    appearance Appearance {   # how is the model going to look
     material Material {      # define the surface color et.c
       diffuseColor 0.50 0.10 0.00
       shininess 0.5
     }                              # end of material
     texture ImageTexture {   # part of the model uses an image texture loaded
      url "../maps/AMBER.GIF"  # from this image file
     }
    }                              # the geometry is an list of faces indexing a
                                    # vertex list
    geometry DEF OFX-MODEL-FACES IndexedFaceSet {
     ccw TRUE                 # the order of vertices in the triangles
     solid FALSE              # for interaction
     coord DEF OFX-MODEL-COORD Coordinate { # vertex coordinates, sets of points
       point [                  # points have 3 coordinates
        -0.780 0.000 -0.490, 0.760 0.000 -0.490, 0.760 0.000 0.520, -0.780 0.000
         0.520]
     }                              # end of vertex coordinates
     # Identilfy which vertex bounds which face. there are
     # three vertices per face. A (-1) tells VRML the end of a face index has
     # occured.
     coordIndex [ 0, 2, 1, -1, 0, 3, 2, -1]
     texCoord DEF OFX-MODEL-TEXCOORD TextureCoordinate {
     # texture coords for vertices
     # lists of 2 coords per vertex
      point [-0.110 0.883, 1.094 0.883, 1.094 0.093, -0.110 0.093]
     }
     texCoordIndex [ 0, 2, 1, -1, 0, 3, 2, -1] # same vertex order as face list
    }
  }    # end of shape
}
```

Listing 5.3. A VRML file describing a textured cube exported from OpenFX's [3] modeler program.

effects, roughness effects and even transparency effects. Integrating moving images from live camera sources and video files may also play a significant role in VR, so we will discuss this separately.

With the exception of simple raw data, most image sources will need to be extracted from a file in one of the commonly used formats. For single photographic-type pictures, this is likely to be the JPEG (Joint Photographic

Experts Group) format. Other formats that frequently crop up are the PNG (Portable Network Graphics) and GIF (Graphics Interchange Format). Readily available software libraries that encode and decode these files into raw 24-bit pixel arrays can be easily integrated with any VR application. We show how to do this in quite a few of the utilities and application programs described in Part II.

Nevertheless, it is useful to have an appreciation of the organization of the file storage, so that it can be adapted to suit special circumstances, such as stereoscopic imaging, which we discuss in Chapters 10 and 16.

5.4.1 JPEG Storage Format

It is generally accepted that for storing photographs the lossy compression algorithm named after the standards body the Joint Photographic Experts Group (JPEG) gives excellent results, and it is thus in common use. The algorithm that underlies the JPEG compression is non-trivial and we won't attempt to describe it in detail here. A good way to obtain the full detail of this algorithm is to first consult [11] and [10], and then read Wallace's [24] paper, which gives a tutorial explanation of the details of the algorithm. A comprehensive book by Pennebaker and Mitchell [10] describes all the implementation steps.

Even if one doesn't want to delve into the inner workings of the JPEG compression process, it is a good idea to have at least a little knowledge of how it works:

> The two-dimensional array of image pixels (3 bytes per pixel representing the red, green and blue color components) is encoded in three stages. First it is separated into luminance and chrominance components. The chrominance components, considered as two difference signals, are scaled down to a lower resolution, the first loss. The data from these three components (luminance and two chrominance signals) are arranged as a single two-dimensional array of $n \times m$ bytes, which is subdivided into blocks of 8×8 bytes that are compressed separately. The 64 pieces of *spatial* data are transformed into the frequency domain using the discrete cosine transform (DCT). This gives 64 frequency components. These are then quantized by multiplying them by a scaling factor that reduces many of them to zero. This is the second loss of information. The remaining frequency components are further compressed using a lossless entropy en-

coding mechanism (Huffman encoding) to produce the final format. After that, they are stored in a file structured to accommodate sufficient information for a decoding process to recover the image, albeit with a little loss of information.

In making use of images encoded in JPEG format, we will need to know how the compressed data is organized in the file. In fact, sometimes it is actually harder to get details like this than details of the encoding algorithms themselves. One thing we can say is that a lot of image file formats, and most 3D mesh description file formats, arrange their data in a series of chucks within the overall file. Each chunk usually has an important meaning, e.g., image size, vertex data, color table etc. In the case of a JPEG file, it conforms to the exchangeable image file format (Exif) [23] specification. This allows extra information to be embedded into the JPEG file. Digital camera manufacturers routinely embed information about the properties of their cameras in what are called application markers (APPx). In Section 16.2, we will see how an APP3 is used to embed stereoscopic information in a JPEG file.

We will now briefly discuss the JPEG file layout. Hamilton [5] gives an extensive outline of the file organization within the Exif framework. Table 5.1 and Figure 5.10 show the key chunk identifiers. Every marker in an Exif file is identified with a double-byte header followed immediately by an unsigned 16-bit integer, stored in *big-endian*[5] order that gives the size of the chunk.

The first byte in any header is always 0xff,[6] so the 16-bit header takes the form 0xff**. The first two bytes in an Exif file are always 0xffd8, representing the start of image (SOI) marker. Immediately following the opening bytes is the APP0 segment marker 0xffe0, followed by the segment length (always 16 bytes). As Figure 5.10 shows, the function of the APP0 marker is

---

[5]Storing integer numbers in binary form which exceed 255 in magnitude will require more than one byte. A 16-bit integer will have a high-order byte and a low-order byte. In writing these two bytes in a file, one has the option to store either the high-order one first or the low-order one first. When the high-order byte is written first, this is called *big-endian* order. When the low-order byte is written first it is called *little-endian* order. Computers based on Intel processors and the Windows operating system use little-endian storage. Other processors, Sun and Apple systems, for example, use big-endian ordering for binary data storage.

[6]The 0x... notation, familiar to C programmers, specifies a byte value using the hexadecimal numbering system. Two hex digits per 8 bits, e.g., 0xff = 255 (decimal). A 16-bit number requires 4-hex digits, e.g., 0x0100 = 256 (decimal).

Marker Value	Identifier	Function
0xfd	SOI	Start of image
0xe0	APP0	Application marker 0 (image dimensions)
0xe1	APP1	Application marker 1
0xe2	APP2	Application marker 2
0xe3	APP3	Application marker 3 (Stereo Image etc.)
0xef	APP15	Application Marker 15
0xc0	SOF0	Image encoding is baseline standard
0xc4	DHT	Define the Huffman table
0xdb	DQT	Define the quantization table
0xda	SOS	Start of scan—the image data follows
0xd9	EOI	End of image

Table 5.1. The most significant JPEG file segment markers for basic images using DCT lossy encoding and their meaning

to give the dimensions of the image and, for example, whether the file also contains a *thumbnail* impression of the full-sized image.

The key segment which stores the actual image data is the DQT segment with the quantization table containing the coefficients which are used to multiply the frequency components after the DCT stage. A decoder can use these to recover the unscaled frequency components; i.e., it divides by the coefficients to reverse. The DHT segment stores the entries in the Huffman encoding table so that the decoders can recover the scaled frequency component values. The start of scan (SOS) segment contains the actual compressed image data from the 8×8 blocks of pixels. These are used together with the image size values from the APP0 header to rebuild the image for viewing.

Computer codes to implement JPEG encoding and decoding are freely available, and we have found those from [9] to work well in practice. They can also be easily integrated into our VR application programs. Rich Geldreich [4] has also implemented a useful concise library based on a C++ class for decoding only.

5.4.2 Lossless Image-Storage Formats

The most enduring lossless image storage format is undoubtedly the GIF format. It is ideal for small graphics containing text, such as those used on webpages. It has one major drawback, in that it requires a color palette and is only able to reproduce 256 different colors in the same image. At the core of the GIF algorithms is *Lempel-Ziv and Haruyasu* (LZH) [29] encoding. Variants

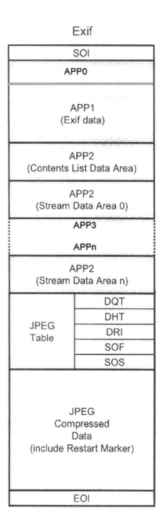

Figure 5.10. The structure of a JPEG file showing the markers and expanded contents of typical APP0, APP1, APP2 and APP3 chunks. More details of the chunks are given in Table 5.1.

of this appear in compressed file archive systems such as ZIP and TGZ. The PNG [20] formats have been relatively successful for losslessly storing 24-bit color images. In a way similar to the Exif file organization, PNG files can have application-specific information embedded with the picture data, and so for example, they can be used to encode stereoscopic formatting information.

5.5   Storing Video

If the high compression ratios attainable with the JPEG algorithm are deemed necessary for single images they become indispensable when working with video. Digital video (DV), digital versatile disc (DVD), Moving Picture Experts Group (MPEG-1/2/3/4...), Xvid and Microsoft's WMV all involve video compression technologies. They do not, however, define how the data should be stored, just how it should be compressed. For storage, the compressed video data will need to be inserted into a container file format of some sort along with synchronized audio and possibly other multimedia content. Apple's QuickTime [1, 7] was one of the first movie container formats to support multiple tracks indexed by time and which was extensible so that it can offer the inclusion of optional and efficient compression technology. Typical of the container file types that have become synonymous with video on the PC platform is Microsoft's AVI (audio video interleave) [14]. It is very flexible, since the video and audio streams it contains may be compressed with a wide selection of encoding methods.

5.5.1   AVI

AVI does not define any new method of storing data. It is an instance of Microsoft's resource interchange file format (RIFF) specifically containing synchronized audio and video data. The RIFF format is typical of many data formats in that it is divided into *chunks*, which in turn are divided into *sub-chunks*. Within a chunk and sub-chunk, a header provides information about the contents of the chunk. In the case of video data, the header will contain such items as the width and height of the video frame, the number of data streams in the chunk, the time between frames and the size of the chunk. For AVI files, the sub-chunk header might indicate that it contains two streams, for example one containing audio samples and the other video frames. Interleaving means that the stream data is arranged so that the picture data for each video frame alternates with the data for a set of audio samples which are to be played back during the time the video frame is onscreen. AVI files allow for any compression method to be applied to the video samples and/or the audio samples. An AVI file might, for example, contain video in uncompressed form, compressed by simple run-length encoding or the H.264/AVC (advanced video coding) [21], which is used for broadcast-quality steaming video with MPEG-4.

5.5.2 Codecs

It would be virtually impossible for every application program that needs to use video sources to include a full spectrum of decoding algorithms. For one thing, new compression algorithms are emerging with regularity. But by separating out, into *loadable* components, those parts of an application program that encode and decode the data, any program can much more easily access all kinds of multimedia content, especially audio and video material. These self-contained units that specifically only fulfill the role of encoding and decoding are the *codecs*. Under Windows, they are implemented in a particular form of dynamic link library (DLL) and can be installed as part of the operating system so that any application program may use them to encode, decode or both encode and decode data streams read from a container file such as AVI.

Marketing codecs has become quite big business itself, and with the popularity of personal music players such as Apple's iPod, a wide variety of audio codecs (e.g., Apple's AAC, now regarded as the successor of the MP3 format, and Sony's ATRAC) are in common use. Providing that a codec's author makes it available for your type of PC system, you should still be able to use a well-written DVD, movie or audio player program to show the latest content encoded with any codec. We shall look (in Chapter 15) at how application programs may be written so that they automatically detect which codec a movie is compressed with, load the code for its decompressor and present the video under a number of guises. However, for video data streams, the encoding method is almost certainly going to be some extension or variant on the standard proposed by the Moving Picture Experts Group (MPEG). Thus we will take a brief look at the concept behind the MPEG compression algorithm in the next section.

The first MPEG standard, MPEG-1, was designed in the late 1980s for data rates of 1.5MBits/s, with video frame sizes[7] of 352 × 240.

The video resolution of MPEG-1 is low by today's standards, even below traditional analog TV. By enhancing the algorithms from MPEG-1 but with-

---

[7]The accompanying compressed audio stream was enhanced in the early 1990s by adding a system that became known as Musicam or Layer 2 audio. The Musicam format defined the basis of the MPEG audio compression format: sampling rates, structure of frames, headers and number of samples per frame. Its technologies and ideas were fully incorporated into the definition of ISO MPEG Audio Layers 1, 2 and later 3, which when used without any video stream gives rise to the familiar MP3 audio encoding format. Uncompressed audio as stored on a compact disc has a bit rate of about 1400 kbit/s. MP3 as a lossy format is able to compress this down to between 128 and 256 kbit/s.

out having to change things radically, it was possible to increase the resolution and compress broadcast-quality video, resolution 720×480 for NTSC video and 720×572 for PAL video. The enhanced standard, MPEG-2, defines the encoding and decoding methods for DVDs, hard-disk video recorders and many other devices. It is so successful that the plans for an MPEG-3 especially for high-definition television were scrapped in favor of MPEG-4, which can accommodate a much broader range of multimedia material within its specification. However, for audio and video, the key concepts of how the encoding is done remain basically the same since MPEG-1. We will look briefly at this now.

5.5.3 Video Data Compression

The success of the JPEG encoding method in compressing single digital images opens the door to ask whether the same high compression ratios can be achieved with digital video. There is one drawback, in that the video decompressor must be able to work fast enough to keep up with the 25 or 30 frames per second required. The MPEG compression standard delivers this by combining the techniques of transformation into the frequency domain with quantization (as used in JPEG) and predictions based on past frames and temporal interpolation. Using these techniques, a very high degree of compression can be achieved. By defining a strategy for compression and decompression rather than a particular algorithm to follow, the MPEG encoding leaves the door open for research and development to deliver higher compression rates in the future.

Figure 5.11 illustrates the ideas that underlie the MPEG compression concept. In Figure 5.11(a), each frame in the video stream is either compressed *as-is* using DCT and entropy encoding, or obtained by prediction from an adjacent frame in the video stream. In Figure 5.11(b), a frame is broken up into macro-blocks, and the movement of a macro-block is tracked between frames. This gives rise to an offset vector. In Figure 5.11(c), a difference between the predicted position of the macro-block and its actual position will give rise to an error. In the decoder (Figure 5.11(d)), a P-frame is reconstructed by applying the motion vector to a macro-block from the most recent I-frame and then adding in the predicted errors.

An MPEG video stream is a sequence of three kinds of frames:

1. I-frames can be reconstructed without any reference to other frames.

2. P-frames are forward predicted from the previous I-frame or P-frame.

3. B-frames are both forward predicted and backward predicted from the previous/next I- or P-frame. To allow for backward prediction, some of the frames are sent out-of-order, and the decoder must have some built-in buffering so that it can put them back into the correct order. This accounts for part of the small delay which digital television transmissions exhibit. Figure 5.11(a) illustrates the relationship amongst the frames.

Figure 5.11(b) shows how prediction works. If an I-frame shows a picture of a moving triangle on a white background, in a later P-frame the same triangle may have moved. A motion vector can be calculated that specifies how to move the triangle from its position in the I-frame so as to obtain the position of the triangle in the P-frame. This 2D motion vector is part of the MPEG data stream. A motion vector isn't valid for the whole frame; instead the frame is divided into small *macro-blocks* of 16×16 pixels. (A macro-block is actually made up from 4 luminance blocks and 2 chrominance blocks of size 8×8 each, for the same reason as in the JPEG image format.)

Every macro-block has its own motion vector. How these motion vectors are determined is a complex subject, and improvements in encoder algorithms have resulted in much higher compression ratios and faster compression speeds, e.g., in the Xvid [28] and DivX [2] codecs. However, as Figure 5.11(c) shows, the rectangular shape (and the macro-block) is not restricted to translation motion alone. It could also rotate by a small amount at the same time. So a simple 2D translation will result in a prediction error. Thus the MPEG stream needs to contain data to correct for the error. In the decoder, an I-frame is recovered in two steps:

1. Apply the motion vector to the base frame.

2. Add the correction to compensate for prediction error. The prediction error compensation usually requires fewer bytes to encode than the whole macro block. DCT compression applied to the prediction error also decreases its size.

Figure 5.11(d) illustrates how the predicted motion and prediction error in a macro-block are recombined in a decoder. Not all predictions come true, and so if the prediction error is too big, the coder can decide to insert an I-frame macro-block instead. In any case, either the I-frame version of the macro-block or the error information is encoded by performing a discrete cosine transform (DCT), quantization and entropy encoding in a very similar way to the way that it is done for still JPEG images.

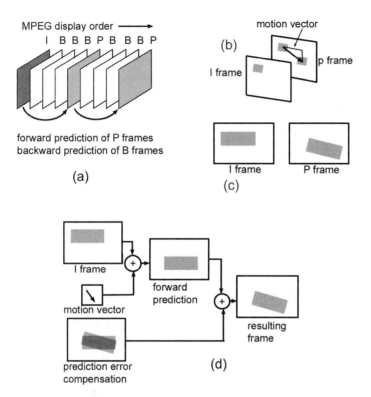

Figure 5.11. The MPEG compression concept. (a) Types of compression applied to successive images. (b) A frame is broken up into macro-blocks and their motion tracked. (c) A position error in the predicted position of the macro-block is obtained. (d) In the decoder, a P-frame is reconstructed.

A similar scheme with macro-blocks is applied to compression of the B-frames, but for B-frames, forward prediction or backward prediction or both may be used, so as to try and obtain a higher compression ratio, i.e., less error.

5.5.4 A Final Comment on Filename Extensions

It is common practice to identify the format of the contents of image and video files with a filename extension: a three or four character string added after a dot at the end of the file's name. Windows normally hides this information when it lists files and folders, in file open dialogs for example. Instead, it substitutes an icon to identify an application which can open this file type. However, as we have seen in the case of movie/video files, the filename extension is of little help in identifying the form of the information contained in the file. With Windows, nearly all video material is contained in an AVI file even if it is an MPEG-2 DVD or Xvid MPEG-4 downloaded HDTV movie.

5.6   Summary

This chapter has looked at the most common data organizations for storing 3D specifications of virtual objects and scenes. We shall be using this information in Chapters 13 and 14, where we look at application programs to render a virtual environment in real time and with a high degree of realism. This chapter also explored the ideas that underlie image storage, first because image-based texture mapping is so important for realism of virtual environments, and second because modifications to the well-known storage formats are needed in Chapter 16 when we discuss software to support stereopsis. The chapter concluded with a look at video and movie file formats, because this impacts on how multimedia content from recorded and live video sources might be used in VR.

Bibliography

[1] Apple Computer Inc. *Inside Macintosh: QuickTime*. Reading, MA: Addison-Wesley, 1993.

[2] DivX, Inc. "DivX Video Codec". http://www.divx.com/.

[3] S. Ferguson. "OpenFX". http://www.openfx.org/, 2006.

[4] R. Geldreich. "Small JPEG Decoder Library". http://www.users.voicenet.com/~richgel/, 2000.

[5] E. Hamilton. "JPEG File Interchange Format". Technical Report, C-Cube Microsystems, 1992.

[6] J. Hartman, J. Wernecke and Silicon Graphics, Inc. *The VRML 2.0 Handbook*. Reading, MA: Addison-Wesley Professional, 1996.

[7] E. Hoffert et al. "QuickTime: An Extensible Standard for Digital Multimedia". In *Proceedings of the Thirty-Seventh International Conference on COMPCON*, pp. 15–20. Los Alamitos, CA: IEEE Computer Society Press, 1992.

[8] M. Hosseni and N. Georganas. "Suitability of MPEG4's BIFS for Development of Collaborative Virtual Environments". In *Proceedings of the 10th IEEE International Workshops on Enabling Technologies: Infrastructure for Collaborative Enterprises,* pp. 299–304. Washington, D.C.: IEEE Computer Society, 2001.

[9] Independent JPEG Group. *Free Library for JPEG Image Compression.* http://www.ijg.org/, 2001.

[10] T. Lane. "Introduction to JPEG". http://www.faqs.org/faqs/compression-faq/part2/section-6.html, 1999.

[11] T. Lane. "JPEG Image Compression FAQ, part 1/2". http://www.faqs.org/faqs/jpeg-faq//part1/, 1999.

[12] Mercury Computer Systems, Inc. "Open Inventor Version 6.0 by Mercury Computer Systems". http://www.mc.com/products/view/index.cfm?id=31&type=software.

[13] Moving Picture Experts Group. "MPEG-4 Description". http://www.chiariglione.org/mpeg/standards/mpeg-4/mpeg-4.htm, 2002.

[14] J. D. Murray and W. vanRyper. *Encyclopedia of Graphics File Formats.* Sebastopol, CA: O'Reilly Media, 1994.

[15] OSG Community. "OpenSceneGraph". http://www.openscenegraph.org, 2007.

[16] H. Peitgen and D. Saupe (editors). *The Science of Fractal Images.* New York: Springer-Verlag, 1988.

[17] H. Peitgen et al. *Chaos and Fractals: New Frontiers in Science.* New York: Springer-Verlag, 1993.

[18] W. Pennebaker and J. Mitchell. *JPEG Still Image Compression Standard.* New York: Van Nostrand Reinhold, 1993.

[19] D. Reiners and G. Voss. "OpenSG". http://www.opensg.org, 2006.

[20] G. Roelofs. "PNG (Portable Network Graphics)". http://www.libpng.org/pub/png/, 2007.

[21] R. Schäfer, T. Wiegand and H. Schwarz. "The Emerging H.264/AVC Standard". *EBU Technical Review* 293 (2003).

[22] Silicon Graphics, Inc. "Open Inventor". http://oss.sgi.com/projects/inventor/, 2003.

[23] Technical Standardization Committee on AV & IT Storage Systems and Equipment. "Exchangeable Image File Format for Digital Still Cameras: Exif Version 2.2". Technical Report, Japan Electronics and Information Technology Industries Association, http://www.exif.org/Exif2-2.PDF, 2002.

[24] G. Wallace. "The JPEG Still Compression Standard". *Communications of the ACM* 34:4 (1991) 31–44.

[25] T. Wegner and B. Tyler. *Fractal Creations*, Second Edition. Indianapolis, IN: Waite Group, 1993.

[26] J. Wernecke and Open Inventor Architecture Group. *The Inventor Mentor: Programming Object-Oriented 3D Graphics with Open Inventor, Release 2*. Reading, MA: Addison-Wesley Professional, 1994.

[27] "X3D". http://www.web3d.org/x3d/, 2006.

[28] "Xvid Codec". http://www.xvid.org/, 2006.

[29] J. Ziv and A. Lempel. "A Universal Algorithm for Sequential Data Compression". *IEEE Transactions on Information Theory* 23:3 (1977) 337–343.

6

A Pocket 3D Theory Reference

For all intents and purposes, the world we live in is three dimensional. Therefore, if we want to construct a realistic computer model of it, the model should be in three dimensions. Geometry is the foundation on which computer graphics and specifically 3D computer graphics, which is utilized in VR, is based. As we alluded to in Chapter 5, we can create a virtual world by storing the locations and properties of primitive shapes. Indeed, the production of the photorealistic images necessary for VR is fundamentally determined by the intersection of straight lines with these primitive shapes.

This chapter describes some useful mathematical concepts that form the basis for most of the 3D geometry that is used in VR. It also establishes the notation and conventions that will be used throughout the book. It will be assumed that you have an appreciation of the Cartesian frame of reference used to map three-dimensional space and that you know the rules for manipulating vectors and matrices; that is, performing vector and matrix arithmetic. In any case, we review the fundamentals of 3D geometry, since it is essential to have a general knowledge in this area when we discuss the more complex topic of rendering, which we cover in Chapter 7. Therefore, in this chapter we explain how to manipulate the numerical description of a virtual world so that we will be able to visualize it with views from any location or angle. Then, in the next chapter, we explain how this jumble of 3D mathematics leads to the beautiful and realistic images that envelop you in a virtual world.

6.1   Coordinate Systems

In order to provide a virtual world, we need some geometric base on which to form our computations. This comes in the form of a coordinate system, one that provides a numerical frame of reference for the 3D universe in which we will develop our ideas and algorithms. Two coordinate systems are particularly useful to us, the ubiquitous Cartesian (x, y, z) rectilinear system and the spherical polar (r, ϑ, φ) or angular system. Cartesian coordinates are the most commonly used, but angular coordinates are particularly helpful when it comes to interacting in 3D, where it is not only important to say where something is located but also in which direction it is pointing or moving.

6.1.1   Cartesian

Figure 6.1 illustrates the Cartesian system. Any point P is uniquely specified by a triple of numbers (a, b, c). Mutually perpendicular coordinate axes are conventionally labeled x, y and z. For the point P, the numbers a, b and c can be thought of as distances we need to move in order to travel from the origin to the point P (move a units along the x-axis then b units parallel to the y-axis and finally c units parallel to the z-axis).

In the Cartesian system, the axes can be oriented in either a *left-* or *right-*handed sense. A right-handed convention is consistent with the vector cross product, and as such all algorithms and formulae used in this book assume a right-handed convention.

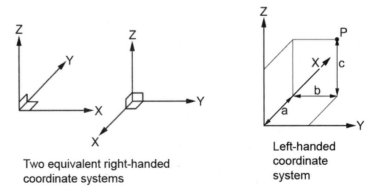

Two equivalent right-handed coordinate systems

Left-handed coordinate system

Figure 6.1. Right-handed and left-handed coordinate systems with the z-axis vertical.

6.1.2 Spherical Polar

Figure 6.2 shows the conventional spherical polar coordinate system in relation to the Cartesian axes. The distance r is a measure of the distance from the origin to a point in space. The angles ϑ and φ are taken relative to the z- and x-axes respectively. Unlike the Cartesian x-, y- and z-values, which all take the same units, spherical polar coordinates use both distance and angle measures. Importantly, there are some points in space that do not have a unique one-to-one relationship with an (r, ϑ, φ) coordinate value. For example, points lying on the positive z-axis can have any value of φ. That is, $(100, 0, 0)$ and $(100, 0, \pi)$ both represent the same point.

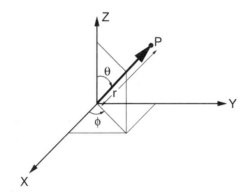

Figure 6.2. The spherical polar coordinate system.

Also, the range of values which (r, ϑ, φ) can take is limited. The radial distance r is such that it is always positive $0 \leq r < \infty$; ϑ lies in the range $0 \leq \vartheta \leq \pi$ and φ takes values $-\pi \leq \varphi < \pi$.

It is quite straightforward to change from one coordinate system to the other. When the point P in Figure 6.2 is expressed as (r, ϑ, φ), the Cartesian coordinates (x, y, z) are given by the trigonometric expressions

$$
\begin{aligned}
x &= r \sin \vartheta \cos \varphi, \\
y &= r \sin \vartheta \sin \varphi, \\
z &= r \cos \vartheta.
\end{aligned}
$$

Conversion from Cartesian to spherical coordinates is a little more tricky. The radial distance is easily determined by $r = \sqrt{x^2 + y^2 + z^2}$ and ϑ follows by trigonometry; $\vartheta = \arccos \frac{z}{r}$. Finally, we can then determine φ, again using simple trigonometry. However, this time it is necessary to separate the calculation for different values of x.

That is,

$$\varphi = \begin{cases} \arctan(\frac{y}{x}) & \text{if } x > 0, \\ \pi - \arctan(\frac{y}{-x}) & \text{if } x < 0, \\ \frac{\pi}{2} & \text{otherwise.} \end{cases}$$

If performing these calculations on a computer, an algorithm is required that tests for the special cases where P lies very close to the z-axis; that is, when ϑ and φ approach zero. A suitable implementation is presented in Figure 6.3.

```
if (x² + y²) < ε  {
    r = z
    ϑ = 0
    φ = 0
}
else {
    r = √(x² + y² + z²)
    ϑ = arccos(z/r)
    φ = ATAN2(y, x)
}
```

Figure 6.3. Algorithm for conversion from Cartesian to spherical coordinates.

The parameter ε is necessary because no computer can calculate with total accuracy. What value is chosen depends on the relative size of the largest and smallest measurements. For example, a 3D animation of atomic and molecular processes would have a very different value of ε from one illustrating planetary dynamics.

In general, a satisfactory result will be obtained when ε is chosen to be about 0.0001% of the smallest unit in use.

The function $ATAN2(y, x)$ is provided in the libraries of many computer languages. It returns a value in the range $(-\pi, \pi)$, and in the first quadrant this is equivalent to $\arctan(y/x)$.

6.2 Vectors

The vector, the key to all 3D work, is a triple of real numbers (in most computer languages, these are usually called floating-point numbers) and is noted in a **bold** typeface, e.g., **P** or **p**. When hand-written (and in the figures of this book), vectors are noted with an <u>underscore</u>, e.g., $\underline{P}$.

Care must be taken to differentiate between two types of vector (see Figure 6.4).

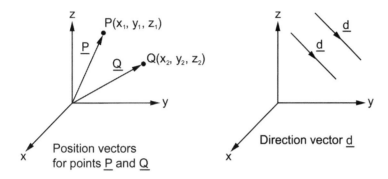

Figure 6.4. Position and direction vectors.

- *Position vectors.* A position vector runs from the origin of coordinates $(0, 0, 0)$ to a point (x, y, z), and its length gives the distance of the point from the origin. Its components are given by (x, y, z). The essential concept to understand about a position vector is that it is anchored to specific coordinates (points in space). The set of points or *vertices* that are used to describe the *shape* of all models in 3D graphics can be thought of as position vectors.

Thus a point with coordinates (x, y, z) can also be identified as the end point of a position vector **p**. We shall often refer to a point as (x, y, z) or **p**.

- *Direction vectors.* A direction vector differs from a position vector in that it is *not* anchored to specific coordinates. Frequently direction vectors are used in a form where they have unit length; in this case they are said to be *normalized.* The most common application of a direction vector in 3D computer graphics is to specify the orientation of a surface or direction of a ray of light. For this we use a direction vector at right angles (*normal*) and pointing away from a surface. Such *normal* vectors are the key to calculating surface shading effects. It is these shading effects which enhance the realism of any computer generated image.

6.3   The Line

There are two useful ways to express the equation of a line in vector form (see Figure 6.5). For a line passing through a point $\mathbf{P_0}$ and having a direction $\hat{\mathbf{d}}$, any point $\mathbf{p}$ which lies on the line is given by

$$\mathbf{p} = \mathbf{P_0} + \mu\hat{\mathbf{d}},$$

where $\mathbf{P_0}$ is a position vector and $\hat{\mathbf{d}}$ is a unit length (*normalized*) direction vector.

Alternatively, any point $\mathbf{p}$ on a line passing through two points $\mathbf{P_0}$ and $\mathbf{P_1}$ is given by

$$\mathbf{p} = \mathbf{P_0} + \mu(\mathbf{P_1} - \mathbf{P_0}).$$

Using two points to express an equation for the line is useful when we need to consider a finite segment of a line. (There are many examples where we need to use segments of lines, such as calculating the point of intersection between a line segment and a plane.)

Thus, if we need to consider a line segment, we can assign $\mathbf{P_0}$ and $\mathbf{P_1}$ to the segment end points with the consequence that any point $\mathbf{p}$ on the line will only be part of the segment if its value for μ in the equation above lies in the interval $[0, 1]$. That is, $(\mu = 0, \mathbf{p} = \mathbf{P_0})$ and $(\mu = 1, \mathbf{p} = \mathbf{P_1})$.

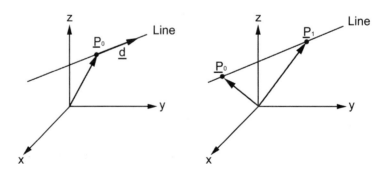

Figure 6.5. Specifying a line.

6.4   The Plane

A plane is completely specified by giving a point on the plane $\mathbf{P_0}$ and the direction $\hat{\mathbf{n}}$ perpendicular to the plane. To write an equation to represent the plane, we can use the fact that the vector $(\mathbf{p} - \mathbf{P_0})$ which lies in the plane must be at right angles to the normal of the plane, thus,

$$(\mathbf{p} - \mathbf{P_0}) \cdot \hat{\mathbf{n}} = \mathbf{0}.$$

Alternatively, a plane could be specified by taking three points $\mathbf{P_2}$, $\mathbf{P_1}$ and $\mathbf{P_0}$ lying in the plane. Provided they are not collinear, it is valid to write the equation of the plane as

$$(\mathbf{p} - \mathbf{P_0}) \cdot \frac{(\mathbf{P_2} - \mathbf{P_0}) \times (\mathbf{P_1} - \mathbf{P_0})}{|\,(\mathbf{P_2} - \mathbf{P_0}) \times (\mathbf{P_1} - \mathbf{P_0})\,|} = \mathbf{0}.$$

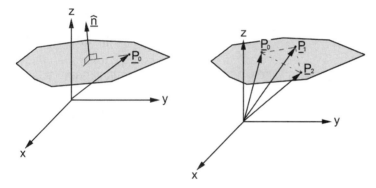

Figure 6.6. Specifying a plane.

Figure 6.6 illustrates these two specifications for a plane. It should be noted that these equations apply to planes that extend to infinity in all directions. We shall see later that the intersection of a line with a *bounded* plane plays a very important role in rendering algorithms.

6.4.1   Intersection of a Line and a Plane

The intersection of a line and a plane will occur at a point $\mathbf{p_i}$ that satisfies the equation of the line and the equation of the plane simultaneously. For the line $\mathbf{p} = \mathbf{P_1} + \mu\hat{\mathbf{d}}$ and the plane $(\mathbf{p} - \mathbf{P_p}) \cdot \hat{\mathbf{n}} = \mathbf{0}$, the point of intersection $\mathbf{p}_I$ is given by the algorithm shown in Figure 6.7.

$$
\begin{aligned}
&\text{if } |\hat{\mathbf{d}} \cdot \hat{\mathbf{n}}| < \varepsilon \;\{\text{ no intersection}\}\\
&\text{else } \{\\
&\qquad \mu = \frac{(\mathbf{P_p} - \mathbf{P_l}) \cdot \hat{\mathbf{n}}}{\hat{\mathbf{d}} \cdot \hat{\mathbf{n}}}\\
&\qquad \mathbf{p}_I = \mathbf{P_l} + \mu\hat{\mathbf{d}}\\
&\}
\end{aligned}
$$

Figure 6.7. Algorithm to determine whether a line and a plane intersect.

Note that we must first test to see whether the line and plane actually intersect. That is, if the dot product of the direction vector of the line and the normal vector to the plane is zero, no intersection will occur because the line and the plane are actually perpendicular to each other. The parameter ε allows for the numerical accuracy of computer calculations, and since $\hat{\mathbf{d}}$ and $\hat{\mathbf{n}}$ are of unit length, ε should be of the order of the machine arithmetic precision.

$$
\begin{aligned}
&\mathbf{a} = \mathbf{P}_0 - \mathbf{P}_p\\
&\mathbf{b} = \mathbf{P}_1 - \mathbf{P}_p\\
&d_a = \mathbf{a} \cdot \mathbf{n}\\
&d_b = \mathbf{b} \cdot \mathbf{n}\\
&\text{if } |d_a| \le \varepsilon_0 \text{ and } |d_b| \le \varepsilon_0 \;\{\\
&\qquad \text{both } \mathbf{P}_0 \text{ and } \mathbf{P}_1 \text{ lie in the plane}\\
&\}\\
&\text{else } \{\\
&\qquad d_{ab} = d_a d_b\\
&\qquad \text{if } d_{ab} < \varepsilon_1 \;\{\\
&\qquad\qquad \text{The line crosses the plane}\\
&\qquad \}\\
&\qquad \text{else } \{\\
&\qquad\qquad \text{The line does not cross the plane}\\
&\qquad \}\\
&\}
\end{aligned}
$$

Figure 6.8. Algorithm to determine whether the line joining two points crosses a plane.

6.4.2 Intersection of a Line Segment and a Plane

Given a line segment joining $\mathbf{P}_0$ to $\mathbf{P}_1$ and a plane $(\mathbf{p} - \mathbf{P_p}) \cdot \hat{\mathbf{n}} = \mathbf{0}$, the algorithm shown in Figure 6.8 determines whether the plane and line intersect. Note that this does not actually calculate the point of intersection. It is a good idea to separate the calculation of an intersection point by first testing whether there will be one before going on to determine the point.

The parameters ε_0 and ε_1 are again chosen as non-zero values because of the numerical accuracy of floating-point calculations.

6.4.3 Intersection of a Line with a Triangular Polygon

Many 3D algorithms *(rendering and polygonal modeling)* require the calculation of the point of intersection between a line and a fundamental shape called a *primitive*. We have already dealt with the calculation of the intersection between a line and a plane. If the plane is bounded, we get a primitive shape called a *planar polygon*. Most 3D rendering and modeling application programs use polygons that have either three sides (triangles) or four sides (quadrilaterals). Triangular polygons are by far the most common, because it is always possible to reduce an *n*-sided polygon to a set of triangles.

This calculation is used time and time again in rendering algorithms (image and texture mapping and rasterization). The importance of determining whether an intersection occurs in the interior of the polygon, near one of its vertices, at one of its edges or indeed just squeaks by outside cannot be overemphasized. We deal with the implications of these intersections in the next chapter.

This section gives an algorithm that can be used determine whether a line intersects a triangular polygon. It also classifies the point of intersection as being internal, a point close to one of the vertices or within some small distance from an edge.

The geometry of the problem is shown in Figure 6.9. The point $\mathbf{P}_i$ gives the intersection between a line and a plane. The plane is defined by the points $\mathbf{P}_0$, $\mathbf{P}_1$ and $\mathbf{P}_2$, as described in Section 6.4.1. These points also identity the vertices of the triangular polygon. Vectors $\mathbf{u}$ and $\mathbf{v}$ are along two of the edges of the triangle under consideration. Provided $\mathbf{u}$ and $\mathbf{v}$ are *not* collinear, the vector $\mathbf{w} = \mathbf{P}_i - \mathbf{P}_0$ lies in the plane and can be expressed as a linear combination of $\mathbf{u}$ and $\mathbf{v}$:

$$\mathbf{w} = \alpha\mathbf{u} + \beta\mathbf{v}. \tag{6.1}$$

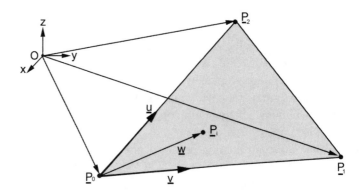

Figure 6.9. Intersection of line and triangular polygon.

Once α and β have been calculated, a set of tests will reveal whether $\mathbf{P}_i$ lies inside or outside the triangular polygon with vertices at $\mathbf{P}_0$, $\mathbf{P}_1$ and $\mathbf{P}_2$.

Algorithm overview:

- To find $\mathbf{P}_i$, we use the method described in Section 6.4.1.

- To calculate α and β, take the dot product of Equation (6.1) with $\mathbf{u}$ and $\mathbf{v}$ respectively. After a little algebra, the results can be expressed as

$$\alpha = \frac{(\mathbf{w} \cdot \mathbf{u})(\mathbf{v} \cdot \mathbf{v}) - (\mathbf{w} \cdot \mathbf{v})(\mathbf{u} \cdot \mathbf{v})}{(\mathbf{u} \cdot \mathbf{u})(\mathbf{v} \cdot \mathbf{v}) - (\mathbf{u} \cdot \mathbf{v})^2};$$

$$\beta = \frac{(\mathbf{w} \cdot \mathbf{v})(\mathbf{u} \cdot \mathbf{u}) - (\mathbf{w} \cdot \mathbf{u})(\mathbf{u} \cdot \mathbf{v})}{(\mathbf{u} \cdot \mathbf{u})(\mathbf{v} \cdot \mathbf{v}) - (\mathbf{u} \cdot \mathbf{v})^2}.$$

Since both expressions have the same denominator, it need only be calculated once. Products such as $(\mathbf{w} \cdot \mathbf{v})$ occur more than once, and therefore assigning these to temporary variables will speed up the calculation. Optimizing the speed of this calculation is worthwhile because it lies at the core of many *time-critical* steps in a rendering algorithm, particularly image and texture mapping functions. C++ code for this important function is available with the book.

The pre-calculation of $(\mathbf{u} \cdot \mathbf{u})(\mathbf{v} \cdot \mathbf{v}) - (\mathbf{u} \cdot \mathbf{v})^2$ is important, because should it turn out to be too close to zero, we cannot obtain values for α or β. This problem occurs when one of the sides of the triangle is of zero length. In practice, a triangle where this happens can be ignored because if one of its sides has zero length, it has zero area and will therefore not be visible in a rendered image.

- Return hit code as follows:

  if $\alpha < -0.001$ or $\alpha > 1.001$ or $\beta < -0.001$ or $\beta > 1.001$ miss polygon
  if $(\alpha + \beta) > 1.001$ miss polygon beyond edge $\mathbf{P}_1 \to \mathbf{P}_2$
  if $\alpha \geq 0.0005$ and $\alpha \leq 0.9995$
      and $\beta \geq 0.0005$ and $\beta \leq 0.9995$
      and $\alpha\beta \leq 0.9995$ inside polygon
  else if $\alpha < 0.0005$ { *along edge* $\mathbf{P}_0 \to \mathbf{P}_1$
     if $\beta < 0.0005$ at vertex $\mathbf{P}_0$
     else if $\beta > 0.9995$ at vertex $\mathbf{P}_1$
     else on edge $\mathbf{P}_0 \to \mathbf{P}_1$ not near vertex
  }
  else if $\beta < 0.0005$ { *along edge* $\mathbf{P}_0 \to \mathbf{P}_2$
     if $\alpha < 0.0005$ at vertex $\mathbf{P}_0$
     else if $\alpha > 0.9995$ at vertex $\mathbf{P}_2$
     else on edge $\mathbf{P}_0 \to \mathbf{P}_2$ not near vertex
  }
  else if $(\alpha + \beta) > 0.9995$ on edge $\mathbf{P}_1 \to \mathbf{P}_2$
  else miss polygon

Note that the parameters -0.001 etc. are not dependent on the absolute size of the triangle because α and β are numbers in the range $[0, 1]$.

6.5 Reflection in a Plane

In many visualization algorithms, there is a requirement to calculate a reflected direction given an incident direction and a plane of reflection. The vectors we need to consider in this calculation are of the *direction* type and assumed to be of unit length.

If the incident vector is $\hat{\mathbf{d}}_{in}$, the reflection vector is $\hat{\mathbf{d}}_{out}$ and the surface normal is $\mathbf{n}$, we can calculate the reflected vector by recognizing that because

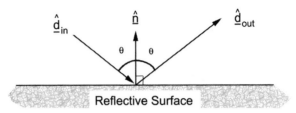

Figure 6.10. Incident and reflection vector.

$\hat{\mathbf{d}}_{out}$ and $\hat{\mathbf{d}}_{in}$ are normalized (the argument would work equally well provided $\mathbf{d}_{out}$ and $\mathbf{d}_{in}$ are the same length), vector $(\hat{\mathbf{d}}_{out} - \hat{\mathbf{d}}_{in})$ is collinear with $\hat{\mathbf{n}}$. Figure 6.10 illustrates this, therefore,

$$\hat{\mathbf{d}}_{out} - \hat{\mathbf{d}}_{in} = \alpha\hat{\mathbf{n}}. \tag{6.2}$$

Where α is a scalar factor. As the incident and reflected angles are equal,

$$\hat{\mathbf{d}}_{out} \cdot \hat{\mathbf{n}} = -\hat{\mathbf{d}}_{in} \cdot \hat{\mathbf{n}}. \tag{6.3}$$

However, we need to have an equation that allows us to compute the reflection vector. To do that, we need to manipulate Equations (6.2) and (6.3). Initially, we take the dot product of both sides of Equation (6.2) with $\hat{\mathbf{n}}$ to give,

$$\hat{\mathbf{d}}_{out} \cdot \hat{\mathbf{n}} = \hat{\mathbf{d}}_{in} \cdot \hat{\mathbf{n}} + \alpha\hat{\mathbf{n}} \cdot \mathbf{n}. \tag{6.4}$$

We then substitute this expression for $\hat{\mathbf{d}}_{out} \cdot \hat{\mathbf{n}}$ into Equation (6.3) to give,

$$(\hat{\mathbf{d}}_{in} \cdot \hat{\mathbf{n}} + \alpha\hat{\mathbf{n}} \cdot \hat{\mathbf{n}}) = -\hat{\mathbf{d}}_{in} \cdot \hat{\mathbf{n}}. \tag{6.5}$$

Since $\mathbf{n}$ is normalized, $\hat{\mathbf{n}} \cdot \hat{\mathbf{n}} = 1$, thus $\alpha = -2(\hat{\mathbf{d}}_{in} \cdot \hat{\mathbf{n}})$ and so after substituting for α in Equation (6.2), we obtain the reflected direction:

$$\hat{\mathbf{d}}_{out} = \hat{\mathbf{d}}_{in} - 2(\hat{\mathbf{d}}_{in} \cdot \hat{\mathbf{n}})\hat{\mathbf{n}}.$$

This is the standard equation used to compute the direction of the reflected vector, and is very useful when studying lighting effects within the virtual world.

6.6   Transformations

Transformations have two purposes in 3D graphics: *to modify the position vector of a vertex* and *to change the orientation of a direction vector.* It is useful to express and use transformations in the form of a matrix. A matrix is simply a two-dimensional array of numbers which can have any number of rows or columns. For example, a 4×4 matrix might be represented by

$$\begin{bmatrix} a_{00} & a_{01} & a_{02} & a_{03} \\ a_{10} & a_{11} & a_{12} & a_{13} \\ a_{20} & a_{21} & a_{22} & a_{23} \\ a_{30} & a_{31} & a_{32} & a_{33} \end{bmatrix}.$$

Each number in the matrix is usually called an *element* of the matrix, e.g., a_{33}. If you are unfamiliar with matrices, vectors or linear algebra in general, it might be useful to consult the books listed in the references at the end of this chapter [1–3].

A matrix that has only one column is called a *vector*,[1] and when it has three rows it conveys the same information and behaves algebraically in the same way as the familiar 3D vector.

The vectors we have used so far in this chapter apply to a 3D universe, and thus have three components. However, we can also imagine an n-dimensional space, in which case our vector will have n components and there is no difference between a vector with n components and a column matrix with n rows. (The n-dimensional space in which these vectors lie might be hard to visualize, but it is no less mathematically useful than the 3D one we use in graphics and VR.)

Thus we can express our familiar 3D vector as having either three components (x, y, z) or as having three elements written in a single column:

$$\begin{bmatrix} x \\ y \\ z \end{bmatrix}.$$

Matrices like these are said to have three rows and one column; they are 3×1 matrices.

Matrices are mathematical objects and have their own algebra, just as real numbers do. You can add, multiply and invert a matrix. You can do things

---

[1]Sometimes this is termed a column vector, and a matrix that has only one row is termed a row vector.

with matrices that have no equivalent for real numbers, such as write their transpose. Even the familiar 3D *dot* and *cross* products of two vectors can be defined in terms of matrix multiplication.

Several different notations are used to represent a matrix. Throughout this text, when we want to show the elements of a matrix, we will group them inside [ ] brackets. Our earlier work used a bold typeface to represent vectors, and we will continue to use this notation with single-column matrices (vectors). A single-row matrix is just the transpose of a vector,[2] and when we need to write this we will use $\mathbf{p}^T$. When we want to represent a rectangular matrix as an individual entity, we will use the notation of an italicized capital letter, e.g., A.

> *It turns out that all the transformations appropriate for computer graphics work (moving, rotating, scaling etc.) can be represented by a square matrix of size 4×4.*

If a transformation is represented by a matrix T, a point $\mathbf{p}$ is transformed to a new point $\mathbf{p}'$ by matrix multiplication according to the rule:

$$\mathbf{p}' = T\mathbf{p}.$$

The order in which the matrices are multiplied is important. For matrices T and S, the product TS is different from the product ST; indeed, one of these may not even be defined.

There are two important points which are particularly relevant when using matrix transformations in 3D graphics applications:

1. How to multiply matrices of different dimensions.

2. The importance of the order in which matrices are multiplied.

The second point will be dealt with in Section 6.6.4. As for the first point, to multiply two matrices, the number of columns in the first must equal the number of rows in the second. For example, a matrix of size 3×3 and 3×1 may be multiplied to give a matrix of size 3×1. However, a 4×4 and a 3×1 matrix cannot be multiplied. This poses a small problem for us, because vectors are represented by 3×1 matrices and transformations are represented as 4×4 matrices.

---

[2]When explicitly writing the elements of a vector in matrix form, many texts conserve space by writing it as its transpose, e.g., $[x, y, z]^T$, which of course is the same thing as laying it out as a column.

The problem is solved by introducing a new system of coordinates called *homogeneous coordinates* in which each vector is expressed with four components. The first three are the familiar (x, y, z) coordinates and the fourth is usually set to "1" so that now a vector appears as a 4×1 matrix. A transformation applied to a vector in *homogeneous* coordinate form results in another *homogeneous* coordinate vector. For all the 3D visualization work in this book, the fourth component of vectors will be set to unity and thus they can be transformed by 4×4 matrices. We will also use transformations that leave the fourth component unchanged. Thus the position vector $\mathbf{p}$ with components (p_0, p_1, p_2) is expressed in homogeneous coordinates as

$$\mathbf{p} = \begin{bmatrix} p_0 \\ p_1 \\ p_2 \\ 1 \end{bmatrix}.$$

Note: The fourth component will play a very important role when we discuss the fundamentals of computer vision in Chapter 8.

The transformation of a point from $\mathbf{p}$ to $\mathbf{p}'$ by the matrix T can be written as

$$\begin{bmatrix} p_0' \\ p_1' \\ p_2' \\ 1 \end{bmatrix} = \begin{bmatrix} t_{00} & t_{01} & t_{02} & t_{03} \\ t_{10} & t_{11} & t_{12} & t_{13} \\ t_{20} & t_{21} & t_{22} & t_{23} \\ t_{30} & t_{31} & t_{32} & t_{33} \end{bmatrix} \begin{bmatrix} p_0 \\ p_1 \\ p_2 \\ 1 \end{bmatrix}.$$

We will now discuss the basic transformations that are utilized in computer graphics: translation, rotation and scaling. These three transformations have one special property in common: they are the only transformations that do not distort their objects. That is, a straight line in an object will remain straight after applying one of these transformations. In fact, all non-distorting transformations can be broken down into a series of rotations, translations and scaling.

6.6.1 Translation

A translation is represented by the general transformation,

$$T_t = \begin{bmatrix} 1 & 0 & 0 & dx \\ 0 & 1 & 0 & dy \\ 0 & 0 & 1 & dz \\ 0 & 0 & 0 & 1 \end{bmatrix}.$$

This transformation will allow us to move the point with coordinates (x, y, z) to the point with coordinates $(x + dx, y + dy, z + dz)$. That is, the translated point becomes $\mathbf{p'} = T_t\mathbf{p}$ or

$$
\begin{bmatrix} x + dx \\ y + dy \\ z + dz \\ 1 \end{bmatrix} = \begin{bmatrix} 1 & 0 & 0 & dx \\ 0 & 1 & 0 & dy \\ 0 & 0 & 1 & dz \\ 0 & 0 & 0 & 1 \end{bmatrix} \begin{bmatrix} x \\ y \\ z \\ 1 \end{bmatrix}.
$$

6.6.2 Scaling

Scaling is represented by the general transformation matrix:

$$
T_s = \begin{bmatrix} s_x & 0 & 0 & 0 \\ 0 & s_y & 0 & 0 \\ 0 & 0 & s_z & 0 \\ 0 & 0 & 0 & 1 \end{bmatrix}.
$$

Essentially this allows us to scale (expand or contract) a position vector $\mathbf{p}$ with components (x, y, z) by the factors s_x along the x-axis, s_y along the y-axis and s_z along the z-axis. Thus, for example, if we scaled the three vertex positions of a polygon, we could effectively change its size, i.e., make it larger or smaller. The scaled point $\mathbf{p'}$ is given by $\mathbf{p'} = T_s\mathbf{p}$ or

$$
\begin{bmatrix} xs_x \\ ys_y \\ zs_z \\ 1 \end{bmatrix} = \begin{bmatrix} s_x & 0 & 0 & 0 \\ 0 & s_y & 0 & 0 \\ 0 & 0 & s_z & 0 \\ 0 & 0 & 0 & 1 \end{bmatrix} \begin{bmatrix} x \\ y \\ z \\ 1 \end{bmatrix}.
$$

6.6.3 Rotation

In two dimensions, there is only one way to rotate, and it is around a point. In three dimensions, we rotate around axes instead, and there are three of them to consider, each bearing different formulae. Thus a rotation is specified by both an axis of rotation and the angle of the rotation. It is a fairly simple trigonometric calculation to obtain a transformation matrix for a rotation about one of the coordinate axes. When the rotation is to be performed around an arbitrary vector based at a given point, the transformation matrix must be assembled from a combination of rotations about the Cartesian coordinate axes and possibly a translation. This is outlined in more detail in Section 6.6.5.

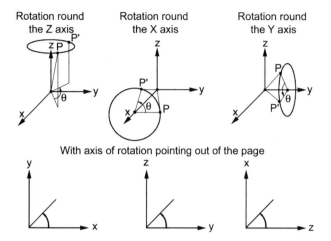

Figure 6.11. Rotations, anticlockwise looking along the axis of rotation, towards the origin.

Rotate About the z-Axis

To rotate round the z-axis by an angle ϑ, the transformation matrix is

$$T_z(\vartheta) = \begin{bmatrix} \cos\vartheta & -\sin\vartheta & 0 & 0 \\ \sin\vartheta & \cos\vartheta & 0 & 0 \\ 0 & 0 & 1 & 0 \\ 0 & 0 & 0 & 1 \end{bmatrix}. \tag{6.6}$$

We can see how the rotational transformations are obtained by considering a positive (anticlockwise) rotation of a point $\mathbf{P}$ by ϑ round the z-axis (which points out of the page). Before rotation, $\mathbf{P}$ lies at a distance l from the origin and at an angle φ to the x-axis (see Figure 6.12). The (x, y)-coordinate of $\mathbf{P}$ is $(l\cos\varphi, l\sin\varphi)$. After rotation by ϑ, $\mathbf{P}$ is moved to $\mathbf{P}'$ and its coordinates are $(l\cos(\varphi+\vartheta), l\sin(\varphi+\vartheta))$. Expanding the trigonometric sum gives expressions for the coordinates of $\mathbf{P}'$:

$$\begin{aligned} P_x' &= l\cos\varphi\cos\vartheta - l\sin\varphi\sin\vartheta, \\ P_y' &= l\cos\varphi\sin\vartheta + l\sin\varphi\cos\vartheta. \end{aligned}$$

Since $l\cos\varphi$ is the x-coordinate of $\mathbf{P}$ and $l\sin\varphi$ is the y-coordinate of $\mathbf{P}$, the coordinates of $\mathbf{P}'$ becomes

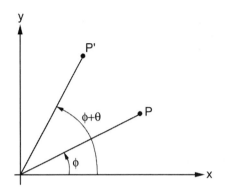

Figure 6.12. Rotation of point **P** by an angle ϑ round the z-axis.

$$
\begin{aligned}
P_x' &= P_x \cos \vartheta - P_y \sin \vartheta, \\
P_y' &= P_x \sin \vartheta + P_y \cos \vartheta.
\end{aligned}
$$

Writing this in matrix form, we have

$$
\begin{bmatrix} P_x' \\ P_y' \end{bmatrix} = \begin{bmatrix} \cos \vartheta & -\sin \vartheta \\ \sin \vartheta & \cos \vartheta \end{bmatrix} \begin{bmatrix} P_x \\ P_y \end{bmatrix}.
$$

There is no change in the z-component of **P**, and thus this result can be expanded into the familiar 4×4 matrix form by simply inserting the appropriate terms to give

$$
\begin{bmatrix} P_x' \\ P_y' \\ P_z' \\ 1 \end{bmatrix} = \begin{bmatrix} \cos \vartheta & -\sin \vartheta & 0 & 0 \\ \sin \vartheta & \cos \vartheta & 0 & 0 \\ 0 & 0 & 1 & 0 \\ 0 & 0 & 0 & 1 \end{bmatrix} \begin{bmatrix} P_x \\ P_y \\ P_z \\ 1 \end{bmatrix}.
$$

Rotation About the y-Axis

To rotate round the y-axis by an angle ϑ, the transformation matrix is

$$
T_y(\vartheta) = \begin{bmatrix} \cos \vartheta & 0 & \sin \vartheta & 0 \\ 0 & 1 & 0 & 0 \\ -\sin \vartheta & 0 & \cos \vartheta & 0 \\ 0 & 0 & 0 & 1 \end{bmatrix}.
$$

Again, it is possible to derive this transformation using the principles described previously for rotation around the z-axis, as can the transformation for the rotation about the x-axis.

Rotation About the x-Axis

To rotate round the x-axis by an angle ϑ, the transformation matrix is

$$T_x(\vartheta) = \begin{bmatrix} 1 & 0 & 0 & 0 \\ 0 & \cos\vartheta & -\sin\vartheta & 0 \\ 0 & \sin\vartheta & \cos\vartheta & 0 \\ 0 & 0 & 0 & 1 \end{bmatrix}.$$

Note that as illustrated in Figure 6.11, ϑ is positive if the rotation takes place in a clockwise sense when looking from the origin along the axis of rotation. This is consistent with a right-handed coordinate system.

6.6.4 Combining Transformations

Section 6.6 introduced the key concept of a transformation applied to a position vector. In many cases, we are interested in what happens when several operations are applied in sequence to a model or one of its points (*vertices*). For example, *move the point* **P** *10 units forward, rotate it 20 degrees round the z-axis and shift it 15 units along the x-axis.* Each transformation is represented by a single 4×4 matrix, and the compound transformation is constructed as a sequence of single transformations as follows:

$$\begin{aligned} \mathbf{p}' &= T_1\mathbf{p}; \\ \mathbf{p}'' &= T_2\mathbf{p}'; \\ \mathbf{p}''' &= T_3\mathbf{p}''. \end{aligned}$$

Where $\mathbf{p}'$ and $\mathbf{p}''$ are intermediate position vectors and $\mathbf{p}'''$ is the end vector after the application of the three transformations. The above sequence can be combined into

$$\mathbf{p}''' = T_3 T_2 T_1 \mathbf{p}.$$

The product of the transformations $T_3 T_2 T_1$ gives a single matrix T. Combining transformations in this way has a wonderful efficiency: if a large model has 50,000 vertices and we need to apply 10 transformations, by combining

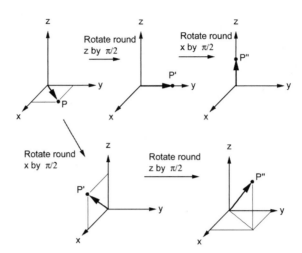

Figure 6.13. Effect of transformations applied in a different order.

the transformations into a single matrix, 450,000 matrix multiplications can be avoided.

It is important to remember that the result of applying a sequence of transformations depends on the order in which they are applied. $T_3 T_2 T_1$ is *not* the same compound transformation as $T_2 T_3 T_1$. Figure 6.13 shows the effect of applying the transformations in a different order.

There is one subtle point about transformations that ought to be stressed. The parameters of a transformation (angle of rotation etc.) are all relative to a *global* frame of reference. It is sometimes useful to think in terms of a local frame of reference that is itself transformed relative to a global frame. This idea will be explored when we discuss keyframe and character animation. However, it is important to bear in mind that when a final scene is assembled for rendering, all coordinates must be specified in the same frame of reference.

6.6.5   Rotation About an Arbitrary Axis

The transformation corresponding to rotation of an angle α around an arbitrary vector (for example, that shown between the two points $\mathbf{P}_0$ and $\mathbf{P}_1$ in Figure 6.14) cannot readily be written in a form similar to the rotation matrices about the coordinate axes.

The desired transformation matrix is obtained by combining a sequence of basic translation and rotation matrices. (Once a single 4×4 matrix has

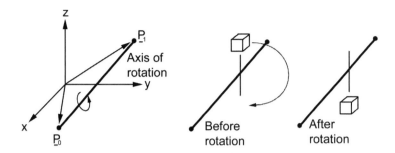

Figure 6.14. Rotation round an arbitrary vector.

been obtained representing the composite transformations, it can be used in the same way as any other transformation matrix.)

The following outlines an algorithm to construct a transformation matrix to generate a rotation by an angle α around a vector in the direction $\mathbf{P}_1 - \mathbf{P}_0$:

1. Translate $\mathbf{P}_0$ to the origin of coordinates

2. Align rotation axis $\mathbf{P}_1 - \mathbf{P}_0$ with the x-axis.

3. Rotate by angle α round x-axis.

4. Make inverse transformation to undo the rotations of Step 2.

5. Translate origin of coordinates back to $\mathbf{P}_0$ to undo the translation of Step 1.

The full algorithm is given in Figure 6.15.

6.6.6 Viewing Transformation

Before rendering any view of a 3D scene, one has to decide from where to view/photograph the scene and in which direction to look (point the camera). This is like setting up a camera to take a picture. Once the camera is set, we just click the shutter. The camera projects the image as seen in the viewfinder onto the photographic film and the image is rendered. Likewise, when rendering a 3D scene, we must set up a viewpoint and a direction of view. Then we must set up the projection (i.e., determine what we can and cannot see), which is discussed in Section 6.6.7.

In mathematical terms, we need to construct a suitable transformation that will allow us to choose a viewpoint (place to set up the camera) and

Let $\mathbf{d} = \mathbf{P}_1 - \mathbf{P}_0$
$T_1 =$ a translation by $-\mathbf{P}_0$
$d_{xy} = d_x^2 + d_y^2$
if $d_{xy} < \varepsilon$ { *rotation axis is in the z-direction*
    if $d_z > 0$ make T_2 a rotation about z by α
    else $T_2 =$ a rotation about z by $-\alpha$
    $T_3 =$ a translation by $\mathbf{P}_0$
    return the product $T_3 T_2 T_1$
}
$d_{xy} = \sqrt{d_{xy}}$
if $d_x = 0$ and $d_y > 0$ $\varphi = \pi/2$
else if $d_x = 0$ and $d_y < 0$ $\varphi = -\pi/2$
else $\varphi = ATAN2(d_y, d_x)$
$\vartheta = ATAN2(d_z, d_{xy})$
$T_2 =$ a rotation about z by $-\varphi$
$T_3 =$ a rotation about y by $-\vartheta$
$T_4 =$ a rotation about x by α
$T_5 =$ a rotation about y by ϑ
$T_6 =$ a rotation about z by φ
$T_7 =$ a translation by $\mathbf{P}_0$
Multiply the transformation matrices to give the final result
$T = T_7 T_6 T_5 T_4 T_3 T_2 T_1$

Figure 6.15. Algorithm for rotation round an arbitrary axis. Points d_x, d_y and d_z are the components of vector $\mathbf{d}$.

direction of view (direction in which to point the camera). Once we have this *view transformation*, it can be combined with any other transformations that needs to be applied to the scene, or to objects in the scene.

We have already seen how to construct transformation matrices that move or rotate points in a scene. In the same way that basic transformation matrices were combined in Section 6.6.5 to create an arbitrary rotation, we can build a single matrix, T_o, that will transform all the points (vertices) in a scene in such a way that the projection of an image becomes a simple standard process. But here is the funny thing. In computer graphics, we do not actually change the position of the camera; we actually transform or change the position of all the vertices making up the scene so that camera is fixed at the center of the universe $(0, 0, 0)$ and locked off to point in the direction $(1, 0, 0)$ (along the x-axis).

There is nothing special about the direction $(1, 0, 0)$. We could equally well have chosen to fix the camera to look in the direction $(0, 1, 0)$, the y-axis, or even $(0, 0, 1)$, the z axis. But we have chosen to let z represent the *up* direction, and it is not a good idea to look directly up, so $(0, 0, 1)$ would be a poor choice for viewing. (Note: The OpenGL and Direct3D software libraries for 3D graphics have their z-axis parallel to the viewing direction.)

Once T_o has been determined, it is applied to all objects in the scene. If necessary, T_o can be combined with other transformation matrices to give a single composite transformation T.

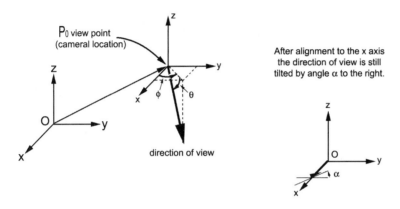

Figure 6.16. Viewpoint and direction of view.

A viewing transformation depends on:

- Position of the viewpoint (camera location). (A vector $\mathbf{p}_o$. See Figure 6.16.)

- The direction in which we wish to look: north, south, east or west. This is measured by an angle φ which is relative to the x-axis and lies in the xy-plane. φ is positive to the right of the x-axis when looking along x from the viewpoint.

- The amount by which we look up or down. The angle ϑ measures this relative to the xy-plane. It is positive when looking down. Note that when determining a viewing transformation, the effect of looking up or down comes into play after the direction of view has been accounted for, and therefore it is equivalent to a rotation around the y-axis.

- The degree to which our head is tilted to the left or right. This is measured by the angle α. To be consistent with the right-handed frame of reference and sense of rotation, α is positive when the camera tilts to the right as it looks from $\mathbf{p}_o$ along the x-axis.

A viewing transformation appears to operate in reverse to ordinary transformations. For example, if you tilt your head to the left, the world appears to tilt to the right. Note carefully that the angle ϑ is positive if we are looking down and negative if we looking up. If you prefer, you can think of φ as the *heading*, ϑ as the *pitch* and α as the degree of *banking*.

The viewing transformations are also combined in the reverse order to the order in which a transformation is assembled for objects placed in a scene. In that case, the rotation around the x-axis is applied first and the translation by $\mathbf{p}_o$ is applied last.

Given the parameters $\mathbf{p}_o$, φ, ϑ and α (illustrated in Figure 6.16), the transformation T_o is constructed by the following algorithm:

Place observer at $(0, 0, 0)$ with the transformation:
T_1 = a translation by $-\mathbf{p}_o$
Rotate the direction of observation into the xz-plane with:
T_2 = a rotation about z by $-\varphi$
Align the direction of observation to the x-axis with:
T_3 = a rotation about y by $-\vartheta$
Straighten the camera up with transformation:
T_4 = a rotation about x by $-\alpha$
Multiply the individual transformation matrices to give
one composite matrix representing the viewing transformation:
$T_0 = T_4 T_3 T_2 T_1$

6.6.7   Projection Transformation

After we set up a camera to record an image, the view must then be projected onto film or an electronic sensing device. In the conventional camera, this is done with a lens arrangement or simply a *pinhole*. One could also imagine holding a sheet of glass in front of the viewer and then having her trace on it what she sees as she looks through it. What is drawn on the glass is what we would like the computer to produce: a 2D picture of the scene. It's even shown the right way up, as in Figure 6.17.

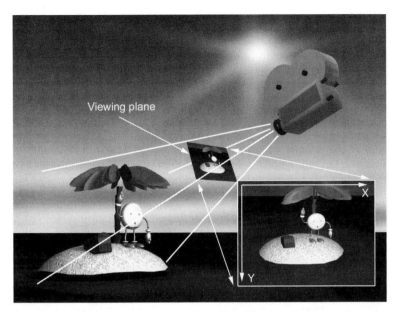

Figure 6.17. Project the scene onto the viewing plane. The resulting two-dimensional image is then recorded or displayed.

It is straightforward to formulate expressions needed to perform this (non-linear) transformation. A little thought must be given to dealing with cases where parts of the scene go behind the viewer or are partly in and partly out of the field of view. The field of view (illustrated in Figure 6.18) governs how much of the scene you see. It can be changed so that you are able to see more or less of the scene. In photography, telephoto and fish-eye lenses have different fields of view. For example, the common 50 mm lens has a field of view of 45.9°. Because of its shape as a truncated pyramid with a regular base, the volume enclosed by the field of view is known as a *frustum*.

One thing we can do with a projective transformation is adjust the *aspect ratio*. The *aspect ratio* is the ratio of height to width of the rendered image. It is 4 : 3 for television work and 16 : 9 for basic cine film. The aspect ratio is related to the vertical and horizontal resolution of the recorded image. Get this relationship wrong and your spheres will look egg shaped.

Before formulating expressions to represent the projection, we need to define the coordinate system in use for the projection plane. It has become almost universal[3] to represent the computer display as a pair of integers in

---

[3]However, there are important exceptions, e.g., the OpenGL 3D library, where floating-point numbers are used.

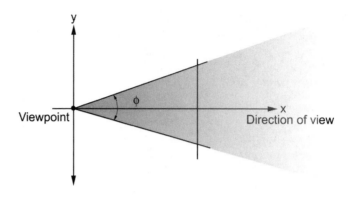

Figure 6.18. The field of view φ governs how much of the scene is visible to a camera located at the viewpoint. Narrowing the field of view is equivalent to using a zoom lens.

the range $0 \to (X_{\max} - 1)$ horizontally and $0 \to (Y_{\max} - 1)$ vertically, where $X_{\max}$ and $Y_{\max}$ refer to the number of pixels in the rendered image in the horizontal and vertical direction, respectively. The coordinate origin $(0, 0)$ is in the top-left corner (see Figure 6.17).

The distance of the projection plane from the viewpoint can be chosen arbitrarily; setting it to a value of 1 simplifies the calculations. Thus, if the plane of projection is located at $(1, 0, 0)$ and oriented parallel to the yz-plane (i.e., the viewer is looking along the x-axis) then the screen coordinates (X_s, Y_s) for the projection of a point (x, y, z) are given by

$$X_s = \frac{X_{\max}}{2} - \frac{y}{x} s_x, \tag{6.7}$$

$$Y_s = \frac{Y_{\max}}{2} - \frac{z}{x} s_y. \tag{6.8}$$

The parameters s_x and s_y are scale values to allow for different aspect ratios and fields of view. This effectively lets us change the *zoom* settings for the camera. Obviously, X_s and Y_s must satisfy $0 \leq X_s < X_{\max}$ and $0 \leq Y_s < Y_{\max}$.

If f_f (measured in mm) is the focal length of the desired camera lens and the aspect ratio is $A_x : A_y$ then

$$s_x = \frac{X_{\max}}{2} \frac{f_f}{21.22} A_x,$$

$$s_y = \frac{Y_{\max}}{2} \frac{f_f}{21.22} A_y.$$

The numerical factor 21.2 is a constant to allow us to specify f_f in standard mm units. For a camera lens of focal length f_f, the field of view ϑ can be expressed as

$$\vartheta \simeq 2ATAN2(21.22, f_f).$$

Any point (x, y, z) for which $x < 1$ will not be transformed correctly by Equations (6.7) and (6.8). The reason for this is that we have placed the projection plane at $x = 1$ and therefore these points will be behind the projection plane. As such, steps must be taken to eliminate these points before the projection is applied. This process is called *clipping*. We talk more about clipping in the next chapter.

6.7 Summary

Let's take a quick review of what we have covered thus far. We know that to construct a virtual environment, we need to be able to connect different primitive shapes together. We then need to be able to assign properties to these shapes; for example, color, lighting, textures etc. Once we have constructed our virtual environment, we then need to be able to produce single images of it, which we do by a process of rendering.

Whilst we have not covered the details of the rendering process yet, we do know that before we render an image, we need to be able to determine what we will actually see of the virtual environment. Some objects or primitives will fall into this view and others will not. In order to determine what we can see, we need to be able to perform some 3D geometry. In this chapter, we covered the basics of this. In particular, we discussed determining the intersection points of planes and lines and setting up viewing transformations.

So now that we have looked at how the viewing transformation is constructed, it is a good idea to look at the process of utilizing this to produce a 2D picture of our 3D environment. It is to the interesting topic of rendering that we now turn our attention.

Bibliography

[1] F. Ayres. *Matrices,* Schaum's Outline Series. New York: McGraw-Hill, 1967.

[2] R. H. Crowell and R. E. Williamson. *Calculus of Vector Functions.* Englewood Cliffs, NJ: Prentice Hall, 1962.

[3] S. Lipschutz. *Schaum's Outline of Linear Algebra,* Third Edition, Schaum's Outline Series. New York: McGraw-Hill, 2000.

7

The Rendering Pipeline

In this chapter, we are going to take a look inside *the* most important element of VR—*the 3D graphics*. Creating realistic-looking 3D graphics (a process called rendering) poses a particular challenge for VR system builders because, unlike making a 3D movie or TV show, we must endow our system with a capability for real-time interactivity. In the past, this has meant that VR graphics *engines*[1] had difficulty in achieving *that really realistic look*. Not any more! For this, we have to thank the computer-game industry. It was the multi-billion dollar market demand for games that accelerated the development of some spectacular hardware rendering engines, available now at knockdown prices. VR system builders can now expect to offer features such as stereopsis, shadowing, high dynamic lighting range, procedural textures and Phong shading—rendered in high resolution and in real time and with full interactivity.

In Chapters 13 and 14 we will look at how to write application programs that use the 3D hardware accelerated rendering engines, but here we intend to look at the algorithmic principles on which they operate. In using any high-tech technology, we hope you will agree that it is of enormous benefit to know, in principle, how it works and what is going on inside, so here goes.

The basic rendering algorithm takes a list of polygons and vertices and produces a picture of the object they represent. The picture (or *image*) is recorded as an array of bytes in the output frame buffer. This array is made up

---

[1] The term *engine* is commonly used to describe the software routines of an application program that carry out the rendering. It originated in the labs of computer-game developers.

of little regions of the picture called fragments.[2] The frame buffer is organized into rows, and in each row there are a number of fragments. Each fragment in the frame buffer holds a number which corresponds to the color or intensity at the equivalent location in the image. Because the number of fragments in the frame buffer is finite (the fragment value represents a small area of the real or virtual world and not a single point), there will always be some inaccuracy when they are used to display the image they represent.

The term *graphics pipeline* is used to describe the sequence of steps that the data describing the 3D world goes through in order to create 2D images of it. The basic steps of the graphics pipeline are:

1. *Geometry and vertex operations.* The primary stage of the graphics pipeline must perform a number of operations on the vertices that describe the shape of objects within a scene or the scene itself. Example operations include transformation of all vertex coordinates to a real-world frame of reference, computation of the surface normal and any attributes (color, texture coordinate etc.) associated with each vertex etc.

2. *Occlusion, clipping and culling.* Occlusion attempts to remove polygons quickly from the scene if they are completely hidden. (Occlusion is different from clipping or culling because algorithms for occlusion operate much faster, but sometimes make a mistake, so it is typically applied only to scenes with enormous polygon counts, greater than 1×10^6 to 2×10^6.) Clipping and culling again remove polygons or parts of polygons that lie outside a given volume bounded by planes (called *clipping planes*), which are formed using a knowledge of the viewpoint for the scene. This allows the rasterizer to operate efficiently, and as a result, only those polygons that we will actually see within our field of view will be rendered.

3. *Screen mapping.* At this stage in the graphics pipeline, we have the 3D coordinates of all visible polygons from a given viewpoint for the scene. Now we need to map these to 2D screen or image coordinates.

4. *Scan conversion or rasterization.* The rasterizer takes the projections of the visible polygons onto the viewing plane and determines which fragments in the frame buffer they cover. It then applies the Z-buffer

---

[2]An individual fragment corresponds to a pixel that is eventually displayed on the output screen.

algorithm to determine, for each fragment, which polygon is closest to the viewpoint and hence visible.

5. *Fragment processing.* For each fragment, this stage calculates what is visible and records it in the output frame buffer. This typically includes an illumination model, a shading model (Gouraud or Phong) and surface texturing, either procedural or via image mapping. It is interesting to note that the term *fragment* is used to describe the output from the rasterizer, whilst the term *pixel* is used to describe the output from the fram ebuffer, i.e., what is displayed on screen. The information stored in the fragment is actually used to calculate the color of its corresponding pixel.

Once rendering is complete, the contents of the frame buffer can be transferred to the screen. So, in simple terms, if we have a 3D model of an environment then in order to produce an image of it from a certain viewpoint, we need to decide what we will actually see from that viewpoint. Sounds simple? Well, as with most things, the easier it sounds the harder it is to do! Therefore, in this chapter we hope to give you some insight into this fascinating subject by outlining in much more detail those steps of the graphics pipeline we summarized previously.

7.1   Geometry and Vertex Operations

At this stage in the graphics pipeline, we need to perform some basic operations at each vertex within the scene. As such, the amount of work that has to be done is in proportion to the number of vertices in the scene, or the scene complexity. Essentially, these operations can be summarized as:

- *Modeling transformation.* When creating a 3D environment, we will almost always require some 3D objects or models to place within that environment. These models will have been created according to a local frame of reference. When we insert them into our 3D environment, we must transform all their vertex coordinates from that local reference frame to the global frame of reference used for our 3D environment. This transformation can of course include some scaling, translation and rotation, all of which were outlined in Section 6.6.

- *Viewing transformation.* Once our global 3D environment has been created, we have basically a 3D model described by a list of polygons and their vertex coordinates. For this to be useful, we need to decide where within the model we want to view it from. Having determined the viewpoint or the position of the virtual camera, we need to apply another transformation to each of the vertices within the 3D model. The reason for this is that we need to transform the virtual camera from its location (which can be anywhere) within the 3D world to the origin of the global reference frame $(0, 0, 0)$. Typically within a right-handed coordinate system, the direction of the viewpoint will be taken along the *x-axis*. (Note: this can vary depending on the graphics rendering engine you are using. For example, in OpenGL, the direction of view is taken along the *z-axis*.) As such, we need to apply the same transformation (which will only involve translation and rotation) to all other vertices within the model.

Some hardware rendering processors also do lighting calculations on a per-vertex basis, but to achieve realistic lighting, it should be done on a per-fragment basis. And since most accelerated graphics hardware allows us to do this now, we will delay our description of the lighting process until later.

7.2   Culling and Clipping

In complex environments, it is unlikely that you will be able to see every detail of the environment from a given viewpoint. For example, an object might be in complete shadow, it might be occluded by another object or perhaps it is outside the viewing area. Since rendering is very processor-intensive, it seems natural to try to avoid even thinking about rendering those objects that you cannot see. Therefore, this stage of the graphics pipeline aims to reject objects which are not within the field of view of the virtual camera and that are occluded by other objects within the environment.

We will use the term *culling* to apply to the action of discarding complete polygons; i.e., those polygons with all their vertices outside the field of view. Clipping will refer to the action of modifying those polygons (or edges) that are partially in and partially out of the field of view. Culling should always be performed first, because it is likely to give the biggest performance gain. Figure 7.1 illustrates the principle of culling, clipping and occluding. In Figure 7.1(c), the importance of occlusion becomes evident in high-complexity scenes. A building plan is depicted with a large number of objects that fall inside the field of view but are occluded by the walls between rooms. These

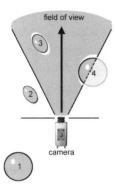

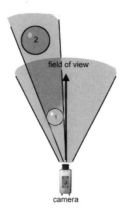

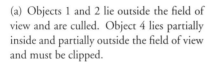

(a) Objects 1 and 2 lie outside the field of view and are culled. Object 4 lies partially inside and partially outside the field of view and must be clipped.

(b) Object 2 is occluded from view behind object 1 and so may be culled.

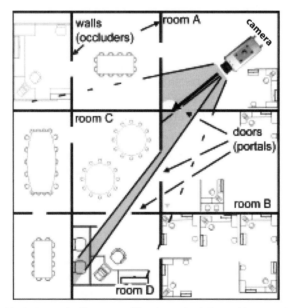

(c) An internal office scene. The viewpoint is located in room A and so the objects in other rooms are hidden beyond the walls. Open doors (portals) allow a few objects in room D to be visible. A special step in the rendering pipeline may be used to determine whether objects are likely to be visible or not.

Figure 7.1. Culling, clipping and occulsion.

should obviously be culled, but the situation is complicated by the presence of portals (the doors): some objects in distant rooms may still be visible from the viewpoint. In computer games and other real-time VR applications, special pre-processing is sometimes applied to determine the visibility status of objects in such scenes.

7.2.1 Culling

For 3D graphics, there are three important types of cull:

1. Remove polygons behind the viewpoint. If the x-coordinate (again, the assumption being that the viewing direction is along the x-axis) of *all* vertices for polygon k are such that $x \leq 1$ then cull polygon k. Remember, in Section 6.6.7 we made the assumption that the viewing plane[3] was located at $x = 1$. So this is equivalent to removing any polygons that lie behind the projection plane.

2. Cull any *backfaces*. A *backface* is any polygon that faces away from the viewpoint. It is called a backface because if the model were solid then any polygons facing away from the viewer would be obscured from view by polygons closer to the viewer. For example, in a street scene, there would be no point rendering the polygons on the sides of the buildings facing away from the virtual camera, since they will be occluded by those sides of the building facing the camera.

   In order to say whether a polygon faces towards or away from an observer, a convention must be established. The most appropriate convention to apply is one based on the surface normal vector. Every polygon (which is essentially a bounded plane) has a surface normal vector as defined in Section 6.4. Since it is possible for a surface normal to point in two opposite directions, it is usually assumed that it is directed so that, in a general sense, it points away from the inside of an object. Figure 7.2 shows a cube with the normal vectors illustrated. If the dot product of the normal vector with a vector connecting the viewpoint to a point on the polygon is negative, the polygon is said to be a *frontface*; if the dot product is positive, the polygon is a *backface*.

3. Cull any polygons that lie completely outside the field of view. This type of cull is not so easy to carry out, and it is usually combined with the clipping operation described next.

---

[3]also called the plane of projection.

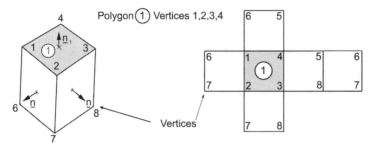

Figure 7.2. Surface normal vectors for a cube and a plan of the vertices showing a consistent ordering of the vertices round each polygonal face. For example, $\mathbf{n}_1 = (\mathbf{P}_2 - \mathbf{P}_1) \times (\mathbf{P}_3 - \mathbf{P}_1)$.

7.2.2 Clipping

After the culling stage, we will probably have rejected a large number of polygons that could not possibly be seen from the viewpoint. We can also be sure that if all of a polygon's vertices lie inside the field of view then it might be visible and we must pass it to the rasterization stage. However, it is very likely that there will still be some polygons that lie partially inside and outside the field of view. There may also be cases where we cannot easily tell whether a part of a polygon is visible or not. In these cases, we need to *clip* away those parts of the polygon which lie outside the field of view. This is non-trivial, because we are essentially changing the shape of the polygon. In some cases, the end result of this might be that *none* of the polygon is visible at all.

Clipping is usually done against a set of planes that bound a volume. Typically this takes the form of a pyramid with the top end of the pyramid at the viewpoint. This bounded volume is also known as a frustum. Each of the sides of the pyramid are then known as the *bounding planes*, and they form the edges of the field of view. The top and bottom of the frustum form the front and back clipping planes. These front and back clipping planes are at right angles to the direction of view (see Figure 7.3). The volume contained within the pyramid is then known as the viewing frustum. Obviously, polygons and parts of polygons that lie inside that volume are retained for rendering. Some polygons can be marked as lying completely outside the frustum as the result of a simple test, and these can be culled. However, the simple test is often inconclusive and we must apply the full rigor of the clipping algorithm.

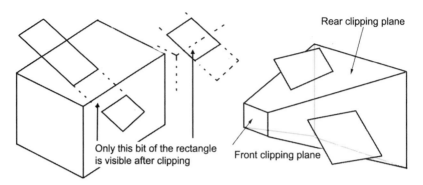

Figure 7.3. Cubic and truncated pyramidal clipping volumes penetrated by a rectangular polygon. Only that portion of the rectangle which lies inside the clipping volume would appear in the rendered image.

To use these multiple clipping planes, it is sufficient to apply them one at a time in succession. For example, clip with the back plane, then the front plane, then the side plane etc.

Clipping a Triangular Polygon with a Clipping Plane

To see how 3D clipping is achieved, we consider the following. Look at Figure 7.4; it shows a triangular polygon *ABC* which is to be clipped at the plane *PP′* (seen end on). Clipping is accomplished by splitting triangle *ABC* into two pieces, *BDE* and *ACED*. The pieces are then either removed from or added to the database of polygons that are to be rendered. For example, when *PP′* is the back clipping plane, triangles *ADE* and *AEC* are added to the database, whilst triangle *BDE* is removed. Points *D* and *E* are determined by finding the intersection between the lines joining vertices of the polygon and the clipping plane.

This calculation is straightforward for both the front or back clipping planes because they are perpendicular to the *x*-axis (that is, the direction of view in the right-handed coordinate system convention). That means they form a plane in the *yz*-axis at points x_f and x_b respectively. That is, x_f refers to the *x*-coordinate of the front clipping plane and x_b to the *x*-coordinate of the back clipping plane. Because clipping at either plane is commonly used, it is worth writing expressions specifically for these particular circumstances. Let's look at the back plane specifically.

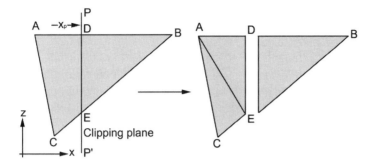

Figure 7.4. Clipping a triangular polygon, *ABC*, with a *yz* plane at *PP'*, at $(x_p, 0, 0)$. Clipping divides *ABC* into two pieces. If the polygons are to remain triangular, the piece *ADEC* must be divided in two.

The intersection point, **p**, between a line joining points $\mathbf{P}_1$ and $\mathbf{P}_2$ and the *yz* plane at x_b is given by

$$
\begin{aligned}
p_x &= x_b, \\
p_y &= P_{1y} + \frac{(x_b - P_{1x})}{(P_{2x} - P_{1x})}(P_{2y} - P_{1y}), \\
p_z &= P_{1z} + \frac{(x_b - P_{1x})}{(P_{2x} - P_{1x})}(P_{2z} - P_{1z}).
\end{aligned}
$$

Likewise for the front viewing plane, except x_b is replaced with x_f in the calculations. Finding the intersection point between polygons intersecting the side viewing planes is more complicated, since these planes are not at right angles to the *x*-axis. As such, a more general calculation of the intersection of a line and a plane needs to be performed. This is shown in Section 6.4.1.

However, regardless of the plane we clip on, once the coordinates of the points *D* and *E* are known, the triangular polygon *ABC* is divided into two or three triangular polygons and stored in the polygon database. This is in essence how several of the most common clipping algorithms work.

7.3   Screen Mapping

At this stage, we know the 3D vertex coordinates of every visible polygon within the viewing volume. Now we need to convert these to 2D image or screen coordinates. For example, in the real world, when we use a camera, we create a 2D picture of the 3D world. Likewise, in the virtual world, we now need to create a 2D image of what our virtual camera can see based on its viewpoint within the 3D environment. 3D coordinate geometry allows us to do this by using a projective transformation, where all the 3D coordinates are transformed to their respective 2D locations within the image array. The intricacies of this are detailed in Section 6.6.7. This process only affects the 3D y- and z-coordinates (again assuming the viewing direction is along the x-axis). The new 2D (X, Y)-coordinates, along with the 3D x-coordinate, are then passed on to the next stage of the pipeline.

7.4   Scan Conversion or Rasterization

At the next stage, the visible primitives (points, lines, or polygons) are decomposed into smaller units corresponding to pixels in the destination frame buffer. Each of these smaller units generated by rasterization is referred to as a fragment. For instance, a line might cover five pixels on the screen, and the process of rasterization converts the line (defined by two vertices) into five fragments. A fragment is comprised of a frame buffer coordinate, depth information and other associated attributes such as color, texture coordinates and so on. The values for each of these attributes are determined by interpolating between the values specified (or computed) at the vertices of the primitive. Remember that in the first stage of the graphics pipeline, attributes are only assigned on a per-vertex basis.

So essentially, this stage of the pipeline is used to determine which fragments in the frame buffer are covered by the projected 2D polygons. Once it is decided that a given fragment is covered by a polygon, that fragment needs to be assigned its attributes. This is done by linearly interpolating the attributes assigned to each of the corresponding 3D vertex coordinates for that polygon in the *geometry and vertex operations* stage of the graphics pipeline.

Now, if we simply worked through the list of polygons sequentially and filled in the appropriate fragments with the attributes assigned to them, the resulting image would look very strange. This is because if the second sequential polygon covers the same fragment as the first then all stored information about the first polygon would be overwritten for that fragment. Therefore, as the second polygon is considered, we need to know the 3D x-coordinate of the relevant vertex and determine if it is in front or behind the previous x-coordinate assigned to that fragment. If it is in front of it, a new attribute value for the second polygon can be assigned to that fragment and the associated 3D x-coordinate will be updated. Otherwise, the current attributes will remain unchanged. This procedure determines which polygon is visible from the viewpoint. Remember, we have defined the viewing plane to be at $x = 1$. Therefore, the 3D x-coordinate of each vertex determines how far it is from the viewing plane. The closer it is to the viewing plane, the more likely that the object connected to that vertex will not be occluded by other objects in the scene. Essentially, this procedure is carried out by the Z-buffer algorithm, which we will now cover in more detail.

7.4.1   The Z-Buffer Algorithm

The Z-buffer algorithm is used to manage the visibility problem when rendering 3D scenes; that is, which elements of the rendered scene will be visible and which will be hidden. Look at Figure 7.5; it shows a polygon and its projection onto the viewing plane where an array of fragments is also illus-

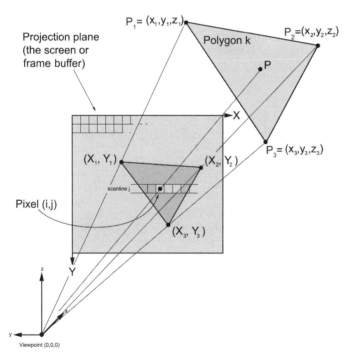

Figure 7.5. Projecting back from the viewpoint through pixel (i, j) leads to a point in the interior of polygon k.

trated. Imagine adding a matching array of similar dimension that is capable of recording a real number for each fragment. Let us call this the Z-buffer. The main action of the Z-buffer algorithm is to record the depth coordinate (or the x-coordinate) associated with the 3D point $\mathbf{P}$ into the buffer at the address (i, j). This action is repeated for all fragments inside the projection of polygon k.

When it comes to drawing another polygon, say l, it too will paint into a set of fragments, some of which may overlap with those previously filled with data from polygon k. Now the Z-buffer comes into play. Before information for polygon l is written to the fragment at location (i, j), the Z-buffer is checked to see whether polygon l appears to be in front of polygon k. If l is in front of k then the data from l is placed in the frame buffer and the Z-buffer depth at location (i, j) is updated to take account of l's depth.

The algorithm is known as the *Z-buffer algorithm* because the first programs to use it arranged their frame of reference with the viewpoint at the origin and the direction of view aligned along the z-axis. In these programs, the distance to any point on the polygon from the viewpoint was simply the point's z-coordinate. (OpenGL and Direct3D use this direction for their direction of view.) The Z-buffer algorithm is summarized in Figure 7.6.

To find P, and hence its x-coordinate, we call on the ideas of Section 6.4.1 for the intersection of line and plane. Initially, we need to transform the 2D

> Fill the depth buffer at $Z(i, j)$ with a *"far away"* depth;
> i.e., set $Z(i, j) = \infty$ for all i, j.
> Repeat for all polygons k {
>   For polygon k find pixels (i, j) covered by it. Fill
>   these with the color or texture of polygon k.
>   With each pixel (i, j) covered by k repeat {
>   Calculate the depth (Δ) of P from V (see Figure 7.5)
>   If $\Delta < Z(i, j)$ {
>   Set pixel (i, j) to color of polygon k
>   Update the depth buffer $Z(i, j) = \Delta$
>   }
>   }
> }

Figure 7.6. The basic ideas of the Z-buffer rendering algorithm.

fragment coordinate at (i, j) into a 3D coordinate by using the inverse of the projective transformation previously discussed. The new 3D coordinate, along with the viewpoint, is used to derive the equation of the line intersecting the polygon k at the point P. The vertices, $\mathbf{P_1}, \mathbf{P_2}, \mathbf{P_3}$, of the polygon k can be used to form a bounded plane. Thus, P can be found using standard 3D geometry, as it is simply the point of intersection between a line and bounded plane.

7.5   Fragment Processing

In the early generations of hardware graphics processing chips (GPUs), fragment processing primarily consisted of interpolating lighting and color from the vertex values of the polygon which has been determined to be visible in that fragment. However, today in software renderers and even in most hardware rendering processors, the lighting and color values are calculated on a per-fragment basis, rather than interpolating the value from the polygon's vertices.

In fact, determining the color value for any fragment in the output frame buffer is the most significant task any rendering algorithm has to perform. It governs the shading, texturing and quality of the final rendering, all of which are now outlined in more detail.

7.5.1   Shading (Lighting)

The way in which light interacts with surfaces of a 3D model is the most significant effect that we can model so as to provide visual realism. Whilst the Z-buffer algorithm may determine *what you can see*, it is mainly the interaction with light that determines *what you do see*. To simulate *lighting* effects, it stands to reason that the location and color of any lights within the scene must be known. In addition to this, we also need to classify the light as being one of three standard types, as illustrated in Figure 7.7. That is:

1. A point light source that illuminates in all directions. For a lot of indoor scenes, a point light source gives the best approximation to the lighting conditions.

2. A directional or parallel light source. In this case, the light comes from a specific direction which is the same for all points in the scene. (The illumination from the sun is an example of a directional light source.)

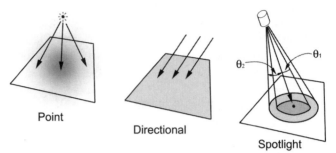

Figure 7.7. Three types of light source.

3. A spotlight illumination is limited to a small region of the scene. The beam of the spotlight is normally assumed to have a graduated edge so that the illumination is a maximum inside a cone of half angle ϑ_1 and falls to zero intensity inside another cone of half angle ϑ_2. Naturally, $\vartheta_2 > \vartheta_1$.

Now that we know a little about the three standard light sources that are available for illuminating our scene, we need to consider how the light from these sources interacts with objects within our scene. To do this, we must consider the spatial relationship between the lights, the camera and the polygons. This is illustrated in Figure 7.8. In the same way that we have three standard light sources, there are also standard lighting components which represent the way light interacts within the scene. These are ambient reflection (I_a), diffuse reflection (I_d), specular reflection (I_s) and depth cuing (I_c). It is further pos-

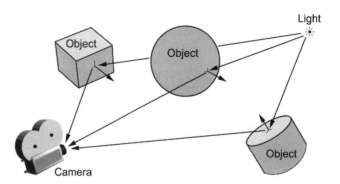

Figure 7.8. Lights, camera and objects. Reflected light finds its way to the observer by being reflected from the object's surfaces.

sible to represent these lighting components as mathematical models, which we will now consider. In these models, we will assume that $\mathbf{p}$ is the point to be illuminated on the visible polygon which has a surface normal $\hat{\mathbf{n}}$ at $\mathbf{p}$.

- *Ambient reflection (I_a).* In a simulated scene, a surface that has no light incident upon it will appear blank. This is not realistic, however, because in a real scene, that surface would be partially illuminated by light that has been reflected from other objects in the scene. Thus, the ambient reflection component is used to model the reflection of light which arrives at a point on the surface from all directions after being bounced around the scene from all other surfaces. In practice, the ambient component is defined to have a constant intensity for all surfaces within the scene. Thus for each surface, the intensity of the ambient lighting is $I_a = k$.

- *Diffuse reflection (I_d).* The term *reflection* is used here because it is light reflected from surfaces which enters the camera. To model the effect, we assume that a polygon is most brightly illuminated when the incident light strikes the surface at right angles. The illumination falls to zero when the direction of the beam of light is parallel to the surface. This behavior is known as the Lambert cosine law and is illustrated in Figure 7.9.

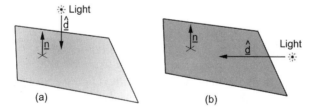

Figure 7.9. Diffuse illumination. (a) The brightest illumination occurs when the incident light direction is at right angles to the surface. (b) The illumination tends to zero as the direction of the incident light becomes parallel to the surface.

The diffuse illumination component I_d is calculated differently for each of the standard types of light source that we have within the scene:

- For a point light source located at $\mathbf{P}_l$:

$$I_d = \frac{\mathbf{P}_l - \mathbf{p}}{|\mathbf{P}_l - \mathbf{p}|} \cdot \hat{\mathbf{n}}.$$

– For a directional light source with incident direction $\hat{\mathbf{d}}$:

$$I_d = -\hat{\mathbf{d}} \cdot \hat{\mathbf{n}}.$$

– For a spotlight located at $\mathbf{P}_l$, pointing in direction $\hat{\mathbf{d}}$ and having light cone angles ϑ_1 and ϑ_2:

$$I_d = \begin{cases} \dfrac{\mathbf{P}_l - \mathbf{p}}{|\mathbf{P}_l - \mathbf{p}|} \cdot \hat{\mathbf{n}} & \text{if } \vartheta < \vartheta_1, \\[2ex] \left(\dfrac{\mathbf{P}_l - \mathbf{p}}{|\mathbf{P}_l - \mathbf{p}|} \cdot \hat{\mathbf{n}} \right) \left(1 - \dfrac{(\vartheta - \vartheta_1)}{(\vartheta_2 - \vartheta_1)} \right) & \text{if } \vartheta_1 \leq \vartheta \leq \vartheta_2, \\[2ex] 0 & \text{if } \vartheta_2 < \vartheta, \end{cases}$$

where ϑ is the angle between the direction of the incident light and the surface normal, and is given by $\vartheta = \cos^{-1}(\hat{\mathbf{d}} \cdot (\mathbf{p} - \mathbf{P}_l))$.

Diffuse lighting is the most significant component of an illumination model, and if there is more than one light source in the scene, the intensity of the diffuse reflection component is summed over all the light sources.

• *Specular reflection (I_s)* . Specular reflection models the light, reflecting properties of shiny or mirror-like surfaces. A perfect mirror reflects all rays of light, and the angle of reflection is equal to the angle of incidence. However, shiny surfaces that are not perfect mirrors introduce small random fluctuations in the direction of the reflected rays. These random fluctuations in the direction of the reflected rays of light tend to be small, so that they all lie within a cone-shaped volume that has its apex at the point of reflection and its axis lying in the direction that a ray would take if the surface were a perfect mirror (see Figure 7.10).

In the lighting model, the specular illumination component I_s is normally modeled with the empirical expression suggested by Phong [10]:

$$I_s = \cos^m \varphi,$$

in which φ is the angle between the reflection vector and the vector leading from the viewpoint at $\mathbf{P}_v$ to $\mathbf{p}$ on the surface. To determine φ from the known geometry shown in Figure 7.10, it is necessary to use a few intermediate steps.

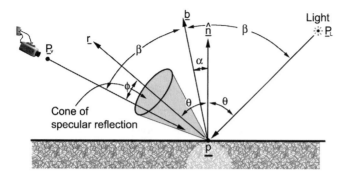

Figure 7.10. Specular reflection.

First calculate the vector $\mathbf{b}$ which bisects the vectors between $\mathbf{p}$ and the viewpoint and between $\mathbf{p}$ and $\mathbf{P}_l$. The vector $\mathbf{b}$ takes an angle β between these vectors. From that, the angle α is easily determined because $\cos\alpha = \hat{\mathbf{b}} \cdot \hat{\mathbf{n}}$ (the surface normal $\hat{\mathbf{n}}$ is known), whilst $\mathbf{b}$ is given by

$$\mathbf{b} = \frac{\mathbf{p} - \mathbf{P}_l}{|\mathbf{p} - \mathbf{P}_l|} - \frac{\mathbf{p} - \mathbf{P}_v}{|\mathbf{p} - \mathbf{P}_v|},$$

$$\hat{\mathbf{b}} = \frac{\mathbf{b}}{|\mathbf{b}|}.$$

In Figure 7.10 we note that

$$\beta = \varphi + \vartheta - \alpha$$

and that

$$\beta = \vartheta + \alpha.$$

Eliminating β and canceling ϑ gives

$$\alpha = \frac{\varphi}{2}.$$

In terms of the known geometry, I_s therefore becomes

$$I_s = \cos^m \varphi, \qquad \text{or}$$

$$I_s = \left(2\cos^2 \frac{\alpha}{2} - 1\right)^m, \qquad \text{or}$$

$$I_s = \left(2(\hat{\mathbf{b}} \cdot \hat{\mathbf{n}})^2 - 1\right)^m.$$

It is the specular component of the light reflected from a surface that makes it look shiny. The cosine power, *m*, is the parameter which governs the *shininess* of a surface. Very shiny surfaces have a high *m*; typical values lie in the range 10 to 99.

Thus, in practice, the effect of specular reflection is to add a *highlight* to parts of a model that have been designated as shiny. The specular highlight takes the color of the light and not that of the surface on which it is visible.

- *Depth cuing (I_c)*. Like any electromagnetic radiation, light is attenuated as it travels away from its source. The reflected light (mainly diffuse) arriving at the viewpoint has traveled from the light source to the object, where it was reflected on to the observer. In theory, the light intensity should be attenuated with distance from its source using an inverse square law, but in practice a linear fall-off looks much more realistic. Quite often, depth cuing can be omitted entirely from the model with only mildly noticeable effect. A linear model is usually written in the form

$$I_c = \frac{1}{d_o + d},$$

where d_o is a constant and d is the distance of **p** from the viewpoint. Assuming that the viewpoint is at $(0, 0, 0)$ and the observer is looking along the *x*-axis, a good assumption is to let $d = p_x$, the *x*-coordinate of **p**.

If depth cuing is to be used, it is also a good idea to perform an *exposure* test by examining all the polygons and scaling the light intensities so that the maximum is always unity. If this is not done then it can require quite a lot of trial and error testing to generate a picture that is neither under- nor overexposed.

Quickly taking stock, we know that when illuminating a scene, we can use three standard light sources and the light can interact within the scene in a number of different ways. It is now time to consider the format for recording a value for the surface color and illumination within a given fragment. Let us record this information using the variable *c*. When recording the color component, it is obvious that we must use a color model. We must use a model to describe color. The simplest color model that we can utilize is the

RGB color model, where a color is stored using three components, one for each of the primary colors, red, green and blue. Any color the eye can perceive can be expressed in terms of an RGB triple. Thus c is recorded as c_R, c_G, c_B, which are usually stored as unsigned 8-bit integers, giving a range of 256 discrete values for each color component. For preliminary calculation, it is usually assumed that c_R, c_G and c_B are recorded as floating-point numbers in the range $[0, 1]$. We will also assume that any light or surface color is also given as an RGB triple in the same range. To determine the color that is recorded in a fragment, the effect of the lights within the scene need to be combined with the surface properties of the polygon visible in that fragment. The simplest way to do this is to break up the mathematical model for light and surface interaction into a number of terms, where each term represents a specific physical phenomenon.

In the following expressions, s_R, s_G, s_B represents the color of the surface, and l_R, l_G, l_B the color of the light. I_a, I_c, I_d and I_s are the four contributions to the lighting model that we have already discussed. Using this terminology, the color c calculated for the fragment in question may be expressed as

$$c_R = I_a s_R + I_c(I_s + I_d s_R)l_R,$$

$$c_G = I_a s_G + I_c(I_s + I_d s_G)l_G,$$

$$c_B = I_a s_B + I_c(I_s + I_d s_B)l_B.$$

To form general expressions for the effect of lighting a surface in a scene, at some point $\mathbf{p}$, with n lights, these equations become

$$c_R \quad = \quad I_a s_R + I_c \sum_{i=0}^{n-1} (I_s(i) + I_d(i)s_R)l_R(i), \qquad (7.1)$$

$$c_G \quad = \quad I_a s_G + I_c \sum_{i=0}^{n-1} (I_s(i) + I_d(i)s_G)l_G(i), \qquad (7.2)$$

$$c_B \quad = \quad I_a s_B + I_c \sum_{i=0}^{n-1} (I_s(i) + I_d(i)s_B)l_B(i). \qquad (7.3)$$

Both I_d and I_s depend on the position of light i and on which type of light it is (spotlight etc.). If $n > 1$, each term in the lighting model must be limited so that it falls in the range $[0, 1]$; otherwise, an overexposed picture will be produced. So now that we have derived expressions to help us compute the surface color due to lighting effects, we can look at the different shading algorithms that are commonly used to implement these calculations.

7.5.2   Gouraud and Phong Shading

Look at Figure 7.11. It shows two pictures of the same faceted model of a sphere. The one on the right looks smooth (apart from the silhouetted edges, which we will discuss later). The one on the left looks like just what it is, a collection of triangular polygons. Although the outline of neither is perfectly circular, it is the appearance of the interior that first grabs the attention. This example highlights the main drawback of the representation of an object with a model made up from polygonal facets. To model a sphere so that it looks smooth by increasing the number (or equivalently decreasing the size) of the facets is quite impractical. Thousands of polygons would be required just for a simple sphere. However, both the spheres shown in Figure 7.11 contain the same number of polygons, and yet one manages to look smooth. How can this happen?

The answer is the use of a trick that fools the eye by smoothly varying the shading within the polygonal facets. The point has already been made that if you look at the outline of both spheres, you will see that they are made from straight segments. In the case of the smooth-looking sphere, it looks smooth because the discontinuities in shading between adjacent facets have been eliminated. To the eye, a discontinuity in shading is much more noticeable than a small angular change between two edges. This smooth shading can be achieved using either the Phong or the Gouraud approach. We shall now discuss these shading methods in more detail.

Figure 7.11. Shading of two similar polygon models of a sphere.

Gouraud Shading

Gouraud shading is used to achieve smooth lighting on polygon surfaces without the computational burden of calculating lighting for each fragment. To do this, a two-step algorithm is used:

1. Calculate the intensity of the light at each vertex of the model; that is, the computed colors at each vertex.

    In Section 7.5.1, we derived equations to calculate the color intensity at a point on the surface of a polygon. We use these equations to calculate the color intensity at vertex i. The normal N used in these equations is taken to be the average of the normals of each of the polygon surfaces which are attached to the vertex. For example, to calculate the light intensity at vertex i, as shown in Figure 7.12, we need to compute the normal at vertex i. To do this, we need to average the surface normals calculated for each of the attached polygons ($j, j + 1, j + 2, j + 3$ and $j + 4$).

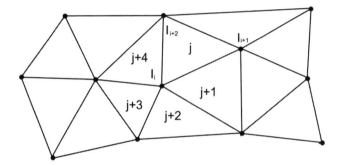

Figure 7.12. Averaging light intensities in polygons j to $j + 4$ adjacent to vertex i will provide an appropriate intensity, I_i, at vertex i. Similarly, light intensities I_{i+1} and I_{i+2} are determined from the light falling on polygons adjacent to those vertices.

2. Use a bilinear interpolation to obtain the intensity at any point within a polygon from the intensity at the vertices of the polygon.

    To interpolate the intensity I_p at the point $\mathbf{p}$ within any surface polygon, as shown in Figure 7.13, there are two alternatives:

    - *Work in three dimensions.* At this stage, we have knowledge of the color intensities (I_0, I_1, I_2) at each of the vertices comprising the polygon ($\mathbf{P_0}, \mathbf{P_1}, \mathbf{P_2}$). We also know the three-dimensional coordinates of these vertices; that is, (x_0, y_0, z_0), (x_1, y_1, z_1) and (x_2, y_2, z_2), respectively. Remembering back to Section 6.4.3, we derived an equation to allow us to determine the point of intersection with a line and a bounded plane. Now if we consider the

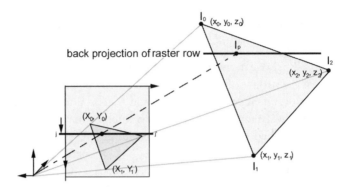

Figure 7.13. Calculate the light intensity at a raster pixel position within the projection of a triangular polygon by interpolating from the intensity values at the polygon's vertices.

surface polygon to be the bounded plane and we can determine a line from a vertex to the point **p**, we can use this equation to obtain values for α and β. These values essentially represent how far the point **p** is from each of the vertices. Hence, we can now use them to bilinearly interpolate the color intensities at each vertex to obtain the color intensity at I_p:

$$I_p = I_0 + (I_2 - I_1)\alpha + (I_1 - I_0)\beta.$$

However, finding I_p in three dimensions is slow. It would be better if we could do the calculation after the polygon's vertices have been projected onto the 2D viewing plane. We can indeed do this, because shading is the last step in the rendering algorithm and all the necessary 3D to 2D conversions have already been carried out. This leads to our second alternative.

- *Work in two dimensions.* Use the two-dimensional coordinates after projection of the polygon's vertices onto the viewing plane, i.e., to points (X_0, Y_0), (X_1, Y_1) and (X_2, Y_2). Further, because the rendering procedure will normally be filling fragments sequentially across a scanline, a few additional arithmetic operations can be saved by calculating an increment in intensity I which is added to the value obtained for the previous fragment on the scanline. For a triangular polygon, k, which spans scanline j (see Figure 7.13) starting at fragment i, the intensity $I_{i,j}$ can be obtained from the following formulae:

$$\Delta = (Y_2 - Y_0)(X_1 - X_0) - (X_2 - X_0)(Y_1 - Y_0),$$

$$\alpha = \frac{(Y_2 - Y_0)(i - X_0) - (X_2 - X_0)(j - Y_0)}{\Delta},$$

$$\beta = \frac{(j - Y_0)(X_1 - X_0) - (i - X_0)(Y_1 - Y_0)}{\Delta},$$

$$I_{i,j} = \alpha(I_1 - I_0) + \beta(I_2 - I_0).$$

The intensity in adjacent fragments $I_{i+1,j}$ etc. is determined by calculating an increment δI from

$$\delta I = \frac{(Y_1 - Y_2)I_0 + (Y_2 - Y_0)I_1 + (Y_0 - Y_1)I_2}{\Delta}.$$

Then for all the remaining fragments on scanline j covered by polygon k, the intensity is given by

$$I_{i+1,j} = I_{i,j} + \delta I.$$

Phong Shading

The Gouraud shading procedure works well for the diffuse reflection component of a lighting model, but it does not give a good result when specular reflection is taken into account. Specular highlights occupy quite a small proportion of a visible surface, and the direction of the normal at precisely the visible point must be known so that the specular reflection can be determined. Thus, Phong's approach was to interpolate the surface normal over a polygon rather than interpolate the light intensity. In the other aspects, the procedure is similar to the Gouraud shading algorithm.

Phong shading is done in the following two stages:

1. Calculate a normal vector for each vertex of the model.

2. Use two-dimensional interpolation to determine the normal at any point within a polygon from the normals at the vertices of the polygon.

In Stage 1, an analogous procedure to the first step of the Gouraud algorithm is used. Figure 7.14 shows a cross section through a coarse (six-sided) approximation of a cylinder. In Figure 7.14(a), the facet normals are illustrated, Figure 7.14(b) shows the vertex normals obtained by averaging normals from facets connected to a particular vertex and Figure 7.14(c) shows the

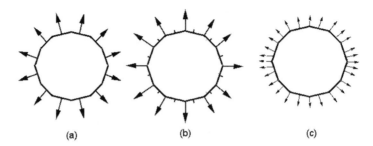

(a)   (b)   (c)

Figure 7.14. In Phong smoothing ,the surface normals are averaged at the vertices and then the vertex normals are interpolated over the flat facets to give an illusion of a continuously varying curvature.

effect of interpolation on normal vectors. A smoothly varying surface normal will interact with the incoming light beam to give smoothly varying shading which will also work correctly with the model for specular highlights.

In Stage 2, expressions very similar to those that occur in the Gouraud model are derived. For a triangular polygon k that spans scanline j at the leftmost fragment i, the normal vector $\hat{\mathbf{n}}_{i,j}$ is given by

$$\Delta = (Y_2 - Y_0)(X_1 - X_0) - (X_2 - X_0)(Y_1 - Y_0),$$

$$\alpha = \frac{(Y_2 - Y_0)(i - X_0) - (X_2 - X_0)(j - Y_0)}{\Delta},$$

$$\beta = \frac{(j - Y_0)(X_1 - X_0) - (i - X_0)(Y_1 - Y_0)}{\Delta};$$

$$\begin{bmatrix} n_x \\ n_y \\ n_z \end{bmatrix}_{i,j} = \begin{bmatrix} \alpha(n_{1x} - n_{0x}) + \beta(n_{2x} - n_{0x}) \\ \alpha(n_{1y} - n_{0y}) + \beta(n_{2y} - n_{0y}) \\ \alpha(n_{1z} - n_{0z}) + \beta(n_{2z} - n_{0z}) \end{bmatrix};$$

$$\hat{\mathbf{n}}_{i,j} = \frac{\mathbf{n}_{i,j}}{|\mathbf{n}_{i,j}|}.$$

For the remaining fragments on scanline j, i.e., fragments $i + 1$, $i + 2$ etc., the normals $(\mathbf{n}_{i+1,j}, \mathbf{n}_{i+2,j})$ are determined by calculating an incremental vector $\delta\mathbf{n}$ given by

$$
\begin{bmatrix} \delta n_x \\ \delta n_y \\ \delta n_z \end{bmatrix} = \frac{1}{\Delta} \begin{bmatrix} (Y_1 - Y_2)n_{0x} + (Y_2 - Y_0)n_{1x} + (Y_0 - Y_1)n_{2x} \\ (Y_1 - Y_2)n_{0y} + (Y_2 - Y_0)n_{1y} + (Y_0 - Y_1)n_{2y} \\ (Y_1 - Y_2)n_{0z} + (Y_2 - Y_0)n_{1z} + (Y_0 - Y_1)n_{2z} \end{bmatrix}.
$$

Normals in the remaining fragments on scanline j covered by polygon k are thus determined sequentially from

$$
\begin{aligned}
\mathbf{n}_{i+1,j} &= \mathbf{n}_{i,j} + \delta\mathbf{n}; \\
\hat{\mathbf{n}}_{i+1,j} &= \frac{\mathbf{n}_{i+1,j}}{|\mathbf{n}_{i+1,j}|}.
\end{aligned}
$$

Once the lighting and shading calculations are completed, each fragment can be written to the output frame buffer, which is then used to output our rendered image onto the screen.

7.6 Texturing

Texturing is a method of adding realism to a computer-generated graphic. It does this by delivering surface detail to the polygonal models within the scene. But where does this fit into the graphics pipeline?

As we mentioned in the previous sections, the Z-buffer algorithm is used to determine which surface fragments are actually visible. Once this is determined, we then need to calculate the surface properties. These were identified by the variables s_R, s_G and s_B in Equations (7.1) to (7.3). These variables could simply represent a single color taken from the color of the polygon which is visible at this fragment. However, if we were only to use color in our scenes, they would not appear realistic. To enhance the appearance of our scene, we can use one of two texturing approaches to create the illusion of almost any type of surface. These are *procedural textures* (often called shaders) and *image maps* (because they drape, or map, a picture stored in a BMP, GIF, JPEG or PNG file over the polygonal mesh).

These textures and maps do not just have to alter the color of a particular fragment; they can be used to change the way the surface normal vector interacts with the direction of the lighting to simulate a rough surface, alter the transparency of the surface at that location or give the illusion that we are looking at an image as if the surface were a mirror. Therefore, if we do add a texture to our scene, the values of s_R, s_G and s_B need to be updated before carrying out the lighting computations. We will now look in more detail at the methods by which we can apply both procedural textures and image maps.

7.6.1 Procedural Textures

Patterns on real-world objects are usually due to variations in geometry across their surface (e.g., thorns, scales, bark etc.). Procedural textures are used to re-create this variation for our computer-generated images. If a procedural texture is applied to a fragment, the color value of that fragment is determined using the relevant mathematical model, that is primarily a function of the position of the fragment. By using this relatively simple technique, you can literally achieve any texture you can imagine. For some practical hints on implementing procedural textures, you might consult Ebert et al. [4].

7.6.2 Image Mapping

An image map performs a similar task to that of a procedural texture. However, image maps facilitate much greater control over the appearance of the surface they are applied to. In essence, any picture may be used as the source for the image map (2D artwork or scanned photographs are common sources). This is particularly useful in enhancing the appearance of outdoor environments and for product design applications where text, manufacturers' logos or labels can be added to a 3D model of their product.

The other major difference between a procedural texture and an image map is that the surface color is not calculated on the basis of a mathematical model. Rather, the surface color is obtained in what one might describe as a data lookup procedure. The color to be used for the surface fragment is obtained from one or a combination of the pixels that make up the image map. It is here that the difficulty lies. We need to determine which pixel within the image map corresponds to the surface fragment and then we need to look it up. More mathematics!

If the surface fragment, centered on point $\mathbf{p}$, lies on the triangular polygon with vertices $\mathbf{P}_0$, $\mathbf{P}_1$ and $\mathbf{P}_2$ then s, the color to be substituted into the lighting model, is determined by executing the following steps. Figure 7.15 illustrates the process.

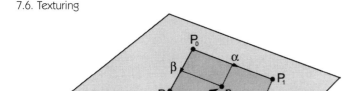

Figure 7.15. Mapping an image into a rectangle with texture coordinates of $(0, 0)$ at $\mathbf{P}_0$, $(1, 0)$ at $\mathbf{P}_1$ and $(0, 1)$ at $\mathbf{P}_2$ respectively. At the point $\mathbf{p}$, an appropriate pixel color is determined from the image using the steps given in the text.

1. Use the coordinates of the vertices of the surface polygon relative to $\mathbf{p}$ to determine parameters α and β, where

$$(\mathbf{p} - \mathbf{P}_0) = \alpha(\mathbf{P}_1 - \mathbf{P}_0) + \beta(\mathbf{P}_2 - \mathbf{P}_0).$$

2. Given texture coordinates of (X_0, Y_0), (X_1, Y_1) and (X_2, Y_2) at $\mathbf{P}_0$, $\mathbf{P}_1$ and $\mathbf{P}_2$ respectively, the texture coordinate (X, Y) for the point $\mathbf{p}$ is determined by

$$
\begin{aligned}
X &= X_0 + \alpha(X_1 - X_0) + \beta(X_2 - X_0), & (7.4) \\
Y &= Y_0 + \alpha(Y_1 - Y_0) + \beta(Y_2 - Y_0). & (7.5)
\end{aligned}
$$

3. Starting with the texture coordinate (X, Y), we need to determine an index (i, j) into the array $A_{n,m}$ (of size $n \times m$) where the image's pixel color values are stored. To calculate the index, we must multiply the values of X and Y by the number of pixels in the horizonal (n) and vertical (m) directions respectively. Thus, a texture coordinate of $(0, 0)$ would give an index of $(0, 0)$, a texture coordinate of $(1, 1)$ would give (n, m) and so on. The general equations to do this are

$$
\begin{aligned}
i &= \text{nearest integer to } (nX), \\
j &= \text{nearest integer to } (mY).
\end{aligned}
$$

4. Copy the pixel color from $A_{i,j}$ to s.

There are two important issues that arise in Step 3. First, to obtain a valid address in A, both i and j must simultaneously satisfy $0 \leq i < n$ and $0 \leq j < m$ or, equivalently, scale to a unit square: $0 \leq X < 1$ and $0 \leq Y < 1$. Thus the question arises as to how to interpret texture coordinates that fall outside the range $[0, 1]$. There are three possibilities, all of which prove useful. They are:

1. Do not proceed with the mapping process for any point with texture coordinates that fall outside the range $[0, 1]$; just apply a constant color.

2. Use a modulus function to *tile* the image over the surface so that it is repeated as many times as necessary to cover the whole surface.

3. Tile the image over the surface, but choose a group of four copies of the image to generate blocks that themselves repeat without any seams, i.e., a mosaic pattern. To generate a mosaic pattern, the α and β values in Equations (7.4) and (7.5) are modified using a floor() function and integer rounding.

The second issue is the rather discontinuous way of rounding the floating-point texture coordinates (X, Y) to the integer address (i, j) ($i = $ truncate(x) and $j = $ truncate(y)) used to pick the color from the image. A better approach is to use a bilinear (or higher order) interpolation to obtain s, which takes into account not only the color in image pixel (i, j) but also the color in the image pixels adjacent to it. Thus, if the texture coordinates at $\mathbf{p}$ are (X, Y), we obtain s by using a bilinear interpolation between color values from $A_{i-1,j}$, $A_{i+1,j}$, $A_{i,j-1}$ and $A_{i,j+1}$ in addition to the color from $A_{i,j}$. Note: Bilinear interpolation give good results when the image map is being magnified, but in cases were the image is being reduced in size, as the object being mapped moves away from the camera, for example, the use of a mipmap (see Möller and Haines [8]) gives a better approximation.

7.6.3   Transparency Mapping

Transparency mapping follows the same steps as basic image mapping to deliver a color value, s, for use at a point $\mathbf{p}$ on polygon k. However, instead of using s directly as a surface color, it is used to control a mixture between the color settings for polygon k and a color derived from the next surface recorded in the transparency buffer. (Remember that the Z-buffer algorithm can be modified to hold several layers so that if any of them were transparent, the underlying surface would show through.) Note that when mixing the proportion of s, due to the underlying surface, *no* lighting effects are applied; that is, when you look through a window from inside a room, it is not the light in the room that governs the brightness of the world outside.

7.6.4 Bump Mapping

This technique, introduced by Blinn [1], uses the color from an image map to modulate the direction of the surface normal vector. To implement a bump map, the change (gradient or derivative) of brightness across the image, from one pixel to another, determines the *displacement vector* added to the surface normal of polygon k at point **p**. Most of the analysis we need in order to calculate the perturbing vector $\Delta\mathbf{n}$ has been covered already. Bump mapping usually uses the same texture coordinates as are used for basic image mapping. In fact, it is essential to do so in cases where an image and bump map are part of the same material. For example, a brick surface can look very good with a bump map used to simulate the effect of differential weathering on brick and mortar.

To determine $\Delta\mathbf{n}$, we need to find incremental vectors parallel to two of the sides of polygon k; call them $\Delta\mathbf{n}_1$ and $\Delta\mathbf{n}_2$. Do this by first finding texture coordinates at points close to **p** (at **p** the texture coordinates are (X, Y) given by Equations (7.4) and (7.5)). The easiest way to do this is to make small increments, say δ_α and δ_β, in α and β and use Equations (7.4) and (7.5) to obtain texture coordinates (X_r, Y_r) and (X_b, Y_b).

Before using the texture coordinates to obtain "bump" values from the map, it is necessary to ensure that the distance *in texture space* between (X, Y) and (X_r, Y_r) and between (X, Y) and (X_b, Y_b) is small. To achieve this, write

$$\begin{aligned}
\Delta_t &= \sqrt{(X_r - X)^2 + (Y_r - Y)^2}, \\
X_r &= X + \delta\frac{(X_r - X)}{\Delta_t}, \\
Y_r &= Y + \delta\frac{(Y_r - Y)}{\Delta_t}.
\end{aligned}$$

and do the same thing to (X_b, Y_b). These equations include an additional scaling factor δ that will be discussed shortly. The next step is to obtain values from the bump image at texture coordinates (X, Y), (X_r, Y_r) and (X_b, Y_b). Using bilinear interpolation, we obtain three bump values s_0, s_1 and s_2. It only remains to construct $\Delta\mathbf{n}$ and add it to the normal, $\hat{\mathbf{n}}$, for polygon k. Thus

$$
\begin{aligned}
\mathbf{d}_1 &= (\mathbf{P}_1 - \mathbf{P}_0), \\
\mathbf{d}_2 &= (\mathbf{P}_2 - \mathbf{P}_0), \\
\Delta\mathbf{n}_1 &= (s_1 - s_0)\left(\frac{\mathbf{d}_1}{|\mathbf{d}_1|}\right), \\
\Delta\mathbf{n}_2 &= (s_2 - s_0)\left(\frac{\mathbf{d}_2}{|\mathbf{d}_2|}\right); \\
\mathbf{n} &= \mathbf{n} + h(\Delta\mathbf{n}_1 + \Delta\mathbf{n}_2), \\
\hat{\mathbf{n}} &= \mathbf{n}/|\mathbf{n}|.
\end{aligned}
$$

Here, h is a parameter that facilitates control of the apparent height of the bumps. Its range should be $[0, 1]$. The choice of δ, δ_α and δ_β is critical to determining how well the algorithm performs.

In the case of δ, it is unlikely that the change in a texture coordinate across a polygon will exceed a value of about 1 or 2. Therefore, $\delta = 0.01$ will work well in most cases, although a little experimentation is well worthwhile. The increments δ_α and δ_β should also be small; values of ≈ 0.05 seem appropriate. Again, a little experimentation may pay dividends.

7.6.5   Reflection Mapping

True reflections can only be simulated with ray tracing or other time-consuming techniques, but for a lot of uses, a *reflection map* can visually satisfy all the requirements. Surfaces made from gold, silver, chrome and even the shiny paint work on a new automobile can all be realistically simulated with a skillfully created reflection map, a technique introduced by Blinn and Newell [2]. An outline of a reflection map algorithm is:

1. With the knowledge that polygon k, at point $\mathbf{p}$, is visible in fragment (i, j) determine the reflection direction $\mathbf{d}$ of a line from the viewpoint to $\mathbf{p}$. (Use the procedure given in Section 6.5.)

2. Since the basic rendering algorithm transformed the scene so that the observer is based at the origin $(0, 0, 0)$ and is looking in direction $(1, 0, 0)$, the reflected direction $\mathbf{d}$ must be transformed with the inverse of the rotational part of the viewing transformation. The necessary inverse transformation can be obtained at the same time. Note: the inverse transformation must not include any translational component, because $\mathbf{d}$ is a direction vector.

3. Get the Euler angles that define the direction of $\mathbf{d}$. These are the heading φ, which lies in the range $[-\pi, \pi]$, and pitch ϑ, which lies in the range $[\frac{-\pi}{2}, \frac{\pi}{2}]$. Scale the range of φ to cover the width X_{max} of the reflection image. Scale the range of ϑ to cover the height Y_{max} of the reflection image:

$$X = \frac{(\pi + \varphi)}{2\pi} X_{max};$$
$$Y = \frac{(\frac{\pi}{2} - \vartheta)}{\pi} Y_{max}.$$

4. Using bilinear interpolation, we may obtain a color value s from the image in the vicinity of (X, Y).

5. Mix s with the other sources contributing to the color seen at pixel (i, j).

You can think of the reflection mapping process as one where the reflected surface is located at the center of a hollow sphere with the image map painted on its inside. Figure 7.16 illustrates the relationship of the object to a reflection map. Another reflection mapping procedure that gives good results is the cubic environment (see Greene [7]).

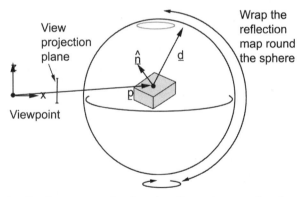

Figure 7.16. A reflection map can be thought of as an image painted on the inside of hollow sphere with the reflective model placed at its center.

7.6.6 GPU Programmable Shaders

The image mapping and procedural textures that we have just been discussing, and indeed the Phong shading and the lighting models, have all been described in terms of algorithms. However, the hardware at the heart of the computer's display is changing dramatically. It is becoming programmable; it is becoming a processor in its own right, a *graphical processing unit* (or GPU). This allows us to implement the algorithms for procedural textures in high performance and dedicated hardware with practically infinite flexibility and degrees of realism.

The term *shader* has become universally accepted to describe these programmable textures and maps. The GPU hardware is dedicated to delivering all stages of the 3D rendering pipeline, and so it appears to the programmer in a somewhat different guise than the well-known microprocessor CPU. There are some restrictions on what operations are permitted and the order in which they are done. Fortunately, the OpenGL and Direct3D software libraries have evolved to work with the GPU by providing what is called a *shader programming language*. We shall look in detail at how this achieved by OpenGL in Chapter 14.

7.7 Summary

In this chapter, we have described the main stages in the process which turns a numerical description of the virtual world into high-quality rendered images. Whole books have been written on this subject; Möller and Haines [8] provide comprehensive detail of all the important aspects of rendering in real-time.

If you need to build into your VR visualization code, special features or effects, it is likely that you will find an algorithm for it in the Graphics Gems series [6], which is a rich source of reference material for algorithms associated with rendering and visualization in general. The new custom processors specifically designed for graphics work (the GPUs) attract their own special algorithms, and the two books in the GPU Gems series [5] provide a fascinating insight into how this hardware can be used to render effects and elements of reality that we previously thought impossible.

The focus of this chapter has been on real-time rendering so that we can bring our virtual worlds and objects to life, make them move and make them interact with us. In the next few chapters, we will be building on this.

Bibliography

[1] J. Blinn. *Jim Blinn's Corner: A Trip Down the Graphics Pipeline*. San Fransisco, CA: Morgan Kaufmann, 1996.

[2] J. Blinn and M. E. Newell. "Texture and Reflection in Computer Generated Images". *Communications of the ACM* 19:10 (1976) 542–547.

[3] R. Bryant and D. O'Hallaron. *Computer Systems: A Programmer's Perspective.* Englewood Cliffs, NJ: Prentice Hall, 2002.

[4] D. S. Ebert et al. *Texturing and Modeling: A Procedural Approach.* Cambridge, MA: Academic Press, 1994.

[5] R. Fernando (editor). *GPU Gems.* Reading, MA: Addison-Wesley Professional, 2004.

[6] A. S. Glassner. *Graphics Gems.* Cambridge, MA: Academic Press, 1990.

[7] N. Greene. "Environment Mapping and Other Applications of World Projections". *IEEE Computer Graphics and Applications* 6:11 (1986) 21–29.

[8] T. Möller and E. Haines. *Real-Time Rendering.* Natick, MA: A K Peters, 1999.

[9] K. Perlin. "An Image Synthesizer". *Computer Graphics* 19:3 (1985) 287–296.

[10] B. T. Phong "Illumination for Computer Generated Pictures". *Communications of the ACM* 18:6 (1975) 311-317.

[11] G. Ward. "A Recursive Implementation of the Perlin Noise Function". In *Graphics Gems II*, edited by J. Arvo, pp. 396–401. Boston, MA: Academic Press, 1991.

8

Computer Vision in VR

Computer vision is, in itself, a useful and interesting topic. It plays a significant role in many applications of artificial intelligence, automated patten recognition. . . a full list of its applications would be too long to consider here. However, a number of the key algorithms which emerged during the development of the science of computer vision have obvious and extremely helpful roles in the implementation and delivery of practical VR applications. The full detail and mathematical basis for all the algorithms we will be looking at in this chapter have been rigorously and exquisitely explored by such notable authors as Faugaras [3] and Hartley and Zisserman [7], so if you want to know more there is no better place to look. But remember, this is not a topic for the faint of heart!

In the context of VR, the geometric aspects of computer vision are the most significant and therefore the ones which we intend to introduce here. For example, in an augmented reality videoconferencing environment [2], the user can see images of the participants on virtual displays. The video images on these virtual displays have to be distorted so as to take on their correct perspective as the viewer moves around the virtual world.

To achieve results like these, one needs to study how the appearance of objects changes as they are viewed from different locations. If you want to make use of interactive video in a virtual world, you also need to consider the effects of distortions in the video camera equipment. And of course all this has to be done in real time. Get it right and you will be able to realistically and seamlessly insert the virtual world into the real world or the real world into the virtual. Get it wrong and you just might end up making your guests feel sick.

In Chapter 6, we examined some of the properties of Euclidian geometry and introduced the concept of homogeneous coordinates in the context of formulating the transformations of translation, scaling and rotation. In that chapter, it appeared that the fourth row and fourth column in the transformation matrices had minimal significance. It was also pointed out that when visualizing a scene, the three-dimensional geometry had to be projected onto a viewing plane. In a simplistic rendering application, this is accomplished with two basic equations (Chapter 6, Equations (6.7) and (6.7)) that implement a nonlinear transformation from a 3D space to a 2D space.

In human vision, the same nonlinear transformation occurs, since it is only a 2D image of the 3D world around us that is projected onto the retina at the back of the eye. And of course, the same projective transformation is at work in all types of camera, both photographic and electronic. It should therefore come as no surprise that in the science of computer vision, where one is trying to use a computer to automatically recognize shapes in two-dimensional images (as seen by the camera feeding its signal to the computer), the result of the projective transformation and the distortions that it introduces are an unwelcome complication. Under a projective transformation, squares do not have to remain square, circles can become ellipses and angles and distances are not generally preserved. Information is also lost during projection (such as being able to tell how far something is away from the viewpoint), and therefore one of main goals in computer vision is to try to identify what distortion has occurred and recover as much of the lost information as possible, often by invoking two or more cameras. (We normally use two eyes to help us do this.)

Of course, this is not a book on computer vision, but some of its ideas such as scene reconstruction using multiple viewpoints can help us build our virtual worlds and interact with them in a realistic manner. So, in this chapter, we will examine some of the basics of computer vision that let us do just this.

In order to make any real progress in understanding computer vision, how to correct for distortion and undo the worst aspects of the projective transformation, it is first a good idea to introduce the mathematical notation of the subject. Then we can use it to express and develop the algorithms that we wish to use for VR.

8.1 The Mathematical Language of Geometric Computer Vision

As we know from Chapter 2, vision is perhaps the most important of the human senses. It provides us with a mechanism for obtaining detailed three-dimensional information about the world around us. And how can we do this? Each of our eyes takes a 2D image of our 3D world. Some processing in the brain combines these images to deliver a stereo picture to us whilst other processing in the brain extracts other useful "intelligent" clues about what we see. Thus we are essentially able to recover depth information, shape and pattern information from two 2D images. Computer vision utilizes a similar concept, and is regarded as the study of the processes which allow the computer to process static images, recover a three dimensional view and help with the first stages in pattern recognition and possibly artificial intelligence. In order to be able to understand how the computer can do this, we need to introduce the principles of projective geometry. Up until now, we have only discussed transforming points in a 3D real-world space to a 2D image plane. Now we have to think about trying to do the reverse, and to do this we need to know the basics of projective geometry.

You are certainly familiar with the two and three dimensional spaces of classical Euclidian geometry (we will represent these symbolically as $\mathbb{R}^2$ and $\mathbb{R}^3$), and one generally uses coordinates to map out the space. 2D space requires two numbers to uniquely specify a point (x, y), and 3D, of course, requires three (x, y, z). Seemingly the origin $(0, 0)$ or $(0, 0, 0)$ takes on special significance, but this is only an illusion, since any point specified in terms of one coordinate system can become the origin in another through a combination of Euclidian transformations, as discussed in Section 6.6.

In two dimensions, the existence of parallel lines can be something of a nuisance in a practical sense, because it introduces a special case that must be excluded when one is calculating, for example, the point of intersection of two lines. Colloquially one says that parallel lines meet at infinity and therefore there is nothing special about parallel lines, because all lines meet at some point. Unfortunately, infinity is not a concept that is particularly amenable to practical calculations performed by a computer, and if possible it should be avoided. One useful device to avoid infinity is to enhance the Euclidian plane by adding the points (at infinity) where parallel lines meet and calling them the ideal points. The ideal points together with the Euclidian plane $\mathbb{R}^2$ deliver a new kind of geometry: *projective geometry*. In the case of two dimensions, $\mathbb{R}^2$ in this enhanced form is written symbolically as $\mathbb{P}^2$. In $\mathbb{P}^2$ all lines intersect and there is no need for special cases because parallel lines now meet at an ideal point.

Still thinking about two dimensions for the moment, how should one specify coordinates to map out $\mathbb{P}^2$? In $\mathbb{R}^2$, we have seen that any point is represented by the coordinate pair (x, y), but it does not change anything if we were to use three numbers $(x, y, 1)$ with the third number fixed at the constant 1. These are called the *homogeneous coordinates*. It is instantly possible to switch back and forth between $(x, y, 1)$ and (x, y); one simply drops or adds the 1.

Now suppose we take the conceptual step of allowing the constant 1 to become a variable, say w, so that we have a triple to represent a point in the 2D plane, (x, y, w). We can still easily recover the original point $(x, y, 1)$ by defining that the triples $(x, y, 1)$, $(2x, 2y, 2)$, $(3x, 3y, 3)$ and $((kx, ky, k)$ for any k) represent the same point. Thus given any homogeneous coordinate (x, y, w), we can recover the original Euclidian point in the 2D plane; that is, $(x/w, y/w)$. So far, we don't seem to have done anything very useful, but then we may ask, what point corresponds to the homogeneous coordinates $(x, y, 0)$? This is certainly a valid homogeneous coordinate, but if you try to obtain the coordinates of the equivalent 2D point, the result $(x/0, y/0)$ is meaningless. In fact, points with $w = 0$ are the points at infinity where anecdotally we claimed parallel lines meet; that is, our ideal points. However homogeneous coordinates pose no practical problems of infinities for our computers, since the points at infinity are now no different from any other points. Thus homogeneous coordinates offer a way forward for performing calculations involving projective geometries and are more than just a convenient device for accommodating translational transformations as 4×4 matrices.

Having seen how the 2D Euclidean space $\mathbb{R}^2$ may be extended to the 2D projective space $\mathbb{P}^2$ by representing points as three-element homogeneous vectors, it is possible to envisage that analogous extensions exist for spaces of n dimensions. A more rigorous analysis confirms our hypothesis that whilst the points at infinity in two dimensions form a line, in three dimensions they form a plane.

We saw in Chapter 6 that a linear transformation may be represented by a matrix multiplication applied to the coordinates of the point. As you might suspect, the same is true of a transformation in a projective space of n dimensions, $\mathbb{P}^n$:

$$\mathbf{X}' = H\mathbf{X},$$

where $\mathbf{X}$ and $\mathbf{X}'$ are points in the projective space and H is the $(n+1) \times (n+1)$ transformation matrix that relates them.

It is interesting to note that under a transformation, points at infinity in $\mathbb{P}^n$ are mapped to other points arbitrarily; that is, they are not preserved as points at infinity. Why are we concerned about these points at infinity? It is because in computer vision, it is convenient to extend the real 3D world into 3D projective space and to consider the images of it, which form at the back of the camera, as lying on a 2D projective plane. In this way, we can deal seamlessly with lines and other artifacts in our image that are parallel, and we do not have to single them out for special treatment.

8.2   Cameras

Understanding the detail of the camera is central to working in computer vision because all the images that a computer can analyze with its recognition algorithms must be subject to a camera projection and any distortions due to the imaging process and specifics of the camera hardware. The projection results in an immediate loss of a dimension, since the projection forms only a 2D representation of the 3D world. If the properties of the camera projection and any distortion can be quantified and measured then, to a certain extent, it should be possible to restore a little of the information lost in going from 3D to 2D. In practice, every camera is slightly different, and it is essential to account for these differences by calibrating the camera before the computer is let loose to analyze the image.

8.2.1   Camera Projections

In Section 6.6.7, we gave simplistic expressions for the act of projecting the 3D world onto the viewing plane using a camera. The equations introduced there are actually quite subtle, and a deeper study can reveal some interesting facts now that we have more information about the homogeneous coordinate system and the projective spaces $\mathbb{P}^2$ and $\mathbb{P}^3$.

For example, when using projective spaces and homogeneous coordinates, it is possible to transform points from $\mathbb{P}^3$ to $\mathbb{P}^2$ using a transformation matrix P:

$$\begin{bmatrix} X \\ Y \\ W \end{bmatrix} = P_{(3 \times 4)} \begin{bmatrix} x \\ y \\ z \\ w \end{bmatrix},$$

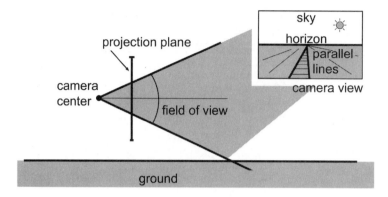

Figure 8.1. Projecting a 2D ground plane onto a 2D viewing plane, where both planes lie in 3D space.

where $[X, Y, W]^T$ are the homogeneous image coordinates and $[x, y, z, w]^T$ are the homogeneous real-world coordinates. If you look at Figure 8.1 you can see the plane $\mathbb{P}^2$ on which the image is formed. The rays that lead from the visible points in the scene back through the projection plane all converge on a single point: *the camera center*. The projection plane is chosen as indicated in Figure 8.1 because it is perpendicular to the principal axis, but

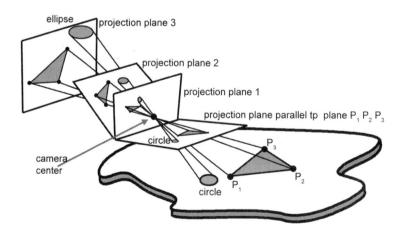

Figure 8.2. Different planes of projection may produce different images, but they all represent the same visible part of a scene. Only if the camera center changes is the view fundamentally altered.

any plane passing through the same point on the principal axis could act as the projection plane. The projection onto any of these planes is still related by a projective transformation. Only if the camera center moves does the view change fundamentally. The resulting planar images will look somewhat different, but in the absence of any other knowledge about the scene, it is hard to argue that one image is more correct than another. For example, if a planar surface in the scene is parallel to the imaging plane then circles on the plane will map to circles in the image, whereas in the case of a different viewing plane, they would map to ellipses. The effect of imaging onto several differently oriented projection planes is illustrated in Figure 8.2. Thus one of the key parameters that describes the properties of a camera is the camera center.

8.2.2 Camera Models

A fairly simplistic view of the imaging process was given in Section 6.6.7, and without stating it explicitly, the camera model used there represents the perfect pinhole camera (see Figure 8.3). In addition, in the previous section we saw that we can transform points from $\mathbb{P}^3$ to $\mathbb{P}^2$. The transformation used actually represents the camera model. In the case of a perfect pinhole camera, the camera model can be represented by the transformation matrix

$$\begin{bmatrix} f(1) & 0 & p_x & 0 \\ 0 & f(2) & p_y & 0 \\ 0 & 0 & 1 & 0 \end{bmatrix}, \tag{8.1}$$

where p_x and p_y are the pixel coordinates of the principal point and $f(1)$ and $f(2)$ are the focal distance expressed in units of horizontal and vertical pixels. The ratio $f(1)/f(2)$ is the *aspect ratio* and is different from 1 if the pixels in the imaging sensor are not square.

If we take a view from the camera center $\mathbf{C}$ along the principal axis, we see that the principal point $\mathbf{P}$ lies at the dead center of the image plane (Figure 8.4). In different types of camera, the principal point may be off-center, and the aspect ratio may not equal 1.

Therefore, if trying to extract accurate 3D information from the 2D image scene, it is necessary to derive an accurate camera matrix through the process of camera calibration.

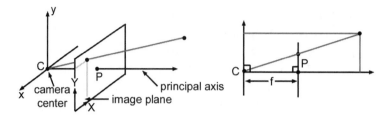

Figure 8.3. The ideal pinhole camera showing the camera (or optical) center **C**, the principal point **P**, the principal axis direction and the image plane. The focal length is f.

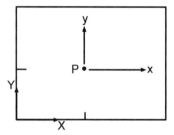

Figure 8.4. The image plane in plan view.

To pursue this idea a little bit further, if we write the action of the projection matrix (Equation (8.1)) for square pixels as

$$\mathbf{X} = K[I|\mathbf{0}]\mathbf{x},$$

where

$$K = \begin{bmatrix} f & 0 & p_x \\ 0 & f & p_y \\ 0 & 0 & 1 \end{bmatrix} \quad \text{and} \quad [I|\mathbf{0}] = \begin{bmatrix} 1 & 0 & 0 & 0 \\ 0 & 1 & 0 & 0 \\ 0 & 0 & 1 & 0 \end{bmatrix},$$

then the square matrix K contains all of the information expressing the parameters of the camera and is called the *camera calibration matrix*. That is, you will need to determine the pixel coordinates of the principal point in addition to the focal distance. In determining these factors, it is important to remember that real optical systems do not act like the perfect pinhole camera. That is, they do not form a point image from a point object; rather they form an intensity distribution of the image. The resulting distortions or aberrations can affect the quality of the image and must be corrected for.

The nonlinear distortions due to the lens systems and associated imperfections in the image sensing technology have implications for some interesting applications in VR, and so we will now examine more advanced ideas in computer vision which involve removing camera distortion and perspective distortion.

8.3   A Brief Look at Some Advanced Ideas in Computer Vision

Using the ideas expressed in Section 8.1, we are in a position to examine a few of the more sophisticated applications of computer vision in VR. We will consider how the effects of lens distortion and perspective distortion may be removed, how we can use a few simple calibrating images to determine the camera matrix for a particular imaging device, how we can create panoramic images from several photographs and how we can reconstruct a 3D scene from a single picture so as to view it from a different viewpoint. Perhaps most useful of all, we'll discuss how we can build a 3D model mesh of a scene out of two, three or more pictures, or a short movie in which the camera moves around the environment.

8.3.1   Removing Camera Distortions

The optical process of digital image formation is prone to some well-known errors or distortions that limit the quality of the image and that if not taken into account, may limit the validity of the derived 3D estimates from the 2D image scene or vice versa.

In essence, geometric distortions change the shape of an image. The two forms of geometric distortions present in the optical systems of image sensors are *pincushion distortion* and *barrel distortion* (see Figure 8.5).

Pincushion distortion magnifies the center of an image less than the edges. This results in the center of an image *bowing in*. Everything will still be in focus, but points in the image will not be in their proper place. Barrel distortion magnifies the center of an image more than the edges. This results in the center of an image *bowing out*. Again, everything will still be in focus, but points in the image will not be in their proper place.

These geometric distortions can be avoided using a well-corrected lens. However, this is expensive, and limits the angle of view since wide-angle lenses

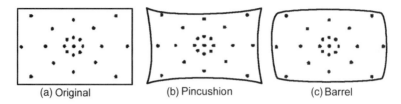

(a) Original   (b) Pincushion   (c) Barrel

Figure 8.5. Effects of pincushion and barrel distortion on the original scene.

are harder and sometimes nearly impossible to correct in this manner. A second method is to correct the distortion using digital image processing techniques. This method is well researched with an abundance of documented distortion correction techniques.

Several researchers have presented various mathematical models of image distortion to find the model parameters required to complete the distortion correction process. Common to most methods is the use of a test chart consisting of equidistant dots. Usually, several adjustments are required to the position of the test chart or optical system so that the central point and orientation of the vertical and horizontal lines on the chart are as intended on the captured image. This is a time-consuming process that has been eliminated by some researchers [14] by creating distortion algorithms that are sensitive to changes in both the horizontal and vertical axis.

Vijayan et al. [14] developed a mathematical model, based on polynomial mapping, which is used to map images from the distorted image space to a corrected image space. The model parameters include the polynomial coefficients, distortion center and corrected center. They also developed an accurate program to extract the dot centers from the test chart. However, Vijayan et al. have assumed that each corrected grid line in the horizontal or vertical direction does not need to have the same slope, and therefore produced line fits that cannot be parallel. Zhang et al. [15] have improved upon this technique so that all lines in any one direction (vertical or horizontal) have the same slope, making them parallel.

We will build on the work in [14] and [15] to illustrate how anti-distortion algorithms can be easily utilized when developing virtual environments to ensure the accuracy of the 3D scene is carried over into the 2D image. In addition, we will only consider barrel distortion, as this is the type most commonly found in cameras.

In general, methods that correct barrel distortion must calculate a distortion center and correct the radial component of the distortion. The distortion correction problem involves two image spaces. These are the distorted space and the corrected space of the image. The goal is to derive a method for transforming the distorted image into a corrected image.

The approach taken to correct for distortion involves a test grid similar to that shown in Figure 8.6(a). Figure 8.6(b) illustrates the original test pattern after imaging. It is evident that distortion is present and needs to be corrected. All the previously straight rows and columns have become curved towards the center of the image. This is typical of barrel distortion, where points on the extremity of the image are moved towards the center of distortion.

Distortion parameters are calculated to fit a line of dots on a captured image. This method offers two advantages, in that the correction coefficients are calculated in an automatic process and the calibration pattern does not need to be exactly horizontal or vertical.

The center of distortion is a fixed point for a particular camera and, once calculated, can be used for all images from that camera for a given focal length. An accurate estimate of the center of distortion is essential for determining the corresponding distortion correction parameters.

The lines of the standard test pattern that remain straight after imaging must lie between adjacent rows and columns of opposite curvatures. The next step is finding where exactly this center of distortion lies within this bounded area. This is shown diagrammatically in Figure 8.7. The intersection of these straight lines should then give the center of distortion.

To estimate the center of distortion, four best-fit polynomials ($f(y) = \sum_{n=0}^{\lambda} a_n x^n$) are defined, passing through the adjacent rows and columns of

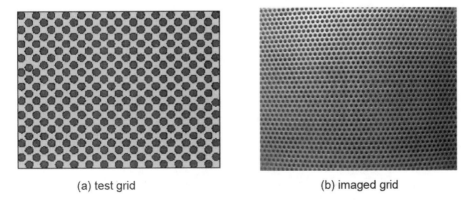

| (a) test grid | (b) imaged grid |

Figure 8.6. Original test pattern and its resulting image.

these opposite curvatures. For each polynomial defined, the (X', Y') data used for the best fit approximation is the pixel locations of the dot centers. It should be noted that the X' and Y' coordinates for each of the two columns need to be interchanged before a polynomial is defined. That is, you need to redefine the columns as rows to simplify the mathematics involved.

Next we need to find the stationary point on each of the curves. The curvature at this stationary point is simply the gradient of the curve, and is found by differentiation. The curvatures are labeled k_1 to k_4. The distortion center is then estimated by interpolating all four curvatures along with the pixel locations of the curvatures. These are shown in Figure 8.7 and are labeled (X'_1, Y'_1) to (X'_4, Y'_4).

That is, the pixel coordinates for the center of distortion (X'_c, Y'_c) are

$$\left(\frac{k_1 X'_1 - k_2 X'_2}{k_1 - k_2}, \frac{k_3 Y'_3 - k_4 Y'_4}{k_3 - k_4} \right).$$

When (X'_c, Y'_c) is found, it is then necessary to relate each pixel in the distorted space (X', Y') to a pixel in the corrected space (X, Y) using a mapping algorithm. Essentially, the aim is to create a new corrected image with the same dimensions as the previous distorted image. If we take the pixel location (X, Y) then we need to compute where that pixel would be within the

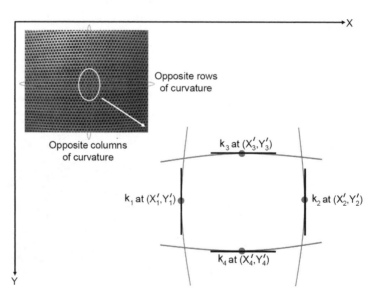

Figure 8.7. Adjacent rows and columns of curvature.

distorted image space (X', Y') and transfer the pixel value or color to the corrected space. In that way, we build up a new image from the distorted image.

The easiest way to do this is to find the absolute distance (r) from (X, Y) to (X_c, Y_c) and the angle (ϑ) that a line joining these two points makes with the x-axis. Note (X_c, Y_c) is the center pixel coordinate for the corrected image. It can be anywhere in the image display frame; however, it is sometimes simpler to assume that it is in the same position as (X'_c, Y'_c). Now we need to convert this absolute distance to the corresponding distance in the distorted image. To do that, we need to characterize the radial component of the distortion. Essentially, a set of N calibration points $(r'_i, i = 0, 1, 2, N - 1)$ needs to be defined from the center of distortion outwards to characterize the radial distortion. The first calibration point r'_0 is taken at the center of distortion. Using these discrete points, a polynomial[1] can be defined to approximate the distortion.

Now the location of the pixel at (r', ϑ') in the distorted image space can be calculated. The polynomial defined for the calibration points is used as an interpolation function to find the pixel location. That is,

$$r' = \sum_{j=0}^{M} a_j \left(\frac{r}{\alpha}\right)^j, \vartheta' = \vartheta,$$

where M is the degree of the polynomial, a_j is the expansion of the coefficients, α is the scale factor of the pixel height versus width and r is the pixel distance in the undistorted image. It is then possible to calculate the distorted pixel coordinates using both the distorted pixel distance and the polar angle:

$$X' = X'_c + r' \cos \vartheta, Y' = Y'_c + r' \sin \vartheta.$$

The brightness or color value of the pixel closest to the location (X', X') can be found directly from the distorted image. This value is then assigned to the pixel in location (X, Y) of the undistorted image. This process can be repeated for each pixel in the undistorted or corrected image space, thereby reproducing the calibrated image without the effects of distortion.

---

[1]The degree of polynomial selected to best represent the distortion should be based upon the error fit and the rate of change of error from the previous smaller degree of polynomial fit. A compromise should always be made between computation time and accuracy, both of which tend to increase with the degree of polynomial.

8.3.2 Removing Perspective Distortion

Section 6.6.7 showed how to implement a projective transformation, but now that we have firmed up on the properties of $\mathbb{P}^2$ and defined the homogeneous coordinates, we can see why a projective transformation or mapping from one 2D plane to another 2D plane embedded in 3D space may be expressed as a nonsingular 3×3 matrix H.

The symbol H is used because in computer vision a planar projective transformation is generally called a *homography*, and although there are nine elements in the matrix, only eight of them are independent. The result can be expressed the other way round: if H is a nonsingular matrix then a projectivity $h()$ exists between a point $\mathbf{P}$ on one projective plane and another point $\mathbf{P}_m = h(\mathbf{P}) = H\mathbf{P}$ on another plane.

Writing out the matrix elements explicitly for $\mathbf{P} = [X, Y, W]^T$ and $\mathbf{P}_m = [X_m, Y_m, W_m]^T$ gives

$$
\begin{bmatrix} X_m \\ Y_m \\ W_m \end{bmatrix} = \begin{bmatrix} h_{00} & h_{01} & h_{02} \\ h_{10} & h_{11} & h_{12} \\ h_{20} & h_{21} & h_{22} \end{bmatrix} \begin{bmatrix} X \\ Y \\ W \end{bmatrix}.
$$

Figure 8.8. Perspective distortion in the view of the side of a building. If four points can be identified that are known to be at the corners of a (known) rectangle (such as a window) then they can be used to determine the elements of H. From that, a perpendicular view of the side can be reconstructed (see Figure 8.9). The four points that we will use to determine H are highlighted.

This relationship is just what is needed to allow us to calculate and hence correct for the projective distortion seen when we look at the photograph of planar surfaces that are not perpendicular to the principal axis of the camera. The idea is illustrated in Figures 8.8, 8.9 and 8.10.

To obtain the elements of H, we need to consider the homogeneous coordinates of matching points. For example, let the point $(X, Y, 1)$ in one plane map to a point (X_m, Y_m, W_m) in another plane. Then, by writing the complete homography for this transformation, the normalized homogenous coordinates for the matched point (X_m, Y_m) can be calculated as follows:

$$X_m = \frac{X_m}{W_m} = \frac{h_{00}X + h_{01}Y + h_{02}}{h_{20}X + h_{21}Y + h_{22}}, \tag{8.2}$$

$$Y_m = \frac{Y_m}{W_m} = \frac{h_{10}X + h_{11}Y + h_{12}}{h_{20}X + h_{21}Y + h_{22}}. \tag{8.3}$$

Remember that whilst there are nine coefficients in the transformation H, only eight are independent. Thus, we only require four sets of matched points in order to determine H up to a scale factor. As we have seen, each matched point generates two equations which are linear in the elements of H, and so using four points, we obtain eight equations. Since there are nine coefficients in H but only eight independent elements, we can choose any value for one

Figure 8.9. Perspective distortion removed by applying H to each pixel in Figure 8.8. The points used to determine H are illustrated in their corrected positions. It should be noted that the correction only applies to those parts of the building which lie in the same plane. If a different plane is used for the correction, a different result is observed (see Figure 8.10).

Figure 8.10. Perspective distortion using points in a different plane results in a different correction.

of them, say $h_{22} = 1$. Thus for points $(X_i, Y_i); 0 \leq i \leq 3$ and their matches $(X_{mi}, Y_{mi}); 0 \leq i \leq 3$, we get the linear equations

$$
\begin{bmatrix}
X_0 & Y_0 & 1 & 0 & 0 & 0 & -X_0 X_{m0} & -Y_0 X_{m0} \\
0 & 0 & 0 & X_0 & Y_0 & 1 & -Y_{m0} X_0 & -Y_{m0} Y_0 \\
X_1 & Y_1 & 1 & 0 & 0 & 0 & -X_1 X_{m1} & -Y_1 X_m \\
0 & 0 & 0 & X_1 & Y_1 & 1 & -Y_{m1} X_1 & -Y_{m1} Y_1 \\
X_2 & Y_2 & 1 & 0 & 0 & 0 & -X_2 X_{m2} & -Y_2 X_{m2} \\
0 & 0 & 0 & X_2 & Y_2 & 1 & -Y_{m2} x_2 & -Y_{m2} Y_2 \\
X_3 & Y_3 & 1 & 0 & 0 & 0 & -X_3 X_{m3} & -Y_3 X_{m3} \\
0 & 0 & 0 & X_3 & Y_3 & 1 & -Y_{m3} X_3 & -Y_{m3} Y_3
\end{bmatrix}
\begin{bmatrix}
h_{00} \\
h_{01} \\
h_{02} \\
h_{10} \\
h_{11} \\
h_{12} \\
h_{20} \\
h_{21}
\end{bmatrix}
=
\begin{bmatrix}
X_{m0} \\
Y_{m0} \\
X_{m1} \\
Y_{m1} \\
X_{m2} \\
Y_{m2} \\
X_{m3} \\
Y_{m3}
\end{bmatrix} .
$$

$$(8.4)$$

This simple-looking (and first-guess) method to determine H, and with it the ability to correct for perspective distortion, is flawed in practice because we cannot determine the location of the four points with sufficient accuracy. The discrete pixels used to record the image and the resulting numerical problems introduce quantization noise, which confounds the numerical solution of Equation (8.4). To obtain a robust and accurate solution, it is necessary to use many more point correspondences and apply one of the estimation algorithms outlined in Section 8.3.4.

8.3.3   Recovering the Camera Matrix

In the previous section, it was shown that it is *in theory* possible to correct for the perspective distortion that accrues when taking photographs. In an analogous way, if we can match several known points in the 3D world scene with their corresponding image points on the projective plane (see Figure 8.11) then we should be able to obtain the camera matrix P. Now why would this be so important? Well, in Sections 8.2.1 and 8.2.2, we saw that we can estimate the position of a real-world point in the image scene if we know the camera matrix. If we cannot assume that the camera we are using is ideal (i.e., the pinhole camera model) then it is very difficult to determine whatthe parameters of the camera matrix are. This enters into the realms of camera calibration, where we need to develop a mathematical model which allows us to transform real-world points to image coordinates. This mathematical model is of course the camera model and is based on three sets of parameters:

1. The extrinsic parameters of the camera which describe the relationship between the camera frame and the world frame; that is, the translation and rotation data needed to align both reference frames.

2. The extrinsic parameters of the camera which describe its characteristics, such as lens focal length, pixel scale factors and location of the image center.

3. Distortion parameters which describe the geometric nonlinearities of the camera. As we described in Section 8.3.1, this can be removed

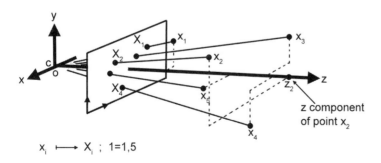

Figure 8.11. Matching image points on the projection plane to points in the 3D world.

separately and does not need to be included in the camera matrix. This is the method we prefer to use.

All these parameters are encompassed in the 3×4 matrix we identified before. That is

$$
\begin{bmatrix} X \\ Y \\ W \end{bmatrix} = P \begin{bmatrix} x \\ y \\ z \\ w \end{bmatrix},
$$

or in full we can write

$$
\begin{bmatrix} X \\ Y \\ W \end{bmatrix} = \begin{bmatrix} p_{00} & p_{01} & p_{02} & p_{03} \\ p_{10} & p_{11} & p_{12} & p_{13} \\ p_{20} & p_{21} & p_{22} & p_{23} \end{bmatrix} \begin{bmatrix} x \\ y \\ z \\ w \end{bmatrix}.
$$

If we multiply out the terms of this matrix and decide, as is normal, that $W = 1$, then we have

$$
X = p_{00}x + p_{01}y + p_{02}z + p_{03}w,
$$

$$
Y = p_{10}x + p_{11}y + p_{12}z + p_{13}w,
$$

$$
1 = p_{20}x + p_{21}y + p_{22}z + p_{23}w.
$$

The third equation in this sequence is interesting because if its value needs to be equal to 1 then the term p_{23} must be dependent on the values of p_{20}, p_{21} and p_{22}. This means that the complete camera matrix contains 11 independent values.

As we know already, every point correspondence affords two equations (one for x and one for y). So $5\frac{1}{2}$ point correspondences are needed between the real-world image coordinates and the image coordinates in order to derive the camera matrix rather than derive it directly from camera calibration methods. If $5\frac{1}{2}$ correspondences are used then the solution is exact.

The problem with this method is that if six or more correspondences are used, we don't find that some correspondences simply drop out of the calculation because they are linear combinations of others. What we find is that we obtain quite wide variations in the numerical coefficients of P depending on which $5\frac{1}{2}$ are chosen for the calculation. *This is due to pixel quantization noise in the image and errors in matching the correspondences. Resolving the difficulty is a major problem in computer vision, one that is the subject of current and ongoing research.*

Intuitively, it would seem evident that if the effect of noise and errors are to be minimized then the more points taken into account, the more accurately the coefficients of the camera matrix could be determined. This is equally true for the coefficients of a homography between planes in $\mathbb{P}^2$ and for the techniques that use two and three views to recover 3D structure. A number of very useful algorithms have been developed that can make the best use of the point correspondences in determining P etc. These are generally referred to as *optimization algorithms*, but many experts who specialize in computer vision prefer the term *estimation algorithms*.

8.3.4 Estimation

So, the practicalities of determining the parameters of the camera matrix are confounded by the presence of noise and error. This also holds true for perspective correction and all the other procedures that are used in computer vision. It is a major topic in itself to devise, characterize and study algorithms that work robustly. By robustly, we mean algorithms that are not too sensitive to changes in noise patterns or the number of points used and can make an intelligent guess at what points are obviously in gross error and should be ignored (outliers).

We will only consider here the briefest of outlines of a couple of algorithms that might be employed in these circumstances. The direct linear transform (DLT) [7, pp. 88–93] algorithm is the simplest; it makes no attempt to account for outliers. The random sample consensus (RANSAC) [5] and the least median squares (LMS) algorithms use statistical and iterative procedures to identify and then ignore outliers.

Since all estimation algorithms are optimization algorithms, they attempt to minimize a *cost function*. These tend to be geometric, statistical or algebraic. To illustrate the process, we consider a simple form of the DLT algorithm to solve the problem of removing perspective distortion. (Using four corresponding points (in image and projection), we obtained eight linear inhomogeneous equations (see Equation (8.4)), from which then the coefficients of H can be estimated.)

If we can identify more than four matching point correspondences, the question is then how to use them. This is where the DLT algorithm comes in. To see how to apply it, consider the form of the matrix in Equation (8.4) for a single point:

$$
\begin{bmatrix}
x_i & y_i & 1 & 0 & 0 & 0 & -x_i x_i' & -y_i x_i' \\
0 & 0 & 0 & x_i & y_i & 1 & -y_i' x_i & -y_i' y_i
\end{bmatrix}
\begin{bmatrix}
h_{00} \\ h_{01} \\ h_{02} \\ h_{10} \\ h_{11} \\ h_{12} \\ h_{20} \\ h_{21}
\end{bmatrix}
=
\begin{bmatrix} x_i' \\ y_i' \end{bmatrix}. \qquad (8.5)
$$

It should be fairly evident that for every corresponding pair of points, there will be an additional two linear equations, and for n corresponding points, the matrix will be of size $2n \times 8$. Since this is an over-determined system, one cannot simply proceed to solve for the h_{ij}, using Gauss elimination, for example. It also turns out that using the inhomogeneous form in Equation (8.5) to compute the coefficients of H in

$$\mathbf{x}' = H\mathbf{x}$$

can lead to unstable results and a better approach is to retain the homogeneous form (which has nine coefficients). Remember, however, that the coefficients are only determined up to a scale factor. Thus for a single point correspondence between (x, y, w) and $(x'y'w')$ (where the w coordinate is retained), we have

$$
\begin{bmatrix}
w_i' x_i & w_i' y_i & w_i' w_i & 0 & 0 & 0 & -x_i x_i' & -y_i x_i' & -x_i' w_i \\
0 & 0 & 0 & w_i' x_i & w_i' y_i & w_i' w_i & -y_i' x_i & -y_i' y_i & -y_i' w_i
\end{bmatrix}
\begin{bmatrix}
h_{00} \\ h_{01} \\ h_{02} \\ h_{10} \\ h_{11} \\ h_{12} \\ h_{20} \\ h_{21} \\ h_{22}
\end{bmatrix}
= \mathbf{0}.
$$

$$(8.6)$$

For n points in this homogeneous form, there is now a $2n \times 9$ matrix, but solutions of the form $A\mathbf{h} = \mathbf{0}$ are less subject to instabilities than the equivalent inhomogeneous formulation. Of course, there still is the little matter of solving the linear system, but the DLT algorithm offers a solution through singular value decomposition (SVD). For details of the SVD step, which is a general and very useful technique in linear algebra, see [12].

Singular value decomposition factorizes a matrix (which does not have to be square) into the form $A = LDV^T$, where L and V are orthogonal matrices (i.e., $V^TV = I$) and D is a diagonal. (D is normally chosen to be square, but it does not have to be.) So, for example, if A is 10×8 then L is 10×10, D is 10×8 and V^T is 10×10. It is typical to arrange the elements in the matrix D so that the largest diagonal coefficient is in the first row, the second largest in the second row etc.

For example, the SVD of

$$
\begin{bmatrix} 1 & 1 \\ 0 & 1 \\ 1 & 0 \end{bmatrix} = \begin{bmatrix} \frac{\sqrt{6}}{3} & 0 & \frac{\sqrt{3}}{3} \\ \frac{\sqrt{6}}{6} & -\frac{\sqrt{6}}{6} & -\frac{\sqrt{3}}{3} \\ \frac{\sqrt{6}}{6} & \frac{\sqrt{2}}{2} & -\frac{\sqrt{3}}{3} \end{bmatrix} \begin{bmatrix} \sqrt{3} & 0 \\ 0 & 1 \\ 1 & 0 \end{bmatrix} \begin{bmatrix} \frac{\sqrt{2}}{2} & \frac{\sqrt{2}}{2} \\ \frac{\sqrt{2}}{2} & -\frac{\sqrt{2}}{2} \end{bmatrix}.
$$

There are two steps in the basic direct linear transform algorithm. They are summarized in Figure 8.12. The DLT algorithm is a great starting point for obtaining a transformation that will correct for a perspective projection, but it also lies at the heart of more sophisticated estimation algorithms for solving a number of other computer vision problems.

We have seen that the properties of any camera can be concisely specified by the camera matrix. So we ask, is it possible to obtain the camera matrix itself by examining correspondences from points (not necessarily lying on a plane) in 3D space to points on the image plane? That is, we are examining correspondences from $\mathbb{P}^3$ to $\mathbb{P}^2$.

Of course the answer is yes. In the case we have just discussed, H was a 3×3 matrix whereas P is dimensioned 3×4. However, it is inevitable that the estimation technique will have to be used to obtain the coefficients of P. Furthermore, the only differences between the expressions for the coefficients of H (Equation (8.6)) and those of P arise because the points in $\mathbb{P}^3$ have four coordinates, and as such we will have a matrix of size $2n \times 12$. Thus the analog of Equation (8.6) can be expressed as $A\mathbf{p} = \mathbf{0}$, and the requirement for the estimation to work is $n \geq 6$. (That is, 6 corresponding pairs with 12 coefficients and 11 degrees of freedom.)

(1) From each point mapping, compute the corresponding
two rows in the matrix A in $A\mathbf{h} = \mathbf{0}$
and assemble the full $2n \times 9$ A matrix.
Note that $\mathbf{h}$ contains all the coefficients in H
written an a vector.

(2) Compute the SVD of A. If the SVD results in
$A = LDV^T$, the elements in H are given by
the last column in V, i.e., the last row of V^T.
(e.g., if the A matrix is dimensioned (64×9) then L is
(64×64), D is (64×9) and L is (9×9))

Figure 8.12. Obtaining the matrix H using the DLT algorithm with n point correspondences. In practice, the DLT algorithm goes through a pre-processing step of normalization, which helps to reduce numerical problems in obtaining the SVD (see [7, pp. 88–93]).

In this case, if we express the correspondence as $(x_i, y_i, z_i, w_i) \mapsto (X_i, Y_i, W_i)$ from $\mathbb{P}^3$ to $\mathbb{P}^2$ and let

$$
\begin{aligned}
\mathbf{x}_i^T &= [x_i, y_i, z_i, w_i], \\
\mathbf{p0}_i &= [p_{00}, p_{01}, p_{02}, p_{03}]^T, \\
\mathbf{p1}_i &= [p_{10}, p_{11}, p_{12}, p_{13}]^T, \\
\mathbf{p2}_i &= [p_{20}, p_{21}, p_{22}, p_{23}]^T,
\end{aligned}
$$

then the 12 coefficients of P are obtained by solving

$$
\begin{bmatrix} W_i\mathbf{x}_i^T & \mathbf{0} & -X_i\mathbf{x}_i^T \\ \mathbf{0} & -W_i\mathbf{x}_i^T & Y_i\mathbf{x}_i^T \end{bmatrix} \begin{bmatrix} \mathbf{p0}_i \\ \mathbf{p1}_i \\ \mathbf{p2}_i \end{bmatrix} = \mathbf{0}. \tag{8.7}
$$

The DLT algorithm is equally applicable to the problem of obtaining P given a set of $n \geq 6$ point correspondences from points in world space (x_i, y_i, z_i, w_i) to points on the image plane, (X_i, Y_i, W_i). However, there are some difficulties in practice, as there are in most numerical methods, that require some refinement of the solution algorithm. For example, the limited numerical accuracy to which computers work implies that some form of scaling (called normalization) of the input data is required before the DLT algorithm is applied. A commensurate de-normalization has also to be applied to the result.

We refer the interested reader to [7, pp. 88–93] for full details of a robust and successful algorithm, or to the use of computer vision software libraries such as that discussed in Section 8.4.

8.3.5   Reconstructing Scenes from Images

One of the most interesting applications for computer vision algorithms in VR is the ability to reconstruct scenes. We shall see (in Section 8.3.7) that if you have more than one view of a scene, some very accurate 3D information can be recovered. If you only have a single view, the opportunities are more limited, but by making some assumptions about parallel lines, vanishing points and the angles between structures in the scene (e.g., known to be orthogonal) some useful 3D data can be extracted.

An example of this is illustrated in Figure 8.13. Figure 8.13(a) is a photograph of a room. Figure 8.13(b) is a view of a 3D model mesh reconstructed from the scene. (Note: some of the structures have been simplified because

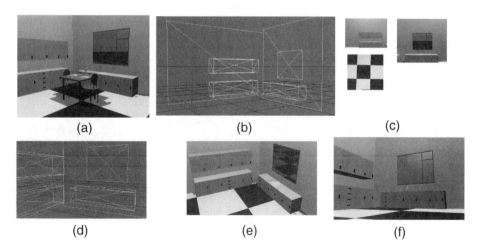

(a)   (b)   (c)

(d)   (e)   (f)

Figure 8.13. Reconstructing a 3D scene from images.

simple assumptions regarding vanishing points and the line at infinity have to be accepted.) Figure 8.13(c) shows image maps recovered from the picture and corrected for perspective. Figure 8.13(d) is a different view of the mesh model. Figure 8.13(e) is a 3D rendering of the mesh model with image maps applied to surfaces. Figure 8.13(f) shows the scene rendered from a low-down viewpoint. (Note: some minor user modifications were required to fix the mesh and maps after acquisition.) The details of the algorithms that achieve these results are simply variants of the DLT algorithm we examined in Section 8.3.4.

In an extension to this theory, it is also possible to match up several adjoining images taken by rotating a camera, as illustrated in Figure 8.14. For example, the three images in Figure 8.15 can be combined to give the single image shown in Figure 8.16. This is particularly useful in situations where you have a wide display area and need to collate images to give a panoramic view (see Figure 8.17). A simple four-step algorithm that will assemble a single panoramic composite image from a number of images is given in Figure 8.18.

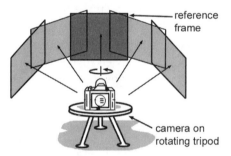

Figure 8.14. A homography maps the image planes of each photograph to a reference frame.

Figure 8.15. Three single photos used to form a mosaic for part of a room environment.

Figure 8.16. The composite image.

Figure 8.17. A composite image used in an immersive VR environment. In this particular example, the image has been mapped to a cylindrical screen.

(1) Pick one image as the reference

(2) Compute the homography H which maps the next
image in the sequence to the reference.
Use the four point correspondence described in Section 8.3.4,
or the RANSAC [5] algorithm can be used with a point
classification procedure to identify matched points [11, Ch. 4].

(3) Projectively align the second image with H
and blend in the non-overlapped part with the current
state of the panoramic image.

(4) Repeat the procedure from Step 2 until all the
images have been mapped and blended into the panorama.

Figure 8.18. The algorithm to assemble a panoramic mosaic from a number of individual images.

8.3.6   Epipolar Geometry and the Fundamental Matrix

So far, we have really only considered scene reconstruction using one camera. However, when a second, third or fourth camera becomes available to take simultaneous pictures of the 3D world, a whole range of exciting opportunities open up in VR. In particular, the realm of 3D reconstruction is very important.

Again, the theory of scene reconstruction from multiple cameras is based on rigorous maths. We will only briefly explore these ideas by introducing *epipolar geometry*. This arises as a convenient way to describe the geometric relationship between the projective views that occur between two cameras looking at a scene in the 3D world from slightly different viewpoints. If the geometric relationship between the cameras is known, one can get immediate access to 3D information in the scene. If the geometric relationship between the cameras is unknown, the images of test patterns allow the camera matrices to be obtained from just the two images alone.

All of the information that defines the relationship between the cameras and allows for a vast amount of detail to be retrieved about the scene and the cameras (e.g., 3D features if the relative positions and orientations of the cameras is known) is expressed in a 4×4 matrix called the *fundamental matrix*. Another 4×4 matrix called the *essential matrix* may be used at times when one is dealing with totally calibrated cameras. The idea of epipolar geometry is illustrated in Figure 8.19(a), where the two camera centers $\mathbf{C}$ and $\mathbf{C}'$ and the camera planes are illustrated.

Any point $\mathbf{x}$ in the world space projects onto points $\mathbf{X}$ and $\mathbf{X}'$ in the image planes. Points $\mathbf{x}$, $\mathbf{C}$ and $\mathbf{C}'$ define a plane S. The points on the image plane where the line joining the camera centers intersect the image planes are called the *epipoles*. The lines of intersection between S and the image planes are called the *epipolar lines*, and it is fairly obvious that for different points $\mathbf{x}$ there will be different epipolar lines, although the epipoles will only depend on the camera centers and the location and orientation of the camera planes (see Figure 8.19(b)).

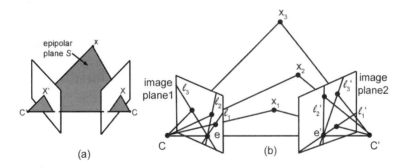

Figure 8.19. (a) Point correspondences and epipolar relationships. (b) Variations of epipolar lines. Point x_1 gives rise to epipolar lines $\mathbf{l}'_1$ and $\mathbf{l}_1$, point x_2 to $\mathbf{l}'_2$ and $\mathbf{l}_2$ etc.

Figure 8.20 shows how the epipolar lines fan out from the epipoles as $\mathbf{x}$ moves around in world space; this is called the *epipolar pencil*.

Now, what we are looking for is a way to represent (parameterize) this two-camera geometry so that information can be used in a number of interesting ways. For example, we should be able to carry out range finding if we know the camera geometry and the points $\mathbf{X}$ and $\mathbf{X}'$ on the image planes, or if we know something about the world point, we can determine the geometric relationship of the cameras.

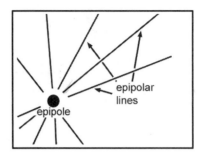

Figure 8.20. Epipole and epipolar lines as they appear in the camera plane view.

In Section 8.3.3, all the important parametric details about a camera were summed up in the elements of the 3×3 matrix, P. It turns out that an exact analog exists for epipolar geometry; it is called the fundamental matrix F. Analogous to the way P provided a mechanism of mapping points in 3D space to points on the camera's image plane, F provides a mapping, not from $\mathbb{P}^3$ to $\mathbb{P}^2$, but from one camera's image of the world point $\mathbf{x}$ to the epipolar line lying in the image plane of the other camera (see Figure 8.21).

This immediately gives a clue as to how to use a two-camera set-up to achieve range finding of points in world space if we know F:

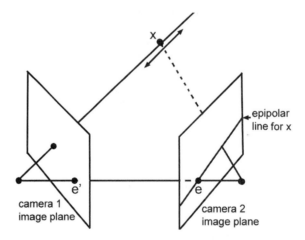

Figure 8.21. A point projected onto the image plane of camera 1 is mapped to some point lying on the epipolar line in the image plane of camera 2 by the fundamental matrix F.

Take the projected image of a scene with two cameras. Use image processing software to identify key points in the picture from image 1. Apply F to these points in turn to find the epipolar lines in image 2 for each of these points. Now use the image processing software to search along the epipolar lines and identify the same point. With the location in both projections known and the locations and orientations of the cameras known, it is simply down to a bit of coordinate geometry to obtain the world coordinate of the key points.

But there is more: if we can *range find*, we can do some very interesting things in a virtual world. For example, we can have virtual elements in a scene appear to interact with real elements in a scene viewed from two video cameras. The virtual objects can appear to move in front of or behind the real objects.

Let's return to the question of how to specify and determine F and how to apply it. As usual, we need to do a bit of mathematics, but before we do, let us recap some vector rules that we will need to apply:

- Rule 1: The cross product of a vector with itself is equal to zero. For example, $\mathbf{X}_3' \times \mathbf{X}_3' = \mathbf{0}$.

- Rule 2: The dot product of a vector with itself can be rewritten as the transpose of the vector matrix multiplied by itself. For example, $\mathbf{X}_3' \cdot \mathbf{X}_3' = \mathbf{X}_3'^T \mathbf{X}_3'$.

- Rule 3: Extending Rule 1, we can write $\mathbf{X}_3' \cdot (\mathbf{X}_3' \times \mathbf{X}_3') = 0$.

- Rule 4: This is not so much of a rule as an observation. The cross product of a vector with another vector that has been transformed by a matrix can be rewritten as a vector transformed by a different matrix. For example, $\mathbf{X}_3' \times T\mathbf{X}_3' = M\mathbf{X}_3'$.

Remembering these four rules and using Figure 8.22 as a guide, let $\mathbf{X_3} = [x, y, z, w]^T$ represent the coordinates in $\mathbb{P}^3$ of the image of a point in the first camera. Using the same notation, the image of that same point in the second camera is at $\mathbf{X}_3' = [x', y', z', w]^T$, also in world coordinates. These positions are with reference to the world coordinate systems of each camera. After projection onto the image planes, these points are at $\mathbf{X} = [X, Y, W]^T$ and $\mathbf{X}' = [X', Y', W]^T$, respectively, and these $\mathbb{P}^2$ coordinates are again relative to their respective projection planes. (Note: $\mathbf{X}_3$ and $\mathbf{X}$ represent the same point, but $\mathbf{X}_3$ is relative to the 3D world coordinate system with its origin at $\mathbf{C}$, whereas $\mathbf{X}$ is a 2D coordinate relative to an origin at the bottom left corner of the image plane for camera 1.)

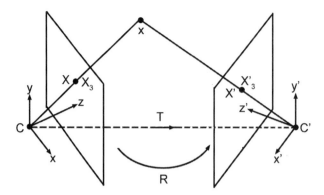

Figure 8.22. Two-camera geometry; the second camera is positioned relative to the first with a rotation R followed by a translation T.

Now, assuming that the properties of the cameras are exactly the same, apply the necessary translation and rotation to the first camera so that it lies at the same location and points in the same direction as the second camera. Then $\mathbf{X}_3$ will have been moved to $\mathbf{X}_3'$, and thus

$$\mathbf{X}_3' = T\mathbf{X}_3,$$

where T is a transformation matrix that represents the rotation and translation of the frame of origin of camera 1 to that of camera 2.

Now if we consider Rule 3, which stated

$$\mathbf{X}_3' \cdot (\mathbf{X}_3' \times \mathbf{X}_3') = 0,$$

then we can replace one of the X_3' terms with TX_3. That is,

$$\mathbf{X}_3' \cdot (\mathbf{X}_3' \times T\mathbf{X}_3) = 0.$$

Utilizing Rule 2, we can rewrite this equation as

$$\mathbf{X}_3'^{T} (\mathbf{X}_3' \times T\mathbf{X}_3) = 0,$$

and bringing in Rule 4, we can simplify this expression to

$$\mathbf{X}_3'^{T} M\mathbf{X}_3 = 0, \tag{8.8}$$

where the matrix M is derived from the translation matrix. Using the camera calibration matrices P and P' to project the world points $\mathbf{X}_3$ and $\mathbf{X}'_3$ onto the viewing planes, $\mathbf{X} = P\mathbf{X}_3$ and $\mathbf{X}' = P'\mathbf{X}'_3$, Equation (8.8) becomes

$$(P'^{-1}\mathbf{X}')^T M(P^{-1}\mathbf{X}) = 0,$$

or

$$\mathbf{X}'^T((P'^{-1})^T MP^{-1}\mathbf{X}) = 0.$$

And finally, by letting $F = (P'^{-1})^T MP^{-1}$, the final result is obtained with an equation that defines the fundamental matrix:

$$\mathbf{X}'^T F\mathbf{X} = 0. \tag{8.9}$$

Equation (8.9) affords a way to calculate the nine coefficients of $\mathbf{F}$ using the same techniques of estimation that were covered in Section 8.3.4. When written in full for some point $\mathbf{x}_i$ projecting to $\mathbf{X}_i$ and $\mathbf{X}'_i$, Equation (8.9) resembles some of the results obtained earlier, e.g., the camera matrix P. The full structure of Equation (8.9) is

$$\begin{bmatrix} X'_i & Y'_i & W'_i \end{bmatrix} \begin{bmatrix} f_{00} & f_{01} & f_{02} \\ f_{10} & f_{11} & f_{12} \\ f_{20} & f_{21} & f_{22} \end{bmatrix} \begin{bmatrix} X_i \\ Y_i \\ W_i \end{bmatrix} = 0.$$

Like the $\mathbb{P}^3$ to $\mathbb{P}^2$ homology H, the fundamental matrix has only eight independent coefficients.

There are a large number of other interesting properties that emerge from a deeper analysis of Equation (8.9), but for the purposes of this introductory chapter and in the context of VR, we simply refer you again to [3] and [7] for full details.

8.3.7 Extracting 3D Information from Multiple Views

Section 8.3.6 hinted that by using the concept of epipolar geometry, two pictures taken from two different viewpoints of a 3D world can reveal a lot more about that world than we would obtain from a single picture. After all, we have already discussed that for VR, work stereopsis is an important concept, and stereoscopic projection and head-mounted displays are an essential component of a high quality VR environment. Nevertheless, one is still entitled to ask if there are any advantages in having an even greater number of views of the world. Do three, four or indeed *n* views justify the undoubted mathematical complexity that will result as one tries to make sense of the data and analyze it?

For some problems, the answer is of course yes, but the advantage is less pronounced from our viewpoint, i.e., in VR. We have two eyes, so the step in going from a single view to two views is simply bringing the computer-based virtual world closer to our own sense of the real world. To some extent, we are not particularly interested in the aspect of computer vision that is more closely allied to the subject of image recognition and artificial intelligence, where these multi-views do help the *stupid* computer recognize things better and reach more human-like conclusions when it looks at something.

For this reason, we are not going to delve into the analysis of three-view or *n*-view configurations. It is just as well, because when three views are considered, matrices (*F* for example) are no longer sufficient to describe the geometry, and one has to enter the realm of the *tensor* and its algebra and analysis. Whilst tensor theory is not an impenetrable branch of mathematics, it is somewhat outside the scope of our book on VR. In terms of computer vision, when thinking about three views, the important tensor is called the *trifocal tensor*. One can read about it from those who proposed and developed it [6, 13].

8.3.8   Extracting 3D Information from Image Sequences

If you think about it for a moment, making a video in which the camera moves from one spot to another effectively gives us *at least* two views of a scene. Of course, we get a lot more than two views, because even if the journey takes as short a time as one second, we may have 25 to 30 pictures, one from each frame of the movie. So on the face of it, taking video or moving footage of a scene should be much more helpful in extracting scene detail than just a couple of camera snapshots.

Unfortunately, there is one small complication about using a single camera to capture scene detail (and we are not talking about the fact that we don't move the camera, or that we simply pan it across the scene, which is no better). We are talking about the fact that elements in the scene get the chance to move. Moving elements will make the recognition of points much more difficult. For example, in using the fundamental matrix to calculate range, we cannot be assured that if we take two pictures of a scene and select a point of interest in one picture (use *F* to find its epipolar line in the second), the image of the point in the second picture will still be lying on the epipolar line.

Nevertheless, video or sequences of images are often so much easier to obtain than two contemporaneously obtained stills that getting information from them has taken on a discipline of its own, the study of *structure from motion (SfM)*. Details of techniques used to solve the SfM problem lie too far away from our theme of VR and so we refer you to the overview [9] (also available online).

One example of SfM which is of interest to us in connection with 3D structure recovery in a noisy environment is illustrated in Figure 8.23, which shows how feature tracking and feature matching are used to build up a 3D model from a sequence of images obtained by moving the viewpoint and/or objects. The three images in Figure 8.23 were acquired from slightly different viewpoints of a standard test model as the camera pans around it. Using a feature-detection algorithm, points of interest are selected in the first and third images. When equivalent points are matched, their predicted positions in the second image can be reconstructed. These are shown on the bottom left. At the bottom right, we see the result of applying the feature detection directly to the second image. The indicated areas show small differences, but the overall impression is that the interpolation is quite good. Consequently, SfM offers high potential as a tool for reconstructing virtual environments from photographs.

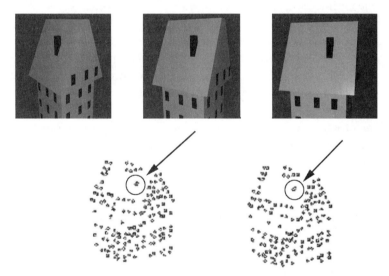

Figure 8.23. Three images taken from a video clip are used by an SfM process to acquire some 3D structure data.

8.4   Software Libraries for Computer Vision

The algorithms we have described in this chapter involve a significant degree of numerical complexity when they are implemented in practice. Practical considerations also affect the format in which images and movies are stored. The length of a computer program written to do a simple image-processing task, such as an edge detection, can be increased by an order of magnitude if the image is stored as a JPEG image as opposed to an uncompressed bitmap.

There are two freely available program function libraries that have proved popular amongst engineers and computer scientists working in VR, image processing and computer vision: Microsoft's Vision SDK [10] and Intel's OpenCV [8].

The Vision SDK reduces much of the drudgery involved in loading images and performing data manipulations. It is a low-level library, intended to provide a strong programming foundation for research and application development; it is not a high-level platform for end-users to experiment with imaging operations. It has a nice interface to Windows, such as shared image memory across processes. It has user-definable pixel types and a device-independent interface for image acquisition. It can be used to create binaries which can be transported and run on a machine with any supported digitizer or camera. It can be easily extended to support new types of digitizers or cameras. Vision SDK is distributed in source form as nine projects for Visual C++ that create libraries and DLLs for use in applications, e.g., the VisMatrix project provides classes for vectors and matrices. Classes CVisVector4, CVisTransform4x4, and CVisTransformChain are specialized to work with three-dimensional vectors and matrices using homogeneous coordinates.

Intel's OpenCV library provides a cross-platform middle- to high-level API that consists of more than 300 C language functions. It does not rely on external libraries, though it can use some when necessary. OpenCV provides a transparent interface to Intel's integrated performance primitives (IPP). That is, it automatically loads IPP libraries optimized for specific processors at run-time, if they are available.

OpenCV has operating-system–independent functions to load and save image files in various formats, but it is *not* optimized for Windows in the way that Vision SDK is. However, it is richly populated with useful functions to perform computer vision algorithms:

- Image processing, including:

    - Gradients, edges and corners

    - Sampling, interpolation and geometrical transforms

    - Filters and color conversion

    - Special image transforms

- Structural analysis, including:

    - Contour processing

    - Computational geometry

    - Planar subdivisions

- Motion analysis and object tracking, including:

    - Accumulation of background statistics

    - Motion templates

    - Object tracking

    - Optical flow

    - Estimators

- Pattern recognition and object detection

- Camera calibration and 3D reconstruction

- Epipolar geometry

By using these two powerful libraries, it is possible to create some user-friendly and sophisticated application programs for fundamental computer vision development and research. In VR applications, the availability of these libraries allows us to craft useful code without having to have a deep understanding of mathematical numerical methods and linear algebra.

One of the most exciting recent applications in VR is augmented reality. We briefly mentioned this in the context of motion tracking (Section 4.3) and earlier in this chapter. A key step in the augmentation is being able to place the virtual object into the real world view with the correct dimension position and orientation. To do this, the image-processing software must be able to recognize one or more predefined planar *marker*(s) and calculate the homography between their planes and the camera image plane. The ARToolKit [1]

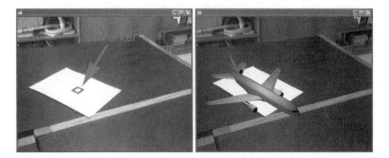

Figure 8.24. A distinctive marker in the field of view of the camera provides a frame of reference that is used to place virtual objects in the scene with a size and orientation calculated to make them look like they are attached to the marker.

provides a neat library of functions for use in a C++/C application program to overlay 3D virtual objects on the video from a digital camera in real time, as illustrated in Figure 8.24. To be able to calculate a transformation that will place the 3D virtual object correctly over the marker, the software has to recognize the marker in the first place and then find its four corners. This requires four steps:

1. Every pixel in the camera's image raster is converted to a binary value, i.e., set to black or white. Above a given threshold, the pixel's intensity is set to "1"; below it, the pixel value is "0".

2. Using the binary image, identify a set of contour lines in the image.

3. Try to find the places where these lines meet to form the corners of the possible image of a marker.

4. Using the known pattern of the markers, attempt to match it to the region bounded by the possible corners and edges of the possible marker.

Once the marker has been correctly identified, the 3D virtual object will be transformed so this it sits on top of it. More details of this process are given in Chapter 18, where we document a project using the ARToolKit.

8.5   Summary

In this chapter, we have given you a quick run through of some of the concepts of computer vision and how they might be applied to enhance and facilitate the VR experience. In particular, a number of the key algorithms which emerged during the development of the geometric theory of computer vision have great utility in the implementation and delivery of practical VR applications.

Two of the most significant applications of computer vision in the context of VR are in augmented reality (for example, *in-place* videoconferencing) and image-based rendering, which is the subject of the next chapter.

Of course, being able to do all this means that the images you take must be geometrically accurate. Therefore, we need to remove any lens distortion and perspective distortion from the images before we try to match them together. We also need to know the geometrical model of the camera i.e., the camera model. This model is difficult to determine, and to do so we entered into the mathematical complexities of the *fundamental matrix etc.* Essentially, we work backwards by identifying similar points from multiple images of the same scene to recover the camera model. This is non-trivial and requires the use of minimization algorithms, all of which were described within this chapter.

Thus, we hope we have given you an insight into the theory that surrounds computer vision and how that theory can be utilized to enhance the VR experience.

Bibliography

[1] *The Augmented Reality Toolkit.* http://www.hitl.washington.edu/artoolkit/.

[2] M. Billinghurst and H. Kato. "Real World Teleconferencing". In *CHI '99 Extended Abstracts on Human Factors in Computing Systems,* pp. 194–195. New York: ACM Press, 1999.

[3] O. Faugeras. *Three-Dimensional Computer Vision: A Geometric Viewpoint.* Cambridge, MA: The MIT Press, 1993.

[4] O. Faugeras and T. Papadopoulo. "A Nonlinear Method for Estimating the Projective Geometry of 3 Views". In *Proceedings of the Sixth International Conference on Computer Vision,* pp. 477–484. Wshington D.C.: IEEE Computer Society, 1998.

[5] M. Fischler and R. Bolles. "Random Sample Consensus: A Paradigm for Model Fitting with Applications to Image Analysis and Automated Cartography". *Communications of the ACM* 24:6 (1981) 381–395.

[6] R. Hartley. "Lines and Points in Three Views: An Integrated Approach". In *Proceedings of the ARPA Image Understanding Workshop,* pp. 1009–1016. San Francisco, CA: Morgan Kaufmann, 1994.

[7] R. Hartley and A. Zisserman. *Multiple View Geometry in Computer Vision,* Second Edition. Cambridge, UK: Cambridge University Press, 2004.

[8] Intel Corp. "Open Source Computer Vision Library". http://www.intel.com/technology/computing/opencv/.

[9] T. Jebara et al. *3D Structure from 2D Motion. IEEE Signal Processing Magazine* 16:3 (1999) 66–84.

[10] Microsoft Research, Vision Technology Group. "The Micorsoft Vision SDK, Version 1.2". http://research.microsoft.com/research/vision, 2000.

[11] M. Nixon and A. Aguado. *Feature Extraction and Image Processing.* Boston, MA: Newnes, 2002.

[12] W. Press et al. (editors). *Numerical Recipes in C++: The Art of Scientific Computing,* Second Edition. Cambridge, UK: Cambridge University Press, 2002.

[13] A. Shashua and M. Werman. "On the Trilinear Tensor of Three Perspective Views and Its Underlying Geometry". In *Proceedings of the International Conference on Computer Vision,* pp. 920–925. Washington, D.C.: IEEE Computer Society, 1995.

[14] A. Vijayan, S. Kumar and D. Radhakrishnan. "A New Approach for Nonlinear Distortion Correction in Endoscopic Images Based on Least Squares Etimation". *IEEE Transactions on Medical Imaging* 18:4 (1999) 345–354.

[15] C. Zhang, J. Helferty, G. McLennan and W. Higgins. "Non Linear Distortion Correction In Endoscopic Video Images". In *Proceedings of International Conference on Image Processing,* pp. 439–442. Washington, D.C.: IEEE Computer Society, 2000.

9 Image-Based Rendering

Image-based rendering (IBR) techniques are a powerful alternative to traditional geometry-based techniques for creating images. Instead of using basic geometric primitives, a collection of sample images are used to render novel views.

In Chapter 5, we discussed these traditional techniques and looked at how 3D scenes constructed from polygonal models were stored in databases. Then, in Chapter 7, which explored the 3D rendering pipeline, we made the implicit assumption that the view one sees in a 3D scene arises as the result of rendering these polygonal models. And for the majority of VR applications, this assumption is probably true. However, IBR offers us an alternative way to set up and visualize a virtual world that does not involve rendering polygons and doesn't require any special 3D accelerated hardware. Fortunately, we have already touched upon the techniques of IBR in Section 8.3.5 when we discussed panoramic image composition, and we also covered one of its most fundamental ideas (that is, removing perspective distortion) in Section 8.3.2.

Of course, there is a lot more to IBR than just allowing one to render views of a virtual world made up of a collection of photographs as opposed to a collection of polygonal models. In this chapter, we will allow ourselves a little indulgence to briefly explore some of these exciting ideas before focusing primarily on those aspects of IBR most relevant to VR. In some situations, IBR is a much more relevant technology to use for VR because it actually allows us to use photographs of real places, and this can make a virtual scene seem anything but virtual!

9.1   General Approaches to IBR

It is generally accepted that there are three different classifications of IBR, which are based on the significance of the geometric component involved in the rendering process [17]. These are:

1. *Rendering with no geometry.* Hypothetically, it is possible to envision storing a complete description of a 3D scene using a seven-dimensional function that gives the light intensity at every point of space (three dimensions x, y, z), looking in every direction (another two dimensions ϑ, φ), at all times t and for every wavelength of light λ. Adelson and Bergen [1] gave this function the name *plenoptic*, $p = P(x, y, z, \vartheta, \varphi, t, \lambda)$. Plenoptic modeling is the general name given to any approach to rendering that may be expressed in these terms. For example, environment mapping as described in Section 7.6.2 is a function of two variables $p = P(\vartheta, \varphi)$, viewed from a single point, at a single instant in time and under fixed lighting conditions. That is, x, y, z, t and λ are constant.

   It is impractical to work with a seven-dimensional scene, but a four-dimensional plenoptic model such as the lumigraph [7] will allow a view in any direction to be determined from any point on a surface surrounding a convex object. The four dimensions are the two surface coordinates (u, v) and two angular directions of view, up/down (ϑ) and left/right (φ). That is, $p = P(u, v, \vartheta, \varphi)$. The lumigraph is illustrated in Figure 9.1. The lumigraph attempts to capture the image that would be seen at every point and by looking in every direction on the surface of a box that surrounds an object. In practice, the surface of the box is divided up into little patches (e.g., 32×32 on each side), and for each patch, a finite number of images (e.g., 256×256) are made by looking in a range of directions that span a wide angle. One of the advantages of this concept is that it allows parallax and stereoscopic effects to be easily and quickly rendered. In addition, the reduction to four degrees of freedom, appropriate discretization and compression reduce the scene description data size even further.

   Apple's QuickTime VR [3] and Shum and Szeliski's [18] method are also examples of the use of no-geometry plenoptic functions. In this case, they are functions of two variables, and so they have manageable data requirements and very high rendering speeds. Shum and Szeliski's

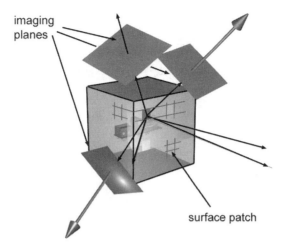

Figure 9.1. The concept of the lumigraph.

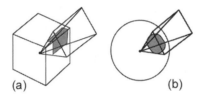

Figure 9.2. The viewing frustum for rendering from an environment map. In (a), a cubic environment is shown, and in (b) a spherical one. Note the distortion on the spherical map and the requirement to use parts of two or three sides of the cube.

method of capturing image panoramas has been used to provide environment maps for a number of the sample programs in Direct3D SDK (which we discuss in Chapter 13). Both these methods render their output by working from a panoramic image stitched together from a collection of photographs. Rendering can be done by traditional spherical or cubic environment maps (see Figure 9.2), which can take advantage of hardware acceleration if it's available, or by a highly optimized image texture lookup function using direction of view input parameters. More interesting perhaps is how the panoramas are produced. We will look at these methods in more detail shortly.

2. *Rendering with implicit geometry.* This class of methods gets its name because it uses inferred geometric properties of the scene which arise

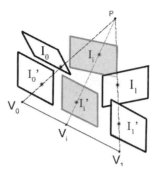

Figure 9.3. Viewmorph. Original images I_0 and I_1 are projected on a plane at I_0' and I_1' respectively. Both these images lie in the same plane, and an interpolation between them will give an image I_i'. If required, a reprojection to the image plane I_i can be made.

from the usual projection calculations of conventional rendering; e.g., using field of view, aspect ratio or frustum. Chen and Williams [5] introduced the idea of image morphing among several photographs taken of a scene. They used range data to obtain the correspondences between the images. Seitz and Dyer [16] show how, given two photographs of a scene and the location and direction of view in each, it is possible to re-create an apparent view from a point between the two locations. Their algorithm consists of three stages, as depicted in Figure 9.3. First the two images are projected onto a common plane, then common points are interpolated to give the desired image and finally the interpolated image is re-projected to give the desired viewpoint.

3. *Rendering with explicit geometry.* Often some range data (depth information along line of sight, possibly obtained with a laser scanner) is available along with a photographic image. In this case, McMillan [12] demonstrated that photographs, apparently taken from locations nearby where no photographic images are available, can be generated synthetically. This is done by projecting the pixels from the images into their 3D locations and then reprojecting them to give a new view. The biggest problems with this are holes due to differences in sampling resolutions in the input and output images and occlusions where an object would be visible in the output image but is hidden in the input images. Figure 9.4 illustrates the idea using only one source image.

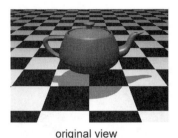

original view

Figure 9.4. Depth rendering. When the distance from the camera to an object's surface is known, a 3D model can be constructed. This can be rendered from a different direction, but note that points on the object's surface which are not visible from one camera location may be visible from another, and since they can't be seen from the location where the range is detected, holes will result in the alternate view.

In VR, the most useful IBR methods are the 2D *no geometry* plenoptic functions of item 1. In the rest of this chapter, we will look at how the panoramic images they use are produced and applied in real-time virtual environment navigation applications.

9.2   Acquiring Images for IBR

One of the trickiest things to do in IBR is acquire the images that are to be used to construct the panoramic view or omnidirectional object views rendered in response to a specified direction of view. Ultimately, one would really like to be able to construct these by using a handheld video camera and making a rough scan of the complete enveloping environment. This would require significant analytical effort in the subsequent image processing stage, termed *mosaicing*. Alternatively, special-purpose imaging hardware has been developed, such as that by Matusik et al. [11]. They use this hardware to acquire images of objects under highly controlled lighting conditions. Equipment of this type is illustrated in Figure 9.5(a).

A camera mounted on a tripod can be used to obtain 360° panoramas (the concept of which is illustrated in Figure 8.14). However, problems arise if one tries to capture images directly overhead. In this case, the production of the complete panorama can be challenging. To increase the vertical field of view, a camera can be mounted sideways. This was suggested by Apple Computer as one way to prepare a panoramic movie for use in QuickTime VR

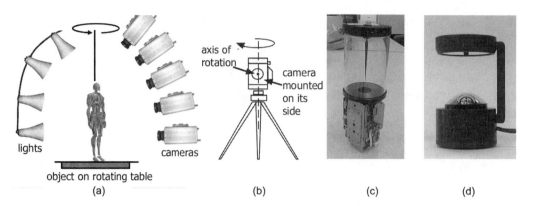

Figure 9.5. (a) Scanning a model with multiple cameras and light sources, after [11]. (b) Panoramic camera mounted on its side for wide vertical field of view. (c) The Omnicam with parabolic mirror. (d) A Catadioptric camera derived from the Omnicam. (Photograph of the Omnicam and Catadioptric camera courtesy of Professor Nayar of Columbia University.)

(see Figure 9.5(b)). QuickTime VR can also be used in *object movie* mode, in which we look inward at an object rather than outward towards the environment. To make a QuickTime VR object movie, the camera is moved in an equatorial orbit round a fixed object whilst taking pictures every 10°. Several more orbits are made at different vertical heights.

Many other ingenious methods have been developed for making panoramic images. In the days before computers, cameras that used long film strips would capture panoramic scenes [14]. Fish-eye lenses also allow wide-angle panoramas to be obtained. Mirrored pyramids and parabolic mirrors can also capture 360° views, but these may need some significant image processing to remove distortion. 360° viewing systems are often used for security surveillance, video conferencing (e.g., Behere [2]) and devices such as the Omnicam (see Figure 9.5(c) and other Catadioptric cameras.) [6] from Columbia or that used by Huang and Trivedi for research projects in view estimation in automobiles [9].

9.3   Mosaicing and Making Panoramic Images

Once acquired, images have to be combined into a large mosaic if they are going to be used as a panoramic source for IBR. If the images are taken from a

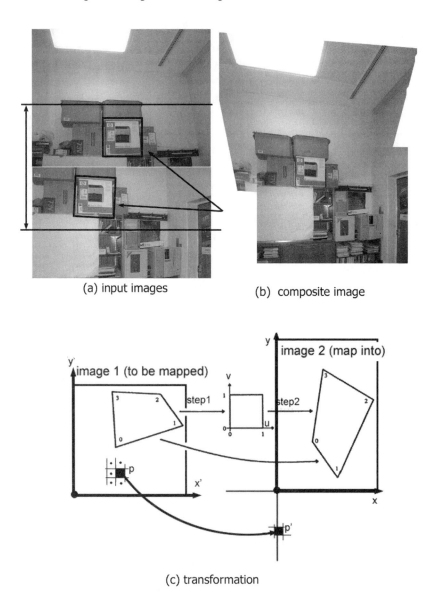

(a) input images   (b) composite image

(c) transformation

Figure 9.6. (a) Input images; (b) combined images; (c) transform in two steps. Image 1 is being blended with image 2. Under the transform defined by the quadrilaterals, point p in image 1 is moved to point p' in the coordinate system used by image 2. Other points map according to the same transformation.

set produced by camera rotation, the term *stitching* may be more appropriate, as sequential images usually overlap in a narrow strip. This is not a trivial task, and ideally one would like to make the process fully automatic if possible.

If we can assume the images are taken from the same location and with the same camera then a panorama can be formed by reprojecting adjacent images and then mixing them together where they overlap. We have already covered the principle of the idea in Section 8.3.2; the procedure is to manually identify a quadrilateral that appears in both images. Figure 9.6(a) illustrates images from two panoramic slices, one above the other, and Figure 9.6(b) shows the stitched result. The quadrilaterals have been manually identified. This technique avoids having to directly determine an entire camera model, and the equation which describes the transformation (Equation (8.4)) can be simplified considerably in the case of quadrilaterals. As shown in Figure 9.6(c), the transformation could be done in two steps: a transform from the input rectangle to a unit square in (u, v) space is followed by a transform from the unit square in (u, v) to the output quadrilateral. This has the advantage of being faster, since Equation (8.4) when applied to step 2 can be solved symbolically for the h_{ij}. Once the h_{ij} have been determined, any pixel p in image 1 can be placed in its correct position relative to image 2, at p'.

Of course, the distortion of the mapping means that we shouldn't just do a naive copy of pixel value from p to p'. It will be necessary to carry out some form of anti-alias filtering, such as a simple bilinear interpolation from the neighboring pixels to p in image 1 or the more elaborate elliptical Gaussian filter of Heckbert [8].

Manually determining the homogeneous mapping between images being mosaiced is fine for a small number of images, but if we want to combine a

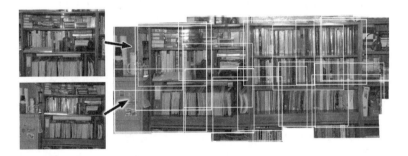

Figure 9.7. The planar mosaic on the right is produced by automatically overlapping and blending a sequence of images. Two of the input images are shown on the left.

large number and they overlap to a significant extent, or we want to build a big library of panoramas, it could get very tedious. Szeliski [19] introduced an algorithm in the context of planar image mosaicing which applies equally well for panoramic images. An example of its application is shown in Figure 9.7.

The algorithm works with two images that should have a considerable overlap, and it attempts to determine the h_{ij} coefficients of Equation (8.4), for the specific case of a 2D transformation, by minimizing the difference in image intensity between the images. The algorithm is iterative. It starts with an assumption for the h_{ij}, uses them to find corresponding pixels (x, y) and (x', y') in the two images and seeks to minimize the sum of the squared intensity error:

$$E = \sum_i \left[I'(x_i', y_i') - I(x_i, y_i) \right]^2$$

over all corresponding pixels i. Once the h_{ij} coefficients have been determined, the images are blended together using a bilinear weighting function that is stronger near the center of the image. The minimization algorithm uses the Levenberg-Marquardt method [15]; a full description can be found in [19]. Once the relationship between the two images in the panorama has been established, the next image is loaded and the comparison procedure begins again.

Szeliski and Shum [20] have refined this idea to produce full-view panoramic image mosaics and environment maps obtained using a handheld digital camera. Their method also resolves the stitching problem in a cylindrical panorama where the last image fails to align properly with the first. Their method will still work so long as there is no strong motion parallax (translational movement of the camera) between the images. It achieves a full view[1] by storing the mosaic not as a single stitched image (for example, as depicted in Figure 9.6), but as a collection of images each with their associated positioning matrices: translation, rotation and scaling. It is then up to the viewing software to generate the appropriate view, possibly using traditional texture mapping onto a polygonal surface surrounding the viewpoint, such as a subdivided dodecahedron or tessellated globe as depicted in Figure 9.8. For each triangular polygon in the sphere, any panoramic image that would overlap it in projection from the center of the sphere is blended together to form a composite texture map for that polygon. Texture coordinates are generated for the vertices by projecting the vertices onto the texture map.

---

[1]The method avoids the usual problem one encounters when rendering from panoramic images, namely that of rendering a view which looks directly up or directly down.

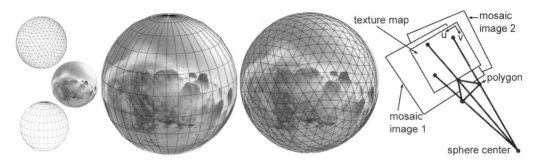

Figure 9.8. Mapping onto polygonal spherical shapes allows a full-view panoramic image to be rendered without any discontinuities.

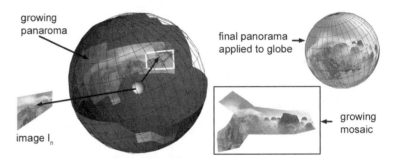

Figure 9.9. Building up a panoramic mosaic by moving the next image I_n into position via a rotation about the optical center. Once in position, I_n is blended with the growing mosaic I_c. A least squares minimization algorithm which matches the overlap between I_n and I_c is used to determine the optimal rotation matrix R_n.

Starting with a sequence of images I_0, I_1, I_2 etc., the refined method seeks to combine the next image, I_n, into the panorama by minimizing the difference between it and the current state of the composite panorama, I_c. A matrix R is used to specify a transformation that will rotate I_n about the view center so that it overlaps part of I_c (see Figure 9.9). The difference between I_n and I_c can be determined as the sum of intensity differences in each pixel where they overlap. A general iterative least squares minimization algorithm [15] will refine R until the alignment between I_n and I_c has minimum difference. Once I_n has been placed in the panorama, it can be blended into I_c by a radial opacity function that decreases to zero at the edge of I_n. The analytic steps in the algorithm are derived in Appendix C.

Of course, getting the iterative algorithm started is no easy matter. When building a panorama, it is likely that image I_n will have a lot of overlap with I_{n-1}, and so starting with the R that placed I_{n-1} in the panorama is probably the best way to generate an initial guess. However, if the overlap between two successive frames is small (less than 50%), an alternative strategy of determining R, or at least getting an initial guess for it, must be followed. The two most successful approaches are:

1. *Hierarchical matching.* Small subsampled and smoothed versions of I_n and I_c are made, e.g., images may be downsampled from 512×512 to 32×32 as a first guess. In these smaller images, the overlap appears greater and, therefore, easier to determine R. Once an R has been determined, a larger subsampled image (e.g., 256×256) is used and the least squares algorithm applied so as to refine the best estimate for R.

2. *Phase correlation* [10]. This is most useful when the overlap is very small. It is generally used to estimate the displacement between two planar images. For example, two aerial photos can be *montaged* using this approach. In it, a 2D Fourier transform of the pair of images is made. The phase difference between them at each frequency is determined. An inverse transform is applied to the difference signal. Back in the spatial domain, the location of a peak in the magnitude shows where the images overlap.

In cases where I_n has no overlap with I_c, I_n will have to be discarded (temporarily) and reconsidered later after other images which may overlap with it have been added into the mosaic. All in all, producing panoramic mosaic images from video footage shot on a basic camcorder is no easy task.

Once panoramic images have been obtained, they can be used in a number of different ways. For example, McMillan and Bishop [13] have demonstrated how two panoramic images acquired from camera centers located about 20 inches apart can generate parallax motion in the viewer.

9.3.1 QuickTime VR

Apple's QuickTime movie file format was introduced in Chapter 5 in the context of a container for storing compressed video and movies. Within a QuickTime movie, there can be multiple tracks, with each one storing some type of linear media indexed with time, video and audio being the most well known. A track can contain compressed data that requires its own player. The QuickTime VR (QTVR) standard adds two new types of track. The first type of track provides an inward-looking (or *object*) view which allows an object to be looked at from any orientation. The second track holds a collection of outward-looking environment *panorama* views. Appropriate interactive software gives the user the impression of control over the view and viewpoint.

In the *object* format, the movie looks conventional, with each frame containing a different view of the photographed object. It is the duty of the player to find and display the correct frame in the movie given the required viewing direction. Extra frames can be added to allow for animation of the object.

In the panoramic format (see Figure 9.10), a number of panoramic images are stored in a conventional movie track. These views give a 360° view from several locations; it could be in different rooms of a virtual house or the same view at different times. The player can pan around these and zoom in and out. To allow the user to select a specific image, possibly by clicking on a hotspot in the image, additional data can be added to the tracks. For example, the panoramic track may contain data called *nodes* which correspond to points in space. The nodes identify which panoramic image is to be used at that point in space and identify how other panoramic images are selected as the user moves from node to node. Hotspot images may also be mixed in to give the appearance of being attached to objects in the panoramic view, say to provide some information about an object, for example exhibits in a virtual museum.

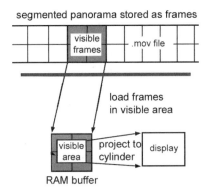

Figure 9.10. QuickTime VR, panoramic display process.

The panoramic images, which tend to be rather large, are usually divided into smaller sub-images that are nearer the size of conventional video frames. When playing the QTVR movie, there is no need for the frames to be stored in any particular order. On playback, only those sub-images that lie in the field of view need to be loaded from the file, so little RAM storage is usually needed to interact with a QTVR movie. In some players, image caching may be used to aid accelerated display. The final stage of rendering is accomplished by a patented cylindrical to planar image warp [4] which renders the view onto the flat screen.

QTVR is not the only panoramic image viewer, but it is the only one tightly integrated into a major operating system. Other panoramic VR viewing image preparation software packages are relatively common and commercially available. Some have been particularly designed for use on the Web. As mentioned before, realtor websites abound with virtual tours of their property portfolios.

9.4   Summary

In this chapter, we looked at the three classical image-based rendering techniques: rendering with no geometry, rendering with implicit geometry and finally rendering with explicit geometry. However, regardless of which technique is used, the main advantage that IBR has to offer over traditional rendering techniques based on polygonal models is its ability to render a scene or object in great detail. The fact that the rendering time is independent of the complexity of the scene is another significant plus. Before 3D-accelerated graphics processors became commonplace, IBR was the only way to achieve high-quality real-time graphics. Parallax effects, virtual tours and varying lighting conditions can all now be simulated with IBR. This has great utility in many different types of applications, one of the most common being archaeological walkthroughs where it is imperative to maintain the detail of the real environment. Here, many people will get to experience a virtual archaeological site, where only a limited number of people would have access to the real site.

The only real negative in IBR is the loss of flexibility in choosing viewing conditions. Where IBR really excels is in producing environmental backgrounds and maps for use as part of a scene described by polygons and rendered in real time with a GPU. So, even if QuickTime VR fades from use, the production of panoramic image maps will still be an essential element of VR work, and as such, IBR will retain its importance to the VR community.

Bibliography

[1] E. Adelson and J. Bergman. "The Plenoptic Function and the Elements of Early Vision". In *Computational Models of Visual Processing*, M. Landy and J. Movshon (editors), pp. 3–20. Cambridge, MA: The MIT Press, 1991.

[2] Be Here Corp. "TotalView". http://www.behere.com/.

[3] S. Chen. "QuickTime VR: An Image-Based Approach to Virtual Environment Navigation". In *Proceedings of SIGGRAPH 95, Computer Graphics Proceedings, Annual Conference Series,* edited by R. Cook, pp. 29–38. Reading, MA: Addison Wesley, 1995.

[4] S. Chen and G. Miller. "Cylindrical to Planar Image Mapping Using Scanline Choerence". United States Patent no. 5,396,583. 1995.

[5] S. Chen and L. Williams. "View Interpolation for Image Synthesis". In *Proceedings of SIGGRAPH 93, Computer Graphics Proceedings, Annual Conference Series,* edited by J. Kajiya, pp. 279–288. New York: ACM Press, 1993.

[6] Columbia University, Computer Vision Laboratory. "Applications of 360 Degree Cameras". http://www1.cs.columbia.edu/CAVE//projects/app_cam_360/app_cam_360.php.

[7] S. Gortler et al. "The Lumigraph". In *Proceedings of SIGGRAPH 96, Computer Graphics Proceedings, Annual Conference Series,* edited by H. Rushmeier, pp. 43–54. Reading, MA: Addison Wesley, 1996.

[8] P. Heckbert. "Filtering by Repeated Integration". *Computer Graphics* 20:4 (1986) 317–321.

[9] K. Huang and M. Trivedi. "Driver Head Pose and View Estimation with Single Omnidirectional Video Stream". In *Proceedings of the 1ˢᵗ International Workshop on In-Vehicle Cognitive Computer Vision Systems,* pp. 44–51. Washington, D.C.: IEEE Computer Society, 2003.

[10] C. Kuglin and D. Hine. "The Phase Correlation Image Alignment Method". In *Proceedings of the IEEE Conference on Cybernetics and Society*, pp. 163–165. Washington, D.C.: IEEE Computer Society, 1975.

[11] W. Matusik et al. "Image-Based 3D Photography using Opacity Hulls". *Proc. SIGGRAPH '02, Transactions on Graphics* 21:3 (2002) 427–437.

[12] L. McMillan. "An Image-Based Approach to Three-Dimensional Computer Graphics". Ph.D. Thesis, University of North Carolina at Chapel Hill, TR97-013, 1997.

[13] L. McMillan and G. Bishop. "Plenoptic Modelling: An Image-Based Rendering System". In *Proceedings of SIGGRAPH 95, Computer Graphics Proceedings, Annual Conference Series,* edited by R. Cook, pp. 39–46. Reading, MA: Addison Wesley, 1995.

[14] J. Meehan. *Panoramic Photography.* New York: Amphoto, 1990.

[15] W. Press et al. (editors). *Numerical Recipes in C++: The Art of Scientific Computing,* Second Edition. Cambridge, UK: Cambridge University Press, 2002.

[16] S. Seitz and C. Dyer. "View Morphing". In *Proceedings of SIGGRAPH 96, Computer Graphics Proceedings, Annual Conference Series,* edited by H. Rushmeier, pp. 21–30. Reading, MA: Addison Wesley, 1996.

[17] H. Shum and S Kang. "A Review of Image-Based Rendering Techniques". In *Proceedings of IEEE/SPIE Visual Communications and Image Processing (VCIP) 2000,* pp. 2–13. Washington, D.C.: IEEE Computer Society, 2000.

[18] H. Shum and R. Szeliski. "Panoramic Image Mosaics". Technical Report MSR-TR-97-23, Microsoft Research, 1997.

[19] R. Szeliski. "Image Mosaicing for Tele-Reality Applications". Technical Report CRL 94/2, Digital Equipment Corp., 1994.

[20] R. Szeliski et al. "Creating Full View Panoramic Image Mosaics and Texture-Mapped Models". In *Proceedings of SIGGRAPH 97, Computer Graphics Proceedings, Annual Conference Series,* edited by T. Whitted, pp. 251–258. Reading, MA: Addison Wesley, 1997.

10   Stereopsis

One of the main things that sets a VR environment apart from watching TV is stereopsis, or depth perception. Some 50 years ago, the movie business dabbled with stereopsis, but it has been only in the last few years that it is making a comeback with the advent of electronic projection systems. Stereopsis adds that *wow* factor to the viewing experience and is a must for any VR system.

Artists, designers, scientists and engineers all use depth perception in their craft. They also use computers: computer-aided design (CAD) software has given them color displays, real-time 3D graphics and a myriad of human interface devices, such as the mouse, joysticks etc. So the next logical thing for CAD programs to offer their users is the ability to draw stereoscopically as well as in 3D. In order to do this, one has to find some way of feeding separate video or images to the viewer's left and right eyes.

Drawing an analogy with stereophonics, where separate speakers or headphones play different sounds into your left and right ears, stereopsis must show different pictures to your left and right eyes. However, stereopsis cannot be achieved by putting two screens in front of the viewer and expecting her to focus one eye on one and the other eye on the other. It just doesn't work. You can of course mount a device on the viewer's head (a head-mounted display or HMD) that shows little pictures to each eye.

Technological miniaturization allows HMDs to be almost as light and small as a rather chunky pair of sunglasses. Figure 10.1 illustrates two commercially available models that can be used for stereoscopic viewing or as part of *very* personal video/DVD players. However, if you want to look at a com-

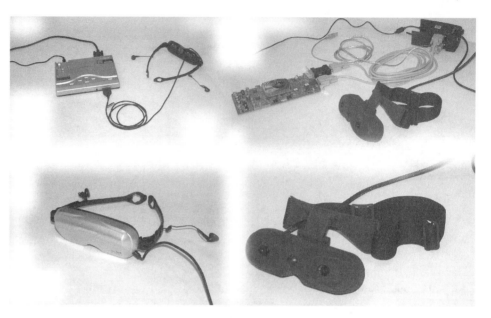

Figure 10.1. Two examples of typical head-mounted display devices for video, VR and stereoscopic work. Small video cameras can be build into the eyewear (as seen on right) so that when used with motion and orientation sensors, the whole package delivers a sense of being immersed in an interactive virtual environment.

puter monitor or watch a big-screen movie, there is no choice but to find a different approach. To date, the most successful approach is to combine images for left and right eyes into a single display that can be separated again by a special pair of eyeglasses worn by the viewer.

Obtaining stereoscopic images or movies requires a different technology again. Traditionally, stereoscopic images were obtained using a custom-built camera or adapters that fit onto the front of conventional single-lens reflex (SLR) cameras. The headset shown on the right of Figure 10.1 has two miniature video cameras built into it which generate video streams from the locations of the left and right eyes. So now we have the option of using two video cameras and a computer to acquire both still and moving stereoscopic images and movies.

In this chapter, we will give a short explanation of the theory and terminology of stereopsis before delving a little deeper into stereoscopic technology. Some example programs for stereoscopic work in a VR context will be presented in Chapter 16.

10.1   Parallax

There are many things we see around us that give clues as to how far away something is or whether one thing is nearer or further away than something else. Not all of these clues require us to have two eyes. Light and shade, inter-position, texture gradients and perspective are all examples of *monocular depth cues*, all of which we detailed in Section 2.1.2. Another monocular depth cue is called *motion parallax*. Motion parallax is the effect we've all seen: if you close one eye and move your head side to side, objects closer to you appear to move faster and further than objects that are behind them. Interestingly, all the monocular cues with the exception of motion parallax can be used for depth perception in both stereoscopic and non-stereoscopic display environments, and they do a very good job. Just look at any photograph; your judgment of how far something was from the camera is likely to be quite reasonable.

Nevertheless, a person's depth perception is considerably enhanced by having two eyes. The computer vision techniques which we discuss in Chapter 8 show just how valuable having two independent views can be in extracting accurate depth information. If it were possible to take snapshots of what one sees with the left eye and the right eye and overlay them, they would be different. Parallax quantifies this difference by specifying numerically the displacement between equivalent points in the images seen from the two viewpoints, such as the spires on the church in Figure 10.2.

Parallax can be quoted as a relative distance, also referred to as the *parallax separation*. This is measured in the plane of projection at the display screen or monitor, and may be quoted in units of pixels, centimeters etc. The value of parallax separation is dependent on two factors. The first factor is the distance between the viewed object and the plane of zero parallax, which is usually the plane of projection. This effect is shown in Figure 10.3. The second factor concerns the distance between the viewpoint and the display. The effect of this is shown in Figure 10.4, where the parallax separation at a monitor screen at a distance s_1 from the viewpoint will be t_1, whilst the parallax separation at a large projection screen a distance s_2 from the viewpoint will be t_2. We can avoid having to deal with this second influence on the value of parallax by using an angular measure, namely the parallax angle which is defined as β. This is also shown in Figure 10.4. For best effect, a parallax (β) of about $1.5°$ is acceptable. Expressing parallax as an angle makes it possible to create stereoscopic images with parallax separations that are optimized for display on a computer monitor or in a theater.

left-eye image

right-eye image

note double vision effect

overlapped left and right images

Figure 10.2. The images we see in our left and right eyes are slightly different, and when overlapped they look like we have just had a bad night on the town.

The spatial relationship between viewed object, projection screen and viewer gives rise to four types of parallax. Figure 10.3 illustrates the four types as the viewed object moves relative to the projection screen and viewer. The parallax types are:

- *Zero.* The object is located at the same distance as the projection screen and so it appears to hover at the screen.

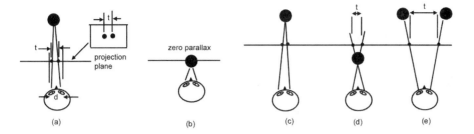

Figure 10.3. Parallax effects: (a) Parallax t is measured in pixel units or inches or centimeters; d is the eye separation, on average 2.5 in. (64 mm) in humans. (b) Zero parallax. (c) Positive parallax. (d) Negative parallax. (e) Divergent parallax (does not occur in nature).

- *Positive.* The object appears to hover behind the projection plane; this effect is rather like looking through a window.

- *Negative.* Objects appear to float in front of the screen. This is the most exciting type of parallax.

- *Divergent.* Normal humans are incapable of divergent vision and therefore this form of parallax warrants no further discussion.

Normally the value of parallax separation (t) should be less than the separation between the two viewpoints (d). This will result in the viewer expe-

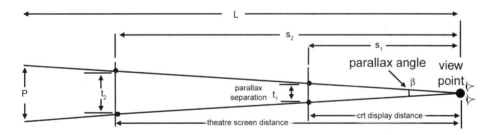

Figure 10.4. Expressing parallax as an angle. For example, if a monitor screen is at a distance of s_1 from the viewpoint, the parallax is t_1. We have the same β for a cinema screen at a distance of s_2 from the viewpoint but the parallax distance is now t_2. In the extreme case where the projection plane is moved back into the scene a distance L, the parallax distance there would be P.

riencing positive, negative or zero parallax. As the value of t approaches d, the viewer can start to feel uncomfortable. As it exceeds d, divergent parallax is being simulated, and the natural viewing perception is lost. For example, sitting in front of a workstation where the average viewing distance is about 18 in., a good and comfortable stereoscopic effect can be obtained with a parallax separation of about $\frac{1}{2}$ in., equivalent to an angular parallax of $3°$.

When rendering images with the intention of providing a parallax separation of $\frac{1}{2}$ in. on a typical monitor, it is necessary to determine the separation between the cameras viewing the virtual scene. This is explained in Section 10.1.1. When setting up the camera geometry, it is also advisable to consider a few simple rules of thumb that can be used to enhance the perceived stereoscopic effect:

- Try to keep the whole object within the viewing window, since partially clipped objects look out of place and spoil the stereoscopic illusion.

- Render the objects against a dark background, especially where you want negative parallax.

- Monocular cues enhance the stereoscopic effects, so try to maximize these by using/simulating wide-angle lenses, which in essence open up the scene to enable these comparisons to be made.

- The goal when creating a stereoscopic effect is to achieve the sensation of depth perception with the lowest value of parallax. Often, the most pleasing result is obtained by placing the center of the object at the plane of the screen.

In practice, to get the best stereo results when viewing the rendered images, you should use as big a monitor as possible or stand as far away as possible by using a projection system coupled with a large-screen display. This has the added benefit of allowing larger parallax values to be used when rendering the images.

10.1.1   A Stereo Camera Model

Since parallax effects play such a crucial role in determining the experience one gets from stereopsis, it seems intuitive that it should be adjustable in some way at the geometric stages of the rendering algorithm (in CAD, animation or VR software). That is, it has an effect in determining how to position the two cameras used to render the scene. Since the required parallax concerns viewpoints and projections, it will make its mark in the geometry stages of rendering, as discussed in Section 6.6.6.

The most obvious geometrical model to construct would simulate what happens in the real world by setting up two cameras to represent the left and right eye viewpoints, both of which are focused on the same target. This target establishes the plane of zero parallax in the scene. Rendering is carried out for the left and right views as they project onto planes in front of the left and right cameras. If the target point is in front of the objects in the scene, they will appear to be behind the screen, i.e., with positive parallax. If the target is behind the objects, they will appear to stand out in front of the screen, i.e., negative parallax. Figure 10.5 illustrates the geometry and resulting parallax. In this approach, we obtain the left view by adding an additional translation by $-d/2$ along the x-axis as the last step (before multiplying all the matrices together) in the algorithm of Section 6.6.6. For the right eye's view, the translation is by $d/2$.

However, when using this form of set-up, the projection planes are not parallel to the plane of zero parallax. When displaying the images formed at these projection planes on the monitor (which is where we perceive the plane of zero parallax to be), a parallax error will occur near the edges of the display. Rather than trying to correct for this distortion by trying to transform the

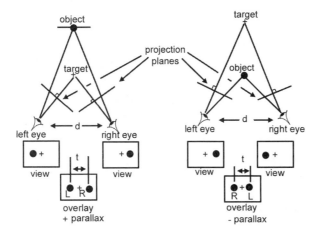

Figure 10.5. The two viewpoints are separated by a distance d, and the views are directed towards the same target. The origin of the coordinate system is midway between the viewpoint with the x-axis towards the right, and the z-axis vertical—out of the plane of the diagram.

projection planes onto the plane of zero parallax, we use an alterative method of projection based on using parallel aligned cameras.

Using cameras with parallel axis alignment is known as *off-axis* rendering, and it gives a better stereoscopic view than the one depicted in Figure 10.5. The geometry depicted in Figure 10.6 sets the scene for this discussion. Two cameras straddling the origin of the coordinate system are aligned so that they both are parallel with the usual direction of view. Unfortunately, the projection transformation is different from the transformation for a single viewpoint, because the direction of view vector no longer intersects the projection plane at its center (see Figure 10.7). Implementing a projective transformation under these conditions will require the use of a nonsymmetric camera frustum rather than a simple field of view and aspect ratio specification. The geometry of this not considered here, but real-time 3D-rendering software libraries such as OpenGL offer functions to set up the appropriate projective transformation.

So far so good, but we have not said what the magnitude of d should be. This really depends on the scale we are using to model our virtual environment. Going back to the point we originally made, we should always have a parallax separation of about $\frac{1}{2}$ in. on the computer monitor and the center of the model at the point of zero parallax. In addition, the farthest point should have a negative parallax of about $-1.5°$ and the nearest a parallax of about $+1.5°$.

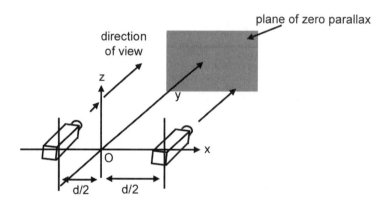

Figure 10.6. Two cameras with their directions of view aligned in parallel, along the *y*-axis. A point halfway between them is placed at the origin of the coordinate system. We need to determine a suitable separation d to satisfy a given parallax condition which is comfortable for the viewer and gives a good stereoscopic effect at all scales.

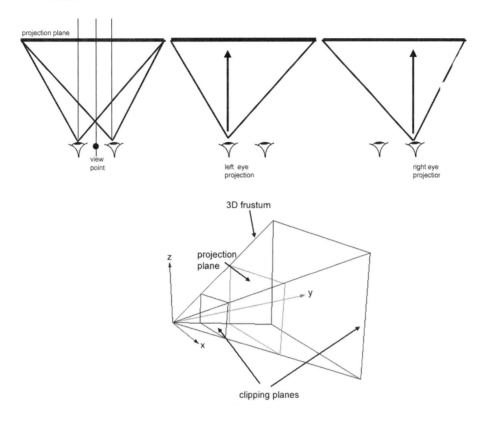

Figure 10.7. Using off-axis (parallel) alignment results in a distorted projection which is different for each of the viewpoints. To achieve this effect, the camera frustum must be made to match the distorted viewing projection. The OpenGL and Direct3D software libraries provide ways to do this.

Consider the geometry in Figure 10.6 and the problem of determining d in the dimensions of the scene given some conditions on the parallax we wish to achieve. Two useful ways to specify parallax are:

1. Require the minimum value of parallax $\beta_{min} = 0$ and set a value on β_{max}, say at about 1.5°. This will give the appearance of looking through a window because everything seems to lie beyond the screen.

2. Specify the point in the scene where we want the zero parallax plane to lie. We shall define the location of this plane by its distance from the viewpoint, D_{zpx}. Therefore, everything that is closer to the cameras than D_{zpx} will appear in front of the screen.

To do this, we will have to ensure that the scene falls within a volume of space bounded by known values: $[x_{min}, x_{max}]$ and $[y_{min}, y_{max}]$. The vertical extent of the scene (between $[z_{min}, z_{max}]$) does not enter the calculation. We will also need to define the required field of view ϑ. This is typically $\vartheta = 40°$ to $50°$. It is useful to define $\Delta x = x_{max} - x_{min}$ and $\Delta y = y_{max} - y_{min}$.

From the angular definition of parallax (see Figure 10.4), we see that for projection planes located at distances s_1 and s_2 from the viewpoint,

$$\beta = 2 \arctan \frac{t_1}{2s_1} = 2 \arctan \frac{t_2}{2s_2} = 2 \arctan \frac{P}{2L}.$$

The measured parallax separation will be t_1 and t_2 respectively. The same perceived angular parallax β gives rise to a measured parallax separation of P at some arbitrary distance L from the viewpoint. With the viewpoint located at the world coordinate origin, if we were to move the projection plane out into the scene then, at a distance L, the measured parallax separation P is given by

$$P = 2L \tan \frac{\beta}{2}. \tag{10.1}$$

In the conventional perspective projection of the scene, using a field of view ϑ and the property that the scene's horizontal expanse Δx just fills the viewport, the average distance L' of objects in the scene from the camera is

$$L' = \frac{\Delta x}{2 \tan \frac{\vartheta}{2}}.$$

For a twin-camera set-up, the parallax separation at this distance is obtained by substituting L' for L in Equation (10.1) to give

$$P = 2L' \tan \frac{\beta}{2} = \frac{\tan \frac{\beta}{2}}{\tan \frac{\vartheta}{2}} \Delta x.$$

Requiring that the scene must have a maximum angular parallax β_{max} and a minimum angular parallax β_{min}, it is equivalent to say that

$$P_{max} = \frac{\tan \frac{\beta_{max}}{2}}{\tan \frac{\vartheta}{2}} \Delta x;$$

$$P_{min} = \frac{\tan \frac{\beta_{min}}{2}}{\tan \frac{\vartheta}{2}} \Delta x.$$

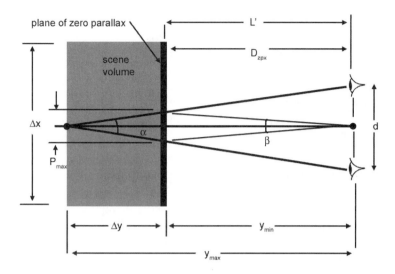

Figure 10.8. Parallax geometry. The plane of zero parallax is at the front of the scene. To determine d, write $\tan \frac{\alpha}{2} = \frac{P_{\max}/2}{\Delta y} = \frac{d/2}{L'+\Delta y}$ then $d = P_{\max}(1 + \frac{L'}{\Delta y})$.

(Note: $P_{\max}$ and $P_{\min}$ are measured in the units of the world coordinate system.)

In the case where we want the minimum angular parallax $\beta_{\min} = 0°$, the distance of zero parallax is $D_{zpx} = y_{\min} = L'$. The equality with L is true because it is usually assumed that the plane of zero parallax lies at the average distance between viewpoint and objects in the scene (see Figure 10.8).

To satisfy this constraint, the cameras must be separated by a distance d (in world coordinate units) given by

$$d = P_{\max}\left(1 + \frac{L'}{\Delta y}\right). \tag{10.2}$$

Alternatively, to choose a specific plane of zero parallax defined by its distance from the viewpoint D_{zpx}, the camera separation d will be given by

$$\begin{aligned}
d_1 &= P_{\min}\left(\frac{L'}{D_{zpx} - y_{\min}} - 1\right); \\
d_2 &= P_{\max}\left(\frac{L'}{y_{\max} - D_{zpx}} + 1\right); \\
d &= \min(|d_1|, |d_2|).
\end{aligned}$$

By using either of these parallax models, it is possible to design a software package's user interface to permit the user to easily find a setting of parallax separation with which they are most comfortable.

> For example, suppose we wish to make a 3D stereoscopic animation of a large protein molecule undergoing enzyme digestion. The scene consists of a polygonal model of the molecule bounded by the cubic volume 1 nm $\times$ 1 nm $\times$ 1 nm. Thus $\Delta x = 1$ nm and $\Delta y = 1$ nm. Using a field of view of $60°$, we must place the viewpoint $L' = \frac{\sqrt{3}}{2}$ nm away from the scene so that it fills the viewport when rendered. To achieve a minimum angular parallax of $0°$ and a maximum angular parallax of $1.5°$, the parallax separation $P_{max} = 0.0225$ nm. By using Equation (10.2), we determine that the viewpoints need to be $d = 0.042$ nm apart.

10.2   Head-Mounted Displays

Head-mounted displays (like those illustrated in Figure 10.1) have been around for quite a long time in specialist applications, such as military training, theme park attractions etc. From the VR perspective, HMDs are an essential component in *augmented reality* applications and have a number of advantages. Principal amongst these is the independence they give to the user. However, there are complications. Unless wireless technology is available, the trailing wires can cause a problem. Another complication comes from the fact that the HMD wearer expects to see a view that matches the way he is moving his head. This involves combining motion- and orientation-sensing equipment, either built into the HMD or attached to the head in some other way. Get this wrong and the HMD wearer can suffer mild to unpleasant motion sickness. Probably the most significant disadvantage of HMDs is that *everything* the wearer sees comes only from the HMD, unlike a VR environment that is presented on a screen. Person-to-person or person-to-equipment/machinery interaction is not possible unless HMDs offer a see-through capability. There are two ways in which this can be achieved, both of which are illustrated in Figure 10.9.

1. An angled glass plate is placed in front of the eyes. The output from two miniature video screens is reflected off the plate and thus it appears

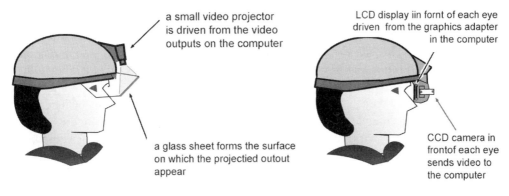

a small video projector
is driven from the video
outputs on the computer

LCD display iin fornt of each eye
driven from the graphics adapter
in the computer

a glass sheet forms the surface
on which the projectied outout
appear

CCD camera in
frontof each eye
sends video to
the computer

Optical see through HMD

Electronic see through HMD

Figure 10.9. Optical and electronic see-through head mounted displays. The device depicted on the right of Figure 10.1 is an example of an electronic stereoscopic HMD.

in the line of sight. Aircraft head-up displays (HUDs) have used this technique for many years and the principle is also used in television autocues and teleprompters. A problem that arises with this type of HMD is that the image may appear *ghostly* or not solid enough.

2. Small CCD video cameras can be placed on the other side of the eye-wear, and video from these can be digitized and mixed back into the video signals sent to each eye. This type of HMD might be subjected to an unacceptable delay between the video being acquired from the cameras and displayed on the video monitors in front of each eye.

The cost of HMDs is falling, and for certain applications they are a very attractive enabling technology. Individual processors could be assigned to handle the stereoscopic rendering of a 3D VR world for each HMD. Position and orientation information from a wide variety of motion-sensing technology can also be fed to the individual processors. The 3D world description and overall system control is easily handled via a master process and network connections. The stereoscopic display is actually the easiest part of the system to deliver. Stereo-ready graphics adapters (Section 16.1.1) are not required for HMD work; any adapter supporting dual-head (twin-video) outputs will do the job. Movies, images or 3D views are simply rendered for left and right eye views and presented on the separate outputs. Under the Windows oper-

ating system, it is a trivial matter to render into a double-width desktop that
spans the two outputs. The left half of the desktop goes to one output and
the right half to the other. Figure 10.10 offers a suggested system diagram
for a multi-user HMD-based VR environment; it shows that each HMD has
a driver computer equipped with a dual-output video adapter. The video
adapter renders the 3D scene specifically from the viewpoint of its user in
stereo. Individual signals are sent to the left and right eye. The driver computers are controlled by a master system that uses position-sensing equipment
to detect the location and orientation of the user's head. The master controller
passes this information to each user processor. It also maintains the 3D scene
description which is available over the shared network.

A project to display stereo images to an HMD is given in Section 18.2,
and a useful, comprehensive and up-to-date list of HMDs is maintained by
Bungert [1].

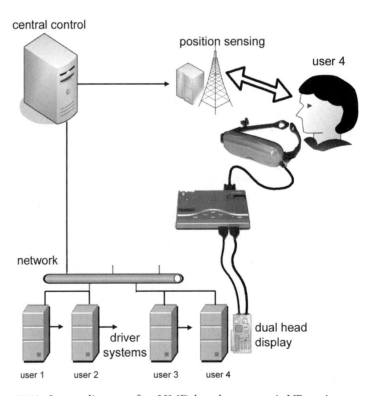

Figure 10.10. System diagram of an HMD-based stereoscopic VR environment.

10.3 Active, Passive and Other Stereoscopic Systems

The alternative to HMDs as a way of presenting different views to the left and right eye is to present them on a monitor or projection screen in the usual way but modify or tag them in some way so that a simple device (which the person looking at the screen uses) can unmodify them just before they enter the eyes. Essentially, this means wearing a pair of glasses, but unlike the HMDs, these glasses can be as simple as a few bits of cardboard and transparent plastic[1].

10.3.1 Anaglyph Stereo Red/Green or Red/Blue

This is probably the simplest and cheapest method of achieving the stereoscopic effect. It was the first method to be used and achieved some popularity in movie theaters during the first half of the twentieth century. A projector or display device superimposes two images, one taken from a left eye view and one from the right eye view. Each image is prefiltered to remove a different color component. The colors filtered depend on the eyeglasses worn by the viewer. These glasses tend to have a piece of red colored glass or plastic in front of the left eye and a similar green or blue colored filter for the right side. Thus, the image taken from the left eye view would have the green or blue component filtered out of it so that it cannot be seen by the right eye, whilst the image taken from the right eye viewpoint would have the red component filtered from it so that it cannot be viewed by the left eye. The advantages of this system are its very low cost and lack of any special display hardware. It will work equally well on CRT and LCD monitors and on all video projectors, CRT, LCD or DLP. The disadvantage is that there is no perception of true color in the pictures or movie. Figure 10.11 illustrates a typical system.

10.3.2 Active Stereo

This system can be used with a single monitor or projector. The viewer wears a special pair of glasses that consist of two remotely controlled LCD shutters working synchronously with the projector or screen. The glasses may be connected to the graphics adapter in the computer or they may pick up

---

[1]This very simple idea can be dressed up in some impressive-sounding technical language: the input signal (the pictures on the screen) is modulated onto a carrier using either time-division multiplexing (active stereo), frequency modulation (anaglyph stereo) or phase shift modulation (passive stereo).

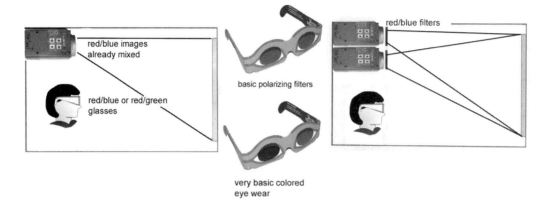

Figure 10.11. Anaglyph stereo, simple paper and plastic glasses with colored filters can give surprisingly good results. This cheap technology is ideal for use with large audiences.

an infrared (IR) signal from an array of emitters placed conveniently in the field of view. The projector or monitor works at twice the normal refresh rate, typically greater than 120 Hz, and displays alternate images for left and right eyes. The LCD shutters are configured so that when the left image is being displayed, the left shutter is open and the right one closed, vice versa when the right image is presented. Figure 10.12 illustrates the principles of active stereo and some examples of active eyewear and emitters. Active stereo systems have the advantage that they can be used interchangeably between workstation monitors, for close up work, and projection-screen systems, for theater and VR suite presentation. The disadvantages are that the eyewear requires batteries, and projector hardware capable of the high refresh rates required tends to be prohibitively expensive, but see Chapter 4 where low-cost options are discussed. LCD projectors cannot operate at a high enough refresh rate to achieve acceptable results.

10.3.3 Passive Stereo

A single CRT or DLP projector operates at double refresh rate and alternates between showing images for left and right eye. A polarizing screen is placed in front of the projector. The screen is synchronized with the graphics adapter to alternate between two states to match the polarizing filters in a pair of glasses worn by the viewer. For example, the glasses may have a horizontal polarizing

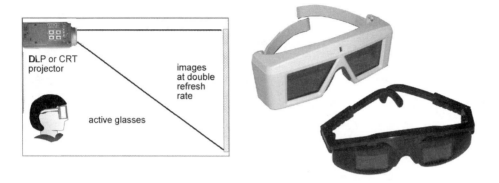

Figure 10.12. Active stereo single CRT–DLP projectors. Top-quality and low-cost eyewear systems.

glass in front of the right eye and a vertical polarizing filter in the glass in front of the left eye. Then, when the polarizing screen is vertically polarized, the light it lets through is also vertically polarized, and will pass through the eyewear to the left eye but be completely blocked from being seen by the right eye. At a high enough refresh rate, greater than 100 Hz, the viewer is unaware of any of this switching.

Unlike active stereo, this system will work equally well with two projectors. One projector has a polarizing screen matching the left eye polarizing filter and the other one's screen matches the right eye. Since there is no need to switch rapidly between left and right images in this case, LCD projectors can be used. CRT projectors can also be used; despite being slightly less bright, they allow for much higher resolutions (greater than 1024×768).

One last comment on passive stereopsis: horizontal and vertical polarization cause a problem as the viewer tilts his head. This effect, known as *crosstalk*, will increase until the tilt reaches $90°$ at which point the views are completely flipped. To avoid crosstalk, circular polarization is used, clockwise for one eye, anticlockwise for the other. For more information on the theory of polarization relating to light and other electromagnetic radiation, browse through Wangsness [4] or any book discussing the basics of electromagnetic theory. Circular polarizing filters must use a tighter match between glasses and polarizing screen and therefore may need to be obtained from the same supplier. Figure 10.13 illustrates typical deployments for passive stereo systems.

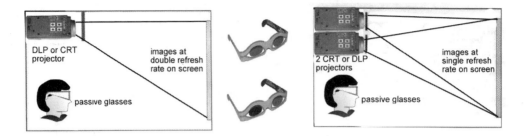

Figure 10.13. Passive stereo can be delivered with one or two projectors and a cheap and simple pair of polarizing glasses.

10.3.4   Autostereopsis

Autostereopsis is the ultimate challenge for stereoscopic technology. Put bluntly, you throw away the eyewear and head-mounted displays, and the display alone gives you the illusion of depth. One approach to autostereopsis requires that the observer's head be tracked and the display be adjusted to show a different view which reflects the observer's head movement. Two currently successful concepts are illustrated in Figures 10.14 and 10.15, both of which have been successfully demonstrated in practical field trials.

In the first approach, the goal is to make the left and right images appear visible to only the left and right eyes respectively by obscuring the other image when seen from a *sweet spot* in front of the screen. There are two ways in which this can be achieved, but both ideas seek to divide the display into two sets of interleaved vertical strips. In set 1 an image for the left eye is displayed,

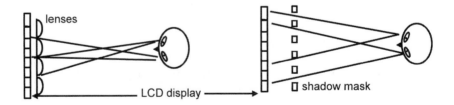

Figure 10.14. Autostereoscopic displays may be built by adding either a special layer of miniature lenses or a grating to the front of a traditional LCD monitor that causes the left and right images to be visible at different locations in front of the monitor. At a normal viewing distance, this corresponds to the separation of the eyes.

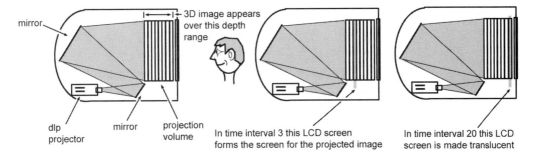

Figure 10.15. LightSpace Technologies' DepthCube [3] creates a volume into which a series of layered 2D images are projected. This builds up a 3D volumetric display that gives the viewer a sense of depth. The display is projected onto one LCD layer in the display volume. When not acting as a translucent screen, the LCD layers are completely transparent.

whilst in set 2 an image for the right eye is displayed. The first approach uses a lenticular glass layer that focuses the vertical strips at two points 30 cm or so in front of the display and approximately at the positions of the left and right eye respectively. The second method uses the grating to obscure those strips that are not intended to be visible from the left eye and do the same for the right eye.

A more user-friendly system developed by LightSpace Technologies [3] builds a layered image by back-projecting a sequence of images into a stack of LCD shutters that can be switched from a completely transparent state into a state that mimics a back projection screen on which the image will appear during a 1 ms time interval. The system uses 20 LCD layers and so can display the whole sequence of depth images at a rate of 50 Hz. A patented antialias technique is built into the hardware so as to blend the layer image, so that they don't look like *layered images*. A DLP projector is used for the projection source.

10.3.5   Crosstalk and Ghosting

In the context of stereopsis, crosstalk is when some of the image that should only be visible by the left eye is seen by the right eye, and vice versa. In active stereo systems, it is primarily a function of the quality of the eyewear. There are three possible sources of crosstalk.

1. Imperfect occlusion, where the polarizing filters fail to block out all the light. Eyewear at the top end of the market is capable of deliver-

ing performance with no perceptible unwanted light leakage. Cheaper systems still do a good job, but they may not go completely opaque. Cheaper eyewear may not give 100% transmissivity when fully open, and this can make the projected output look less bright.

2. Slow switching, where the opaque/clear transition may overshoot, allowing some leakage from one eye's image to the other. This should not be a big problem, since LCD technology can now respond at rates up to 75 Hz, which is quite acceptable for each eye.

3. CRT phosphor afterglow, which applies to screens and CRT projectors, as well as most TVs. The light from CRTs is emitted by a phosphor coating on the front of the CRT after it has been excited by the passing electron beam. The light emission does not stop as soon as the electron beam passes that spot on the screen; it decays over a short time interval. Thus, if the eyewear switches faster than the phosphor light decays, a little crosstalk will be evident.

In passive systems, the quality of the polarizing filters in the glasses and the projector's polarizing screens both play a factor. It is very hard to manufacture perfect polarizing filters, but unlike the active system where switched polarization must be implemented, there will be no crosstalk during transients.

Ghosting is the term used to describe perceived crosstalk. It is a subjective term. It varies with the brightness and color of the image and particularly with the parallax. The larger the value of the parallax the more unpleasant crosstalk will be.

10.3.6 Stereopsis for Multiple Users

In VR applications, there is a good chance that in the larger-scale systems, we will want to be able to accommodate a number of individuals working with, interacting with and viewing the displays. However, everything we have discussed so far makes an important assumption: the display is generated from a single point of view. And the stereoscopic element will have a single parallax too, so everyone has to accept the same depth perception. For CAD work at individual workstations and in a movie theater, this is acceptable. In a VR suite (e.g., a *cave* as described in Chapter 4, where two, three or more participants will be enclosed in an immersive environment), unless HMDs are in use, a screen will be needed. For monoscopic projections onto enclosing screens, there are no problems, but the challenge is to accommodate stereopsis under such conditions. Each individual will have a slightly different viewpoint, point of focus etc., and these effects amplify the differences. Revolving and providing solutions to VR-specific issues such as this are still at the research stage, and no definitive or obvious answer has emerged. Some experiments have been done by using shutter eye systems that display two images for every participant in the room. Unfortunately, this will involve operating the display and polarizing shutters at much higher frequencies.

At present, systems that can accommodate three people have been demonstrated, and in one case even a revolving *mechanical* disk acting as a shutter that sits in front of a projector has been tried. It would seem, however, that the most promising approach involves trying to improve the LCD shutter technology and projection systems. Fortunately, to a first approximation, the effect on individuals of rendering a generic stereoscopic viewpoint from the center of the cave does not introduce a distortion that is too noticeable. The overwhelming sensation of depth perception outweighs any errors due to the actual point of focus not being the one used to generate the stereoscopic view. We will not take this discussion any further here, but we suggest that you keep an eye on the research literature that appears at conferences such as the SPIE [2] annual meeting on electronic imaging.

10.3.7   Front or Back Projection

All the illustrations used so far have depicted front projection. There is no particular reason for this. Back projection works equally well for stereopsis. Sometimes the term used to describe back-projection stereopsis is a *stereo wall*. If there are any drawbacks, they are purely in terms of space, but even this can be overcome by using a mirror with the projectors mounted in a number of different possible configurations. Back projection has the major advantage that anyone working in the stereo environment can go right up to the screen without blocking out the light from the projector.

10.4   Summary

This chapter has introduced the concept, notation and hardware of stereopsis. We have looked at both low-cost (e.g., stereo glasses) and high-cost (e.g., head-mounted displays) hardware and how they work to achieve the stereo effect. So regardless of your VR budget, you should be able to develop a stereoscopic working environment which considerably enhances the VR experience.

What we have not considered in the chapter is the topic of display adaptors and how they are intrinsically linked to the creation of stereo output. However, we shall return to this topic and also examine some practical application programs and how they are constructed in Chapter 16. There we will also look at how to use two simple webcams to record stereoscopic images and play stereoscopic movies.

Bibliography

[1] C. Bungert. "HMD/Headset/VR-Helmet Comparison Chart". http://www.stereo3d.com/hmd.htm.

[2] IS&T and SPIE Electronic Imaging Conference Series. http://electronicimaging.org/.

[3] LightSpace Technologies. "DepthCube". http://www.lightspacetech.com/.

[4] R. Wangsness. *Electromagnetic Fields,* Second Edition. New York: John Wiley & Sons, 1986.

11 Navigation and Movement in VR

So far, we have looked at how to specify 3D scenes and the shape and appearance of virtual objects. We've investigated the principles of how to render realistic images as if we were looking at the scene from any direction and using any type of camera. Rendering static scenes, even if they result in the most realistic images one could imagine, is not enough for VR work. Nor is rendering active scenes with a lot of movement if they result from predefined actions or the highly scripted behavior of camera, objects or environment. This is the sort of behavior one might see in computer-generated movies. Of course, in VR, we still expect to be able to move the viewpoint and objects around in three dimensions, and so the same theory and implementation detail of such things as interpolation, hierarchical movement, physical animation and path following still apply.

What makes VR different from computer animation is that we have to be able to put ourselves (and others) into the virtual scene. We have to appear to interact with the virtual elements, move them around and get them to do things at our command. To achieve this, our software must be flexible enough to do everything one would expect of a traditional computer animation package, but *spontaneously*, not just by pre-arrangement. However, on its own, software is not enough. VR requires a massive contribution from the human interface hardware. To put ourselves in the scene, we have to tell the software where we are and what we are doing. This is not easy. We have to synchronize the *real* with the *virtual*, and in the context of navigation and movement, this requires *motion tracking*. Motion-tracking hardware will allow us to determine where we (and other real objects) are in relation to a *real* coordinate frame of reference. The special hardware needed to do this has its own unique set of practical complexities (as we discussed in Section 4.3). But once we can acquire movement data in real time and feed it into the visualization software, it is not a difficult task to match the real coordinate system with the virtual one so that we can appear to touch, pick up and throw a virtual object. Or more simply, by knowing where we are and in what direction we are looking, a synthetic view of a virtual world can be fed to the display we are using, e.g., a head-mounted display.

In this chapter, we start by looking at those aspects of computer animation which are important for describing and directing movement in VR. Principally, we will look at inverse kinematics (IK), which is important when simulating the movement of articulated linkages, such as animal/human movement and robotic machinery (so different to look at, but morphologically so similar). We will begin by quickly reviewing how to achieve smooth motion and rotation.

11.1   Computer Animation

The most commonly used approach to 3D computer animation is *keyframe in-betweening (tweening),* which is most useful for basic rigid-body motion. The idea of the keyframe is well known to paper-and-pencil animators. It is a "description" of a scene at one instant of time, a key instant. Between key instants, it is assumed that nothing "startling" happens. It is the role of the *key* animators to draw the key scenes (called keyframes), which are used by a team of others to draw a series of scenes filling in the gaps between the keys so that jumps and discontinuities do not appear This is called *tweening* (derived from the rather long and unpronounceable word inbe*tweening*).

The task of *tweening* is a fairly monotonous and repetitive one. Thus, it is ideally suited to some form of automation with a computer. A half-hour animated movie may only need a few thousand keyframes, about four percent the total length. Some predefined and commonly used actions described by library scripts might cut the work of the animators even further. For example, engineering designers commonly need to visualize their design rotating in front of the camera; a script or template for rotation about some specified location at a fixed distance from the camera will cut the work of the animator even further.

Thus, in computer animation, the basic idea is:

> Set up a description of a scene (place models, lights and cameras in three dimensions) for each keyframe. Then use the computer to calculate descriptions of the scene for each frame in between the keyframes and render appropriate images.

Most (if not all) computer-animation application programs give their users the task of describing the state of the action in keyframes, and then they do their best to describe what happens in the snapshots taken during the intervening frames. This is invariably done by interpolating between the description of at least two, but possibly three or four keyframes.

For applications in VR, we cannot use the keyframe concept in the same way, because events are taking place in real time. However, we still have to render video frames, and we must synchronize what is rendered with a real-time clock. So suppose we need to simulate the movement of a flight simulator which represents an aircraft flying at 10 m/s along a path 10 km long over a virtual terrain. The total flight time will take approximately 1,000 seconds. This means that if the graphics card is refreshing at a rate of 100 Hz (or every one hundredth of a second), our visualization software will have to render 100,000 frames during this 1,000-second flight. In addition, if we get external course correction information from joystick hardware every second and use it to update the aircraft's current position, we are still left with the task of having to smoothly interpolate the viewpoint's virtual position and orientation 99 times every second (so that all the video frames can be rendered without looking jerky). This is analogous to what happens in computer animation when *tweening* between two keyframes to generate 100 pictures.[1]

In practical terms, our rendering software has a harder job than that of its equivalent in computer animation because of the need to synchronize to a real-time clock. Going back to our simple flight simulator example, if we cannot render 100 pictures per second, we shall have to make larger steps in viewpoint position. Therefore, instead of rendering 100 frames per second (fps) and moving the viewpoint 0.1 m at each frame, we might render 50 fps and move the viewpoint 0.2 m. Unfortunately, it is not usually possible to know ahead of time how long it is going to take to render a

---

[1]There is a small subtle difference. In the animation example, we know the starting and ending positions and *interpolate*; in the flight-simulator example, we use the current position and the last position to *extrapolate*.

video frame. For one view it might take 10ms, for another 50 ms. As we indicated in Chapter 7, the time it takes to render a scene depends on the scene complexity, roughly proportional to the number of vertices within the scene.

As we shall see in Part II of the book, application programs designed for real-time interactive work are usually written using multiple threads[2] of execution so that rendering, scene animation, motion tracking and haptic feedback can all operate at optimal rates and with appropriate priorities. If necessary, the rendering thread can skip frames. These issues are discussed in Chapter 15.

11.2   Moving and Rotating in 3D

Rigid motion is the simplest type of motion to simulate, because each object is considered an immutable entity, and to have it move smoothly from one place to another, one only has to specify its position and orientation. Rigid motion applies to objects such as vehicles, airplanes, or even people who do not move their arms or legs about. All *fly-by* and *walkthrough*-type camera movements fall into this category.

A point in space is specified by a position vector, $\mathbf{p} = (x, y, z)$. To specify *orientation*, three additional values are also required. There are a number of possibilities available to us, but the scheme where values of heading, pitch and roll ($\varphi, \vartheta, \alpha$) are given is a fairly intuitive description of a model's orientation (these values are also referred to as the Euler angles). Once the six numbers $(x, y, z, \varphi, \vartheta, \alpha)$ have been determined for each object or a viewpoint, a transformation matrix can be calculated from them. The matrix is used to place the object at the correct location relative to the global frame of reference within which all the scene's objects are located.

11.2.1   Moving Smoothly

Finding the position of any object (camera, etc.) at any time instant t, as it moves from one place to another, involves either mathematical interpolation or extrapolation. Interpolation is more commonly utilized in computer animation, where we know the starting position of an object (x_1, y_1, z_1) and its

---

[2]Any program in which activities are not dependent can be designed so that each activity is executed by its own separate and independent thread of execution. This means that if we have multiple processors, each thread can be executed on a different processor.

orientation $(\varphi_1, \vartheta_1, \alpha_1)$ at time t_1 and the final position (x_2, y_2, z_2) and orientation $(\varphi_2, \vartheta_2, \alpha_2)$ at t_2. We can then determine the position and orientation of the object through interpolation at any time where $t_1 \leq t \leq t_2$. Alternatively, if we need to simulate real-time motion then it is likely that we will need to use two or more positions in order to predict or extrapolate the next position of the object at a time in the future where $t_1 < t_2 \leq t$.

Initially, let's look at position interpolation (angular interpolation is discussed separately in Section 11.2.2) in cases where the movement is along a smooth curve. Obviously, the most appropriate way to determine any required intermediate locations, in between our known locations, is by using a mathematical representation of a curved path in 3D space—also known as a *spline*.

It is important to appreciate that for spline interpolation, the path must be specified by at least four points. When a path is laid out by more than four points, they are taken in groups of four. Splines have the big advantage that they are well-behaved and they can have their flexibility adjusted by specified parameters so that it is possible to have multiple splines or paths which go through the same *control* points. Figure 11.1 illustrates the effect of increasing and decreasing the *tension* in a spline.

The equation for any point **p** lying on a cubic spline segment, such as that illustrated in Figure 11.2, is written in the form

$$\mathbf{p}(\tau) = \mathbf{K}_3 \tau^3 + \mathbf{K}_2 \tau^2 + \mathbf{K}_1 \tau + \mathbf{K}_0, \tag{11.1}$$

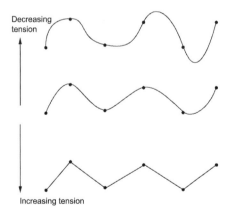

Figure 11.1. Changes in the flexibility (or tension) of a spline allow it to represent many paths through the same control points.

or equivalently (to speed up its calculation),

$$\mathbf{p}(\tau) = ((\mathbf{K}_3 \tau + \mathbf{K}_2)\tau + \mathbf{K}_1)\tau + \mathbf{K}_0.$$

The unknown vector constants $\mathbf{K}_0$, $\mathbf{K}_1$, $\mathbf{K}_2$ and $\mathbf{K}_3$, as well as the parameter τ, have to be determined. To do this, we initially impose four conditions on Equation (11.1):

1. The spline passes through the point $\mathbf{P}_i$ at the start of the curve. At this point, $\tau = 0$.

2. The spline passes through the point $\mathbf{P}_{i+1}$ at the end of the curve. At this point, $\tau = 1$

3. The derivative of the spline at $\mathbf{P}_i$ is given. That is $\mathbf{P}'_i$.

4. The derivative of the spline at $\mathbf{P}_{i+1}$ is given. That is $\mathbf{P}'_{i+1}$.

The parameter τ in the equations for the spline is related to the distance traveled along a given section of the curve. This is shown in Figure 11.2. In addition, in order to determine the vector constants, we only utilize a segment of the spline available.

Obviously, in 3D space, there are three components (x, y, z) to any vector position $\mathbf{p}$, and these conditions apply to each of these components. For example, to obtain the x-component we need to consider the values x_i, x_{i+1}, x'_i, x'_{i+1}.

In addition, differentiating Equation (11.1) with respect to τ gives us a standard equation by which we can specify $\mathbf{p}'(\tau)$:

$$\mathbf{p}'(\tau) = 3\mathbf{K}_3 \tau^2 + 2\mathbf{K}_2 \tau + \mathbf{K}_1. \tag{11.2}$$

Thus, in Equation (11.1), if we substitute τ with 0 to represent $\mathbf{p}(x_i)$ and τ with 1 to represent $\mathbf{p}(x_{i+1})$, and in Equation (11.2) if we substitute τ with 0 to represent $\mathbf{p}'(x_i)$ and τ with 1 to represent $\mathbf{p}'(x_{i+1})$, we will have four simultaneous equations which can be used to solve for the x-component of the vector constants $\mathbf{K}_0, \mathbf{K}_1, \mathbf{K}_2, \mathbf{K}_3$. Written in matrix form, these are

$$\begin{bmatrix} 0 & 0 & 0 & 1 \\ 1 & 1 & 1 & 1 \\ 0 & 0 & 1 & 0 \\ 3 & 2 & 1 & 0 \end{bmatrix} \begin{bmatrix} K_{3_x} \\ K_{2_x} \\ K_{1_x} \\ K_{0_x} \end{bmatrix} = \begin{bmatrix} x_i \\ x_{i+1} \\ x'_i \\ x'_{i+1} \end{bmatrix}.$$

On solution, the following expressions are obtained:

$$
\begin{aligned}
K_{3_x} &= 2x_i - 2x_{i+1} + x'_i + x'_{i+1}, \\
K_{2_x} &= -3x_i + 3x_{i+1} - 2x'_i - x'_{i+1}, \\
K_{1_x} &= x'_i, \\
K_{0_x} &= x'_{i+1}.
\end{aligned}
$$

However, as you can see, determination of the vector constants is dependent on finding the gradient of the spline (or its derivative) at the two control points, $\mathbf{P}_i$ and $\mathbf{P}_{i+1}$. Remember that we can only define a spline using a minimum of four values. Thus we can use the knowledge we have of the other two points ($\mathbf{P}_{i-1}$ and $\mathbf{P}_{i+2}$) on the spline in order to estimate the derivative at $\mathbf{P}_i$ and $\mathbf{P}_{i+1}$. We do this using finite differences, where $x'_i = \frac{x_{i+1} - x_{i-1}}{2}$ and $x'_{i+1} = \frac{x_{i+2} - x_i}{2}$. That is, we are really finding the gradient of the spline at these two points.

This results in the following sequence of equations:

$$
K_{3_x} = -\frac{1}{2}x_{i-1} + \frac{3}{2}x_i - \frac{3}{2}x_{i+1} + \frac{1}{2}x_{i+2}, \tag{11.3}
$$

$$
K_{2_x} = x_{i-1} - \frac{5}{2}x_i + 2x_{i+1} - \frac{1}{2}x_{i+2}, \tag{11.4}
$$

$$
K_{1_x} = -\frac{1}{2}x_{i-1} + \frac{3}{2}x_{i+1}, \tag{11.5}
$$

$$
K_{0_x} = x_i. \tag{11.6}
$$

Similar expressions may be written for the y- and z-components of the K terms, and thus the constant vectors $\mathbf{K_c}$ become

$$
\mathbf{K_c} = \begin{bmatrix} K_{c_x} \\ K_{c_y} \\ K_{c_z} \end{bmatrix},
$$

for $c = 0, 1, 2, 3$.

At this stage, we need to draw a distinction about how we intend to use our spline. When we are animating camera movement, for example, $\mathbf{P}_{i-1}$ to $\mathbf{P}_{i+2}$ are all predetermined positions of the camera. That is, they are key positions. Then we can simply interpolate between these predetermined positions in order to estimate how the camera moves so that the transition will appear smooth, with no sudden changes of direction. We can do this by using Equation (11.1) to determine any position $\mathbf{p}$ at time t along the spline. Of course, in order to use this equation, we need to insert a value for τ. Assuming that an object following the spline path is getting its known locations (the $\mathbf{P}_i$) at equal time intervals Δt, we can parameterize the curve by time and obtain τ from

$$\tau = \left(\frac{t - t_{-2}}{\Delta t} \right).$$

This arises because the spline has been determined using four control positions. In Figure 11.2, these control points are $\mathbf{P_{i-1}}$, $\mathbf{P_i}$, $\mathbf{P_{i+1}}$ and $\mathbf{P_{i+2}}$. To determine the constant vectors, we set $\tau = 0$ at $\mathbf{P_i}$. If we wished to interpolate between control points $\mathbf{P_{i+1}}$ and $\mathbf{P_{i+2}}$, we would use values of τ in the range $1 < \tau < 2$.

Of course, as we mentioned earlier, spline interpolation is only useful when we have predetermined key positions. Extrapolation is required when there is no knowledge of future movement. Take for example our flight simulator, where we obtain information about position from external hardware every second. However, suppose we must render our frames every hundredth of a second. Assuming that $\mathbf{P_{i+2}}$ is the last known position obtained from the

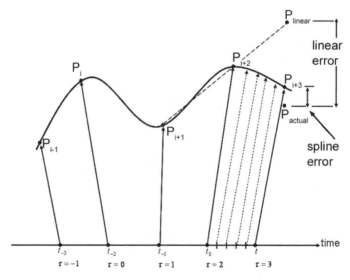

Figure 11.2. A spline segment used for interpolation and extrapolation. The parameter τ lies in the interval $[0, -1]$ between the points $\mathbf{P}_{i-1}$ and $\mathbf{P}_i$, and $\mathbf{p}$ is a position on the curve at any time $t > t_0$

hardware, our software needs to extrapolate the position of our aircraft every one hundredth of a second until the next position, $\mathbf{P_{i+3}}$, is available from the hardware.

Again, we may use Equation (11.1) to determine the position $\mathbf{p}$ at any time greater than t. Thus we try to predict how the spline will behave up until our system is updated with the actual position. This may or may not be the same as our extrapolated position. Typically, parameterization by time and extrapolation can lead to a small error in predicted position. For example, the point labeled P_{actual} in Figure 11.2 is the actual position of the next point along the true path, but it lies slightly off the predicted curve.

So we need to recalculate the equation of the spline based on this current actual position and its previous three actual positions. Thus, $\mathbf{P_i}$ becomes $\mathbf{P_{i-1}}$ and so on. Using the four new control positions, we recompute the spline, and this new spline will then be used to predict the next position at a given time slice. And so the procedure continues.

It is also worth noting at this stage that whilst spline extrapolation is not without error, linear extrapolation is usually much more error-prone. For the example in Figure 11.2, if linear extrapolation of the positions $\mathbf{P_{i+1}}$ and $\mathbf{P_{i+2}}$ is used to predict the new position at time t, the associated positional error is much greater than with spline interpolation.

11.2.2   Rotating Smoothly

In this section, we turn our attention to interpolating angles of orientation. Angles *cannot* be interpolated in the same way as position coordinates are interpolated. For one thing they are periodic in the interval $[0, -2\pi]$. It is now generally agreed that the best way to obtain smooth angular interpolation is by using *quaternions*. Appendix A provides the background on quaternions. It gives algorithms for converting between Euler angles, quaternions and rotation matrices, and defines the function specifically for solving the problem of orientation tweening; that is, the *slerp()* function. It also demonstrates how to calculate the transformation matrix used to set the orientation for an object or direction of view T_{ak} at time t_k, obtained by interpolation of the orientations at times t_l and t_m.

T_{ak} cannot be obtained by directly interpolating the matrices expressing the orientation at times t_l and t_m (see Section A.2). At times t_l and t_m, the matrices T_{al} and T_{am} are actually determined from the known values of $(\varphi_l, \vartheta_l, \alpha_l)$ and $(\varphi_m, \vartheta_m, \alpha_m)$ respectively. Therefore, whilst it may not be possible to interpolate matrices, it is possible to interpolate a quaternion associated with a rotation matrix using the *slerp()* function in the following three steps:

1. Given an orientation that has been expressed in Euler angles at two time points, l and m, calculate equivalent quaternions q_l and q_m, using the algorithm given in section A.2.1.

2. Interpolate a quaternion q_k that expresses the orientation at time t_k using:

$$\mu = \frac{t_k - t_l}{t_m - t_l};$$

$$\rho = \cos^{-1} q_l \cdot q_l;$$

$$q_k = \frac{\sin(1 - \mu)\rho}{\sin \rho} q_l + \frac{\sin \mu \rho}{\sin \rho} q_m.$$

See Appendix A for details.

3. Use the expressions from Section A.2.2 to obtain T_{ak} given the quaternion q_k.

And there we have it: T_{ak} is a matrix representing the orientation at time t_k so that the orientation of any object or the viewpoint changes smoothly during the interval t_l to t_m.

11.3   Robotic Motion

Consider the following scenario:

> You are wearing a haptic feedback glove and a head-mounted stereoscopic display (HMD). The headset and the glove contain sensors that feed their position and orientation into a VR simulator with computer-generated characters. It should be possible for you to reach forward and shake hands with one of the synthetic characters or take the synthetic dog for a walk. The stereoscopic HMD should confuse your eyes into believing the character is standing in front of you; the haptic glove should give you the illusion of a firm handshake.[3] Now for the movement

---

[3] We haven't seen this actually done yet, but all the hardware components to achieve it are already commercially available.

problem: how does the rendering software animate the virtual
figure as it reaches out to grasp your hand and shake it?

The software will know a few things about the task. The character will
be standing on the virtual floor, so the position of its feet will be well defined.
It also must grasp your hand at the point in virtual space where the motion
sensor thinks it is located. This leaves a lot of freedom for the software, but
there are other constraints which should be adhered to if realism is to be
maintained. For example, the virtual character's arms cannot change length
or pass through the character's body. The angle of the character's joints, elbow
and shoulders must also conform to some limits; otherwise, it could look like
a very contorted handshake indeed! To overcome these difficulties, we will
need to do the same thing that computer animators do. We need to build a
skeleton into the virtual character and animate it using inverse kinematics.

A large number of animal species have their movements primarily controlled by some form of *skeleton* that is in essence hierarchical. For example,
a finger is attached to a hand, which is connected to a lower arm, etc. The
skeleton imposes constraints on how an animal behaves (it cannot suddenly
double the length of its legs, for example). For animation purposes, the idea
of a skeleton is very useful. In traditional clay animation, a rigid wire skeleton is embedded in the clay, and this allows the animator to manipulate the
model in a realistic way. (The Oscar-winning Wallace and Gromit are excellent examples of clay characters with a wireframe skeleton.)

In computer animation, the skeleton fulfills two important functions:

1. It provides a rigid framework which can be pivoted, twisted and rotated. Vertices and polygons are assigned to follow a specific *bone* in
   the skeleton, and thus the model will appear to take up various poses,
   just as the clay model does.

2. The hierarchical nature of the skeleton allows for natural behavior. For
   example, pivoting the upper arm about the shoulder in a model of a
   human figure will cause the lower arm and hand to execute the same
   pivot without the animator having to do it explicitly.

Consider the example shown in Figure 11.3, where the model is pictured
on the left, in the center its skeleton is shown as a thick black line and on the
right is a diagrammatic representation of the hierarchical links in the skeleton.

Using the skeleton, one has the option to pivot parts of the model about
the end points (nodes) of any of its bones. Taking the model in Figure 11.3 as

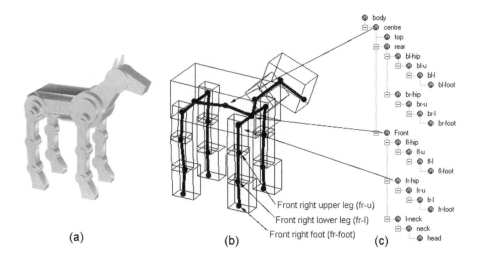

Figure 11.3. A model with its skeleton: In (a), the model is shown; (b) shows the skeleton (the thick lines). A pivot point is shown at the end of each *bone*. The thin lines represent boxes that contain all the parts of the model attached to each bone. In (c), a hierarchical representation of all the *bones* and how they are connected is illustrated.

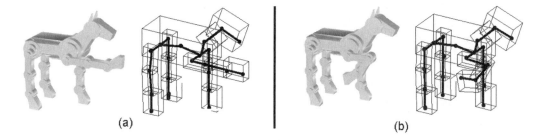

Figure 11.4. Pivoting the leg of the model shown in Figure 11.3 into two poses.

an example, a rotation of the front right upper leg around the hip joint moves the whole front right leg (see Figure 11.4(a)). If this is followed by rotations of the lower leg and foot, the pose illustrated in Figure 11.4(b) results.

11.3.1 Specifying Robotic Positions

Any skeleton has what we can term a *rest* pose, which is simply the form in which it was created, before any manipulation is applied. The skeleton illustrated in Figure 11.3 is in a rest pose. With the knowledge of hierarchical *connectivity* in a skeleton, a position vector giving the location of the end (the node or the joint) of each bone uniquely specifies the skeleton's rest position. To be able to set the skeleton into any other pose, it must satisfy the criteria:

1. Be obtained with a rotational transformation about an axis located at one of the nodes of the skeleton.

2. Of a transformation is deemed to apply to a specific bone, say i, then it must also be applied to the descendent (child) bones as well.

For example, consider the simple skeleton illustrated in Figure 11.5. It shows four bones; bones 2 and 3 are children to bone 1, and bone 4 is a child to bone 3. Each bone is given a coordinate frame of reference (e.g., (x_3, y_3, z_3) for bone 3). (For the purpose of this example, the skeleton will be assumed to lie in the plane of the page.) $\mathbf{P}_0$ is a node with no associated bone; it acts as the base of the skeleton and is referred to as the *root*.

Suppose that we wish to move bone 3. The only option is to pivot it around a direction vector passing through node 1 (to which bone 3 is attached). A rotational transformation is defined as a 4×4 matrix, and we can combine rotations around different axes to give a single matrix M that encodes information for any sequence of rotations performed at a point. This matrix takes the form shown in Equation (11.7), whilst the general theory

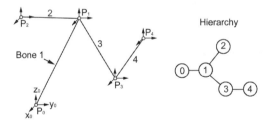

Figure 11.5. Specifying a rest pose of a skeleton with four bones 1 to 4. In addition to the positions $\mathbf{P}_i$ of the nodes at the end of the bone, a local frame of reference $(\mathbf{x}_i, \mathbf{y}_i, \mathbf{z}_i)$ is attached to each node. On the right, a hierarchical diagram of the skeleton is shown. $\mathbf{P}_0$ is a root node which has no bone attached and acts as the base of the skeleton.

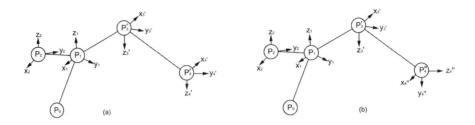

Figure 11.6. (a) Posing for the skeleton of Figure 11.5 by a rotation of bone 3 around the node at the end of bone 1. (b) Co-positional nodes do not imply that two poses are identical. The nodes here are at the same location as in (a), but the local frames of reference do not take the same orientation.

of transformation matrices and homogeneous coordinates is given in Section 6.6:

$$
M = \begin{bmatrix} a_{00} & a_{01} & a_{02} & 0 \\ a_{10} & a_{11} & a_{12} & 0 \\ a_{20} & a_{21} & a_{22} & 0 \\ 0 & 0 & 0 & 1 \end{bmatrix}.
\tag{11.7}
$$

Once M has been calculated, its application to $\mathbf{P}_3$ and $\mathbf{P}_4$ will move them to appropriate locations for the new pose. The example pose in Figure 11.6(a) was obtained by a rotation of π round axis y_2.

There are a few important observations that emerge from this simple example:

- Node 4 is affected by the transformation because it is descended from node 3.

- Nodes 0, 1 and 2 are unaffected and remain at locations $\mathbf{p}_0$, $\mathbf{p}_1$ and $\mathbf{p}_2$. Importantly, the node about which the rotation is made is not disturbed in any way.

- The coordinate frames of reference attached to nodes 3 and 4 are also transformed; they become $(\mathbf{x}_3', \mathbf{y}_3', \mathbf{z}_3')$ and $(\mathbf{x}_4', \mathbf{y}_4', \mathbf{z}_4')$.

- Although node 4 (at $\mathbf{p}_4$) is moved, its position is unchanged in the local frame of reference attached to node 3.

- When M is applied to nodes 3 and 4, it is assumed that their coordinates are expressed in a frame of reference with origin at $\mathbf{P}_1$.

It is possible to apply additional rotational transformations until any desired pose is achieved. To attain the pose depicted in Figure 11.4(b), several successive transformations from the rest position of Figure 11.3 were necessary.

Note that even if the position of the nodes in two poses are identical, it does not mean that the poses are identical. Look at Figure 11.6(b); this shows four bones (four nodes) co-positional with the nodes in the pose of Figure 11.6(a). Close inspection reveals that the local frames of reference for the joint at P_4 are very different. Therefore, the parts of the object attached to P_4 will appear different. For example, if P_4 represents the joint angle at the wrist of a virtual character then the character's hand will have a different orientation between Figure 11.6(a) and Figure 11.6(b).

As we have already mentioned, if it is necessary to interpolate between robot poses then quaternions offer the best option. Full details are given in [1]. This is especially true in animation, where knowledge of the starting and final position is available. However, for real-time interactive movement, it is more appropriate to use inverse kinematics. This is another specialized branch of mathematics that deals specifically with determining the movement of interconnected segments of a body or material such that the motion is transmitted in a predictable way through the parts.

11.4   Inverse Kinematics

A lot of fundamental work on inverse kinematics (IK) has been developed as a result of ongoing research in the area of robotics. In robotics, an important task is to move the working part of the robot (or more technically, its effector) into a position and orientation so as to be able to perform its task. As we have seen previously in this chapter, the way robots move is not dissimilar to the way animals move, and so the same IK can be applied to their movement. IK is the ideal way to program the motion of the virtual components of our *handshake* scenario.

A rigorous theory of IK calculations is *not* trivial. It involves solving nonlinear systems of equations in which there are always more unknowns than equations. Thus, the system is either poorly constrained or not constrained at all. As we look at IK, we will see how these problems are solved, and we will finish off this section with a short and simple practical algorithm for constrained IK motion in 3D.

It is much simpler to understand IK in 2D, so this is where we will begin. The solution of the 3D IK problem follows exactly the same conceptual argument as that used in 2D. It is only in setting up the problem that differences occur; thus, we will use a 2D example to present a rigorous method of the solution.

11.4.1   The IK Problem

For a robot used as a surrogate human on a production line, an important design feature is that the hand can approach the working materials at the correct angle. To do this, there must be enough linkages from the place where the robot is anchored to its hand and each link must have sufficient degrees of freedom. If the robot can put its hand in the correct place then the next question is how does it get there? It may be starting from a rest position or the place in which it has just finished a previous task. From the operator's point of view, it would be helpful if she only had to move the robot's hand, not adjust every link from base to hand. However, the hand is obviously at the end of the articulated linkage, and it is the relative orientation of these links that dictate where the hand is.

Let's assume that each link's orientation is given by the set $(\vartheta_1, \vartheta_2, \ldots \vartheta_n)$. To simplify matters, we can group these linkages into a vector given by Θ, where

$$\Theta = \begin{bmatrix} \vartheta_1 & \vartheta_2 & \ldots & \vartheta_n \end{bmatrix}^T.$$

Similarly, we can say that the position of each linkage within the articulation is given by $\mathbf{X}$, where

$$\mathbf{X} = \begin{bmatrix} x_1 & x_2 & \ldots & x_n \end{bmatrix}^T.$$

We can then further say that the location of the end of the articulation $\mathbf{X}$ is given by some function,

$$\mathbf{X} = f(\Theta). \tag{11.8}$$

Thus, if we know the orientation of each of the links connected to the effector, we can compute its position. This process is called *forward kinematics*. However, in a typical environment, the operator of the articulated linkage is really only concerned with setting the position of the effector and not with deriving it from linkage orientation values. Thus we enter the realm of inverse kinematics. That is, knowing the position we require the effector $\mathbf{X}$ to take, can we determine the required position and orientation of the adjoining linkages to take the effector to this position and orientation?

To solve this problem, we must find the inverse of the function $f()$ so that we can write Equation (11.8) as

$$\Theta = f^{-1}(\mathbf{X}). \tag{11.9}$$

It should be stressed that it may not be possible for the effector to reach the desired position $\mathbf{X}$, and hence we call it a *goal*.

Since $\mathbf{X}$ and Θ are both vectors, it may be possible to rewrite Equation (11.8) in matrix form assuming the function $f()$ is linear:

$$\mathbf{X} = A\Theta, \tag{11.10}$$

and hence we can solve it for Θ by calculating the inverse of A:

$$\Theta = A^{-1}\mathbf{X}.$$

Unfortunately, except for the simplest linkages, $f()$ is a nonlinear function and therefore it is not possible to express Equation (11.8) in the form given by Equation (11.10). To solve this problem, we must use an iterative technique which is essentially the same as that used for finding the solution of a set of simultaneous nonlinear equations. McCalla [2] (for example) describes the usual procedure.

The key to inverse kinematics is to try to linearize Equation (11.8) by expressing it in differential form:

$$\Delta\mathbf{X} = f'(\Theta)\Delta\Theta, \tag{11.11}$$

and then incrementally stepping towards the solution. Since $f()$ is a multi-variable function, $f'()$, the Jacobian matrix of partial derivatives, is given by

$$J(\Theta) = \begin{bmatrix} \dfrac{\partial X_1}{\partial \vartheta_1} & \dfrac{\partial X_1}{\partial \vartheta_2} & \cdots & \dfrac{\partial X_1}{\partial \vartheta_n} \\ \dfrac{\partial X_2}{\partial \vartheta_1} & \dfrac{\partial X_2}{\partial \vartheta_2} & \cdots & \dfrac{\partial X_2}{\partial \vartheta_n} \\ \cdots & \cdots & \cdots & \cdots \\ \dfrac{\partial X_m}{\partial \vartheta_1} & \dfrac{\partial X_m}{\partial \vartheta_2} & \cdots & \dfrac{\partial X_m}{\partial \vartheta_n} \end{bmatrix}.$$

By calculating an inverse for J, Equation (11.11) may be written:

$$\Delta\Theta = J^{-1}\Delta\mathbf{X}.$$

The Jacobian J is of course dependent on Θ, so an initial guess must be made for the values ϑ_1, ϑ_2 .. ϑ_n. In the case of the IK problem, the initial guess is simply the current configuration. After the first step, we proceed to find a new Θ by iterating:

$$\Theta_{n+1} = \Theta_n + \Delta\Theta = \Theta_n + J^{-1}\Delta\mathbf{X},$$

where n is a count of the number of iterations, and at each step J^{-1} is recalculated from the current Θ_n.

11.4.2   Solving the IK Problem in Two Dimensions

We will look at an example of a three-link articulation. Because it's restricted to two dimensions, it moves only in the plane of the page. When equations have been obtained for this specific case, we will see how they may be extended to a system with n links. The links are fixed at one end, which we will designate as the origin $(0, 0)$. The other end, at point (x, y), is moved towards a goal point, which it may or may not reach. We hope a solution to the IK problem will reveal where it gets to.

A solution of Equation (11.9) gives us the orientation of each link in the chain, and from that we can find out how close to the goal the endpoint actually gets. Figure 11.7(a) illustrates several possible configurations for a three-link articulation, and Figure 11.7(b) gives the notation used to specify and solve the three-link IK problem. The first link is anchored at $(0, 0)$ and lies at an angle of ϑ_1 to the x-axis. The lengths of the links are l_1, l_2 and l_3, and link 2 makes an angle of ϑ_2 from link 1. The endpoint $\mathbf{P}_4$ lies at coordinate (x, y).

Figure 11.7(b) shows us that if we know the values of $\mathbf{P}_1$ (anchor point), l_1, l_2, l_3 (lengths of the links), ϑ_1, ϑ_2 and ϑ_3 (the relative orientation between one link and the next) then the current position of every joint in the link is fully specified (including the endpoint $\mathbf{P}_4$).

Since $\mathbf{P}_1$, l_1, l_2 and l_3 are constant and independent of the orientation of the linkage, only the ϑ_i are the unknowns in our IK function, Equation (11.9). For our specific example,

$$\mathbf{P}_4 = \mathbf{f}(\vartheta_1, \vartheta_2, \vartheta_3).$$

Note carefully how ϑ_i is specified as an angle measured anti-clockwise from the direction in which link $(i - 1)$ is pointing to the direction in which

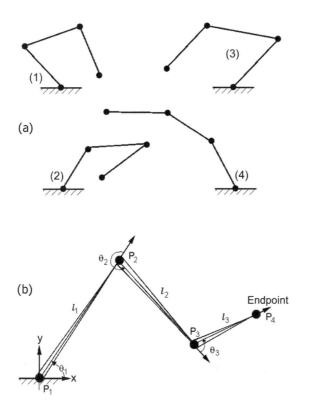

Figure 11.7. (a) Several possible orientations for a three-link two-dimensional artic-ulated figure. In (1) and (2), the articulation reaches to the same endpoint. (b) A three-link articulation.

link i is pointing. The first link angle ϑ_1 is referenced to the x-axis. Using this information, we can write an expression for the (x, y)-coordinate of $\mathbf{P}_4$:

$$\left[\begin{array}{c} x \\ y \end{array} \right] = \left[\begin{array}{c} l_1 \cos(\vartheta_1) + l_2 \cos(\vartheta_1 + \vartheta_2) + l_3 \cos(\vartheta_1 + \vartheta_2 + \vartheta_3) \\ l_1 \sin(\vartheta_1) + l_2 \sin(\vartheta_1 + \vartheta_2) + l_3 \sin(\vartheta_1 + \vartheta_2 + \vartheta_3) \end{array} \right].$$

(11.12)

Equation (11.12) shows that $f(\vartheta_1, \vartheta_2, \vartheta_3)$ is, as expected, a nonlinear function. Therefore, we will need to solve the IK problem by linearizing Equation (11.12) and iterating towards the desired goal position of $\mathbf{P}_4$. The first step in this procedure is to obtain the Jacobian. For the 2D IK problem of three links, this is the 2×3 matrix:

$$J = \left[\begin{array}{ccc} \dfrac{\partial x}{\partial \vartheta_1} & \dfrac{\partial x}{\partial \vartheta_2} & \dfrac{\partial x}{\partial \vartheta_3} \\[2mm] \dfrac{\partial y}{\partial \vartheta_1} & \dfrac{\partial y}{\partial \vartheta_2} & \dfrac{\partial y}{\partial \vartheta_3} \end{array} \right].$$

The terms $\frac{\partial}{\partial \vartheta_i}$ are obtained by differentiating Equation (11.12) to give

$$J = \left[\begin{array}{ccc} J_{11} & J_{12} & J_{13} \\ J_{21} & J_{22} & J_{23} \end{array} \right], \tag{11.13}$$

where

$$
\begin{aligned}
J_{11} &= -l_1 \sin(\vartheta_1) - l_2 \sin(\vartheta_1 + \vartheta_2) - l_3 \sin(\vartheta_1 + \vartheta_2 + \vartheta_3); \\
J_{12} &= -l_2 \sin(\vartheta_1 + \vartheta_2) - l_3 \sin(\vartheta_1 + \vartheta_2 + \vartheta_3); \\
J_{13} &= -l_3 \sin(\vartheta_1 + \vartheta_2 + \vartheta_3); \\
J_{21} &= l_1 \cos(\vartheta_1) + l_2 \cos(\vartheta_1 + \vartheta_2) + l_3 \cos(\vartheta_1 + \vartheta_2 + \vartheta_3); \\
J_{22} &= l_2 \cos(\vartheta_1 + \vartheta_2) + l_3 \cos(\vartheta_1 + \vartheta_2 + \vartheta_3); \\
J_{23} &= l_3 \cos(\vartheta_1 + \vartheta_2 + \vartheta_3).
\end{aligned}
$$

Once J has been calculated, we are nearly ready to go through the iteration process that moves $\mathbf{P}_4$ towards its goal. An algorithm for this will be based on the equations

$$
\begin{aligned}
\Theta_{n+1} &= \Theta_n + J(\Theta_n)^{-1} \Delta \mathbf{X}; \\
\mathbf{X}_{n+1} &= f(\Theta_{n+1}).
\end{aligned}
$$

Unfortunately, because J is a non-square matrix, we cannot just look to the conventional methods such as Gauss elimination to obtain its inverse J^{-1}. However, non-square matrices can have what is called a *generalized inverse*. Of course, for underdetermined systems, there can be no unique solution, but a generalized inverse of J is quite sufficient for our uses of IK. Appendix B gives a little background about generalized inverses.

Iterating Towards the Goal

Having found a way to invert J, we develop an algorithm to iterate from one configuration towards the goal. A suitable algorithm is given in Figure 11.8.

Step 4 of the algorithm provides the mechanism to test for convergence of the IK solution procedure. It is based on ensuring that the norm of the

Start with the linkage configuration defined by the set of angles:
$\vartheta_1, \vartheta_2, \vartheta_3...\vartheta_n$ (which we will write as Θ), and endpoint located at $\mathbf{P}$,
i.e., at coordinate (x, y).
Apply the steps below to move the endpoint towards its goal
at $\mathbf{P}_g$ (i.e., at :(x_g, y_g))

Step 1:
Calculate the incremental step $\Delta\mathbf{X} = \mathbf{P}_g - \mathbf{P}$

Step 2:
Calculate $J(\vartheta_1, \vartheta_2, \vartheta_3, ...\vartheta_n)$
(use the current values of ϑ_1 etc.)

Step 3:
Find the inverse of J, which we will call J^-
$J^- = J^T(JJ^T)^{-1}$
(if J is a $2 \times n$ matrix, J^- is a $n \times 2$ matrix)

Step 4:
Test for a valid convergence of the iteration:
if $\|(I - JJ^-)\Delta\mathbf{X}\| > \varepsilon$ then the step towards the
goal (the $\Delta\mathbf{X}$) is too large, so set $\Delta\mathbf{X} = \dfrac{\Delta\mathbf{X}}{2}$
and repeat Step 4 until the norm is less than ε.
If the inequality cannot be satisfied after a certain number
of steps then it is likely that the goal cannot be reached and
the IK calculations should be terminated
(This step is discussed in more detail in the text)

Step 5:
Calculate updated values for the parameters ϑ_1 etc.
$\Theta = \Theta + J^- \Delta\mathbf{X}$
Θ is the vector of angles for each link $[\vartheta_1, \vartheta_2, ...]^T$

Step 6:
Calculate the new state of the articulation from ϑ_1, ϑ_2 etc.
Check the endpoint $\mathbf{P}_4$ to see if is close enough to the goal.
It is likely that $\Delta\mathbf{X}$ will have been reduced by Step 4
and thus the endpoint will be somewhat short of the goal.
In this case, *go back and repeat the procedure from Step 1.*
Otherwise, we have succeeded in moving $\mathbf{P}_4$ to the goal point $\mathbf{P}_g$

Figure 11.8. Iterative algorithm for solving the IK problem to determine the orientation of an articulated linkage in terms of a number of parameters ϑ_i given the goal of moving the end of the articulation from its current position $\mathbf{P}$ towards $\mathbf{X}_g$.

vector $\Delta \mathbf{X} - J(\Theta)\Delta\Theta$ is smaller than a specified threshold:

$$\|\Delta \mathbf{X} - J(\Theta)\Delta\Theta\| < \varepsilon.$$

If we substitute for $\Delta\Theta$ and call J^- the generalized inverse of J, then in matrix form,

$$\|\Delta \mathbf{X} - JJ^- \Delta\mathbf{X}\| < \varepsilon,$$

or simplifying we have

$$\|(I - JJ^-)\Delta\mathbf{X}\| < \varepsilon,$$

where I is the identity matrix such that $I(\Delta\mathbf{X}) = \Delta\mathbf{X}$.

We use this criterion to determine a $\Delta\mathbf{X}$ that satisfies the condition on the norm. We also use this to determine whether the iteration can proceed, or whether we must accept that the goal cannot be reached. A simple test on the magnitude of $\Delta\mathbf{X}$ will suffice: when it is less than a given threshold, then either the goal has been reached or it is so small that the end will never get there.

Equations for a General Two-Dimensional Articulated Linkage

To extend the solution technique to n links in 2D, we essentially use the same algorithm, but the expressions for (x, y) and the Jacobian terms $\frac{\partial x}{\partial \vartheta_i}$ become

$$x = \sum_{i=1}^{n} l_i \cos(\sum_{j=i}^{i} \vartheta_j),$$

$$y = \sum_{i=1}^{n} l_i \sin(\sum_{j=i}^{i} \vartheta_j),$$

$$\frac{\partial x}{\partial \vartheta_k} = -\sum_{i=k}^{n} l_i \sin(\sum_{j=1}^{i} \vartheta_j),$$

$$\frac{\partial y}{\partial \vartheta_k} = \sum_{i=k}^{n} l_i \cos(\sum_{j=1}^{i} \vartheta_j);$$

$$k = 1, \cdots, n.$$

11.4.3 Solving the IK Problem in Three Dimensions

All the steps in the solution procedure presented in Figure 11.8 are equally applicable to the 3D IK problem. Calculation of generalized inverse, iteration towards a goal for the end of the articulation and the termination criteria are all the same. Where the 2D and 3D IK problems differ is in the specification of the orientation of the links, the determination of the Jacobian and the equations that tell us how to calculate the location of the joints between links.

Although the change from 2D to 3D involves an increase of only one dimension, the complexity of the calculations increases to such an extent that we cannot normally determine the coefficients of the Jacobian by differentiation of analytic expressions. Instead, one extends the idea of iteration to write expressions which relate infinitesimal changes in linkage orientation to infinitesimal changes in the position of the end effector. By making small changes and iterating towards a solution, it becomes possible to calculate the Jacobian coefficients under the assumptions that the order of rotational transformation does not matter, and that in the coefficients, $\sin \vartheta \mapsto 1$ and $\cos \vartheta \mapsto 1$. We will not consider the details here; they can be found in [1]. Instead, we will move on to look at a very simple algorithm for IK in 3D.

A Simple IK Algorithm for Use in Real Time

The *Jacobian* IK algorithm is well defined analytically and can be rigorously analyzed and investigated, but in practice it can sometimes fail to converge. In fact, all iterative algorithms can sometimes fail to converge. However, when the Jacobian method fails to converge, it goes wrong very obviously. It does not just grind to a halt; it goes *way over the top!* Hence, for our VR work, we need a *simple* algorithm that can be applied in real time and is stable and robust.

A heuristic algorithm called the *cyclic coordinate descent* (CCD) method [3] works well in practice. It is simple to implement and quick enough to use in real time even though it doesn't always give an optimal solution. It consists of the five steps given in Figure 11.9.

The algorithm is best explained with reference to the example given in Figures 11.10 and 11.11. The steps are labeled (a) to (h), where (a) is the starting configuration. A line joining pivot point 2 to the target shows how the last link in the chain should be rotated about an axis perpendicular to the plane in which links 1 and 2 lie. After this rotation, the linkage will lie in the position shown in (b). Next, the linkage is rotated about point 1 so that the end effector lies on the line joining point 1 to the target point. This results in configuration (c), at which time a rotation about the base at point 0 is made to give configuration (d). This completes round 1 of the animation. Figure 11.11 configuration (e) is the same as Figure 11.11(d), and a pivot

Step 1:
Start with the last link in the chain (at the end effector).
Step 2:
Set a loop counter to zero. Set the max loop count to some upper limit.
Step 3:
Rotate this link so that it points towards the target. Choose
an axis of rotation that is perpendicular to the plane defined
by two of the linkages meeting at the point of rotation.
Do not violate any angle restrictions.
Step 4:
Move down the chain (away from the end effector) and go back to Step 2.
Step 5:
When the base is reached, determine whether the target has
been reached or the loop limit is exceeded. If so, exit;
if not, increment the loop counter and repeat from Step 2.

Figure 11.9. The cyclic coordinate descent iterative algorithm for manipulating an IK chain into position. It works equally well in two or three dimensions.

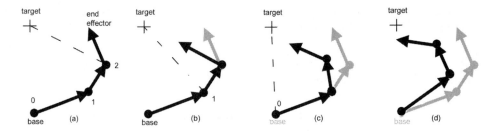

Figure 11.10. The cyclic coordinate descent algorithm. Iteration round 1 (loop count=1). The original position of the linkage is shown in light grey.

at point 2 is made to give configuration (f). Further pivots about point 1 result in configuration (g). Finally, a pivot about point 0 leads to configuration (h), by which time the end effector has reached the target and the iteration stops.

Of course, there are many other configurations that would satisfy the goal of the end effector touching the target, but the configuration achieved with the CCD algorithm is the one that *feels* natural to the human observer. It is also possible that in certain circumstances, the end effector may never reach

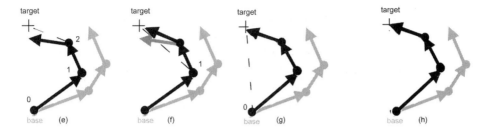

Figure 11.11. The cyclic coordinate descent algorithm. Iteration round 2 (loop count=2) completes the cycle because the end effector had reached the target.

the goal, but this can be detected by the loop count limit being exceeded. Typically, for real-time work, the loop count should never go beyond five. The Jacobian method has the advantage of responding to pushes and pulls on the end effector in a more natural way. The CCD method tends to move the link closest to the end effector. This is not surprising when you look at the algorithm, because if a rotation of the last linkage reaches the target, the algorithm can stop. When using these methods interactively, there are noticeable differences in response; the CCD behaves more like a loosely connected chain linkage whilst the Jacobian approach gives a configuration more like one would imagine from a flexible elastic rod. In practical work, the CCD approach is easy to implement and quite stable, and so is our preferred way of using IK in real-time interactive applications for VR.

11.5 Summary

This chapter looked at the theory underlying two main aspects of movement in VR, rigid motion and inverse kinematics (IK). Rigid motion is performed by using some form of interpolation or extrapolation, either linear, quadratic, or cubic in the form of a 3D parametric spline. Interpolation is also important in obtaining smooth rotation and changes of direction. In this case, the most popular approach is to use quaternion notation. That was explained together with details of how to convert back and forward from the more intuitive Euler angles and apply it in simulating hierarchical motion.

IK is fundamentally important in determining how robots, human avatars and articulated linkages move when external forces are applied at a few of their joints. We introduced two algorithms that are often used to express IK behavior in VR systems, especially when they must operate in real time.

There is a close connection between the use of IK and the motion capture and motion-tracking hardware discussed in Chapter 4. This is because IK can predict the motion of parts of a linkage that have not been explicitly tracked. It is especially useful in those cases where the tracked points relate to human actions. For example, some dance steps might be captured by tracking five markers on a human figure: two on the feet, two on the hands and the other on the torso. A well-thought-out IK model will allow the motion of ankles, knees, hips, wrists, elbows and shoulders to be simulated with a very satisfying degree of realism.

And so, with all the theoretical detail explained with regards to creating your own VR environment, it is now time to summarize all we have reviewed before going on to the practical aspects of VR implementation.

Bibliography

[1] R. S. Ferguson. *Practical Algorithms for 3D Computer Graphics.* Natick, MA: A K Peters, 2001.

[2] T. R. McCalla. *Introduction to Numerical Methods and FORTRAN Programming.* New York: John Wiley and Sons, 1967.

[3] C. Welman. *Inverse Kinematics and Geometric Constraints for Articulated Figure Manipulation.* MSc Thesis, Simon Fraser University, 1993.

Summing Up

We have now reached the halfway point of the book. Back in Chapter 3, we offered the opinion that VR has many useful applications with benefits for mankind, but it also has its drawbacks, can be misused and can be applied in situations where the original non-VR alternatives are still actually better. We suggested that you (the reader) should make up your own mind whether VR is appropriate or will work satisfactorily in an area that you may be considering it for.

In the intervening chapters, we have explored the theory and technology that delivers the most up-to-date virtual experiences. We have seen that a VR system must interact with as many senses as possible to deliver a virtual experience. To date, most research has been dedicated to delivering a realistic visual experience. This is because sight is regarded as the most meaningful human sense, and therefore is the sense that VR primarily tried to stimulate. In addition, the process of sight is well understood, and most modern technology makes use of this knowledge, e.g., cameras and visual display devices. Visualization technology is also very advanced because of its importance to the entertainment industry (e.g., computer gaming). As a result, we can deliver highly realistic graphics in real time to the users of the VR environment. Our ability to do this is also thanks to the huge revolution that graphics processors have undergone. Now we are able to program these directly, and as such we can achieve much more realism in our graphic displays.

However, when we try to stimulate even more senses in VR, e.g., sound and touch, the experience exceeds the sum of its parts. This is a very important point and is one reason why our guide covers such a broad spectrum of ideas, from computer graphics to haptics. Indeed, we try to hint that the term *rendering*, so familiar in graphics, is now being used to present a 3D audio scene as well as painting haptic surfaces that we feel rather than see. Sadly, this is difficult, and it may be a long time before it is possible to render acoustically or haptically as realistic a scene as one can do graphically.

Having described all the relevant technology involved in VR, we then tried to give you an understanding of the internals of delivering the VR experience; that is, the software processing that must be undertaken to deliver this experience. This is not a trivial topic, but if you want to build your own virtual environment then you should be able to grasp the basics of this, e.g., graphics rendering, the Z-buffer, lighting etc.

To complete our first section, we also introduced the concept of computer vision. For some time, the sciences of computer vision (with its applications in artificial intelligence and robotics) and computer graphics have been considered separate disciplines, even to the extent of using different notations for expressing exactly the same thing. But, for VR and the increasingly important *augmented reality* (AR), we need to bring these together and use both as if they are one. For example, we can utilize computer vision to create VR from pictures rather than build it graphically from mesh models. This is an ideal way of creating the surrounding environment for certain VR experiences, e.g., walkthroughs.

Now, at the end of Part I, we hope that you have read enough on the concepts of VR to help you answer whether *VR is appropriate and suitable to help enhance your own application.*

The Future for VR

VR still has a lot of undiscovered potential for use in the future, but we should not expect too much of it.

Ultimately, we would like to achieve a VR that is indistinguishable from reality. But is this possible? Virtual humans whom you can touch and who can touch you are always going to be figments of the imagination. Artificial intelligence may one day offer virtual actors that can act autonomously and intelligently, but they can't be made to take physical form using any type of projection system.

Is it possible to predict the future for VR? We think the answer has to be *no,* but we do believe that today's uses as summarized in Chapter 3 will only form a small portfolio of the areas in which VR can play a part. Provided we know our technological limitations and use the possibilities wisely, VR does offer a benefit to mankind, one that should equal the communications revolution of the Internet. In fact, VR is really just a part of it.

Current Research Focus

In the short term, a high proportion of research effort in VR appears focused along two pathways, haptics and distributed reality. The main technical challenges are in haptics. h e physical process of touch and feel is still not fully understood, and our attempts to build equipment to interface with some of its facets are bulky, look decidedly crude and frankly don't, in many cases, work very well either. When compared to the current graphics engines that deliver startlingly realistic graphics, the realism of haptic rendering *should* be a major focus for research.

Distributed reality is a popular topic in many VR research labs. For one thing, it facilitates collaboration, with its obvious advantages in taking a conference call to a new level of reality. The Internet is touching life in many ways that could not have been imagined, and using it to link VR environments which appear simultaneously across the earth is easily possible and has benefits yet to be quantified.

To say much more on either of these two topics now could make our book look dated very quickly. We just urge you to keep an eye on these two areas of research and development in VR, as they are likely to be the source of the next big thing, whatever that might be.

Putting VR into Practice

To be of any real use, VR must be put into practice. This is where the second part of the book and its CD resources come into play. At the center of any VR application is a computer. It controls all the interface hardware and provides a medium in which to represent the virtual. The computer is now probably the cheapest and most readily available component in any VR system. Even in its most basic form, the computer hardware is capable of sensing in real time whilst simultaneously delivering graphics, audio and even haptic control to almost any level of detail. What you can deliver in terms of reality just depends on how much you are prepared to pay for the human interface components and having the right software. We cannot (in a book) offer you a low-cost and easy way of constructing the interface hardware, but we can offer you some exciting and useful software. In many cases, the same software can be used to drive low-cost *bespoke*/experimental VR technology (as illustrated in Chapter 4), right up to the most lavishly endowed commercial systems. What we can offer you are some pieces of the software jigsaw for VR, which you can place together with some basic interface components to achieve your own VR experience. This is where Part II comes in! So please read on.

Part II   Practical Programs for VR

12

Tools, Libraries and Templates for VR

This part of the book provides practical solutions and small projects for a wide variety of VR requirements such as image and video processing, playing movies, interacting with external devices, stereoscopy and real-time 3D graphics.

The ideas, algorithms and theoretical concepts that we have been looking at in Part I are interesting, but to put them into practice, one must turn them into computer programs. It is only in the memory of the computer that the virtual world exists. It is through VR programs and their control over the computer's hardware that the virtual (internal) world is built, destroyed and made visible/touchable to us in the real world. This part of the book will attempt to give you useful and adaptable code covering most of the essential interfaces that allow us to experience the virtual world, which only exists as numbers in computer files and memory buffers.

To prepare the computer programs, it is necessary to use a wide variety of software tools. No one would think of trying to use a computer these days if it didn't have an operating system running on it to provide basic functions. In this chapter, we will start by looking at the development tools needed to write the VR application programs. We will also discuss the strategy used in the examples: its goal is to make the code as simple and adaptable as possible. And we will look at the special problems and constraints posed by VR applications, especially how hardware and software must collaborate to deliver the interfaces and speed of execution we require. VR applications, more often than not, are required to operate in a real-time environment. There is a definite symbiotic relationship between hardware platforms and application software. The ultimate expression of this relationship is in the custom integrated circuits of the real-time image and video processors or the embedded software applications running in customizable integrated circuits (ICs).

Today, the hardware/software dichotomy has become even more blurred with the introduction of the programmable graphics processing unit (the GPU), so that almost every PC sold today has the potential to deliver performance capable of driving a magnificently realistic VR environment. From the graphics point of view, there is little that can't be done on a domestic PC. It is only the physical interaction, such as the sense of touch and possibly the sense of depth perception, that are still too expensive to simulate exactly in consumer-level hardware.

In the remainder of this chapter, we will look at the application frameworks within which our sample VR programs will execute, but first we begin by discussing the hardware requirements for getting the best out of the example programs.

12.1   The Hardware Environment

In order to build and execute all the sample applications described in the next few chapters, you will need to have a basic processor. Because Windows is the dominant PC desktop environment, all our applications will be targeted for that platform. A few of the examples can be compiled on other platforms; when this is possible, we will indicate how. The most important hardware component for creating a VR system is the graphics adapter. These are discussed in Section 12.1.1.

For the video applications, you will also need a couple of sources. This can be as low-tech as hobbyist's USB (universal serial bus) webcams, digital camcorder with a FireWire (or IEEE 1394) interface or a TV tuner card. Unfortunately, if you try to use two digital camcorders connected via the IEEE 1394 bus then your computer is unlikely to recognize them both. Applications making use of DirectShow[1] (as we will be doing in Chapter 15) will only pick up one device. This is due to problems with the connection

---

[1]DirectShow, DirectInput, Direct3D and DirectDraw are all part of Microsoft's comprehensive API library for developing real-time interactive graphical and video applications called DirectX. Details of these components will be introduced and discussed in the next few chapters, but if you are unfamiliar with them, an overview of their scope can be found at [5].

management procedure (CMP) and bandwidth issues. A full discussion of this and other problems is given in [8]. The problem is not an easy one to resolve, and for the examples we will create that need two video sources, we will assume two USB-type webcams are available.

Where the applications have stereoscopic output, not only is the choice of graphics cards important, but the choice of whether to use active or passive glasses can affect the way your application has to be written. However, the good news is that if you program your 3D stereoscopic effects using the OpenGL[2] graphics API (application programmer interface), the internal operation of OpenGL offers a device-independent programming interface.

12.1.1   Graphics Adapters

Different display adapter vendors often provide functions that are highly optimized for some specific tasks, for example hardware occlusion and rendering with a high dynamic range. While this has advantages, it does not make applications very portable. Nevertheless, where a graphics card offers acceleration, say to assist with Phong shading calculations, it should be used. Another interesting feature of graphics cards that could be of great utility in a practical VR environment is support for multiple desktops or desktops that can span two or more monitors. This has already been discussed in Chapter 4, but it may have a bearing on the way in which the display software is written, since two outputs might be configured to represent one viewport or two independently addressable viewports. One thing the graphics card really should be able to support is *programmability*. This is discussed next.

12.1.2   GPUs

The graphics display adapter is no longer a passive interface between the processor and the viewer—it is a processor in its own right! Being a processor, it can be programmed, and this opens up a fantastic world of opportunities, allowing us to make the virtual world more realistic, more detailed and much more responsive. We will discuss this in Chapter 14. It's difficult to quantify the outstanding performance the GPU gives you, but, for example, NVIDIA's GeForce 6800 Ultra chipset can deliver a sustained 150 Gflops, which is about 2,000 times more than the fastest Pentium processor. It is no

---

[2]OpenGL is a standard API for real-time 3D rendering. It provides a consistent interface to programs across all operating systems and takes advantage of any hardware acceleration available.

wonder that there is much interest in using these processors for non-graphical applications, particularly in the field of mathematics and linear algebra. There are some interesting chapters on this subject in the GPU Gems book series, volumes 1 and 2 [3, 9].

The architectural design of all GPUs conforms to a model in which there are effectively two types of subprocessor, the vertex and fragment processors. The phenomenal processing speed is usually accomplished by having multiple copies of each of these types of subprocessor: 8 or 16 are possible. Programming the vertex processor and fragment processor is discussed in Chapter 14.

12.1.3   Human Interface Devices (HIDs)

For VR application programs, interaction with the user is vital. In addition to the display hardware, we need devices to acquire input from the user. Devices that make this possible come under the broad heading of *human interface devices* (HIDs). For most normal computer applications, the keyboard and/or mouse is usually sufficient. Some drawing applications can benefit from a stylus tablet, and of course computer games link up with a variety of interesting devices: joysticks, steering consoles etc. For gaming, many joysticks are not simply passive devices; they can kick back. In other words, they provide force feedback or *haptic* responses. In VR, these devices may be significant, and in application programs we should be able to use them. From a programmer's perspective, these devices can be tricky to access, especially if they have very different forms of electrical connection to the computer. Fortunately, most commercially available joysticks and game consoles have an accompanying operating system driver that makes the device behave like a standard component and which acts as an interface between application program and device. With a system driver installed as part of Windows, a program can access a broad range of devices in the same way and without the need to implement system driver code itself. This is easily done using the DirectInput API and is discussed in detail in Chapter 17.

With even a rudimentary knowledge of electronic circuits, it is not too difficult to design and build a device interface to either the PC's serial port or the new standard for serial computer communication: USB. Section 17.5 discusses how one can use a programmable interrupt controller (PIC) to interface almost any electrical signal to a PC via the USB.

12.2   Software Tools

At the very least, we need a high-level language compiler. Many now come with a friendly *front end* that makes it easier to manage large projects with the minimum of fuss. (The days of the traditional *makefile* are receding fast.) Unfortunately, this is only half the story. For most applications in VR, the majority of the code will involve using one of the system API libraries. In the case of PCs running Windows, this is the platform software developers kit (SDK). For real-time 3D graphics work, a specialist library is required. Graphics APIs are introduced and discussed in Chapter 13, where we will see that one of the most useful libraries, OpenGL, is wonderfully platform-independent, and it can often be quite a trivial job to port an application between Windows, Linux, Mac and SGI boxes. We shall also see in Chapter 17 that libraries of what is termed *middleware* can be invaluable in building VR applications.

12.2.1   Why C and C++

Almost without exception, VR and 3D graphics application programs are developed in either the C and/or C++ languages. There are many reasons for this: speed of program execution, vast library of legacy code, quality of developer tools—all of the operating systems on the target platforms were created using C and C++. So we decided to follow the majority and write our code in C and C++. We do this primarily because all the appropriate real-time 3D graphics and I/O libraries are used most comfortably from C or C++. Since OpenGL has the longest history as a 3D graphics library, it is based on the functional approach to programming, and therefore it meshes very easily with applications written in C. It does, of course, work just as well with application programs written in C++. The more recently specified API for Windows application programs requiring easy and fast access to graphics cards, multimedia and other I/O devices, known as DirectX, uses a programming interface that conforms to the component object model (COM). COM is intended to be usable from any language, but in reality most of the examples and application programs in the VR arena use C++. Sadly, programming with COM can be a nightmare to do correctly, and so Microsoft's Visual Studio development tools include a library of advanced C++ template classes to help. Called the ATL, it is an enormous help to application developers, who sometimes get *interface reference counting* object creation and destruction wrong. We talk more about this topic in Section 12.4.2.

12.2.2   Compilers

If one is developing for Windows, without question Microsoft's Visual C++ compiler is the most commonly used tool. In fact, the compiler itself is offered free of charge by Internet download from Microsoft [7]. The Application and Class "Wizards" in Visual C++ make for rapid and relatively pain-free development, and all our working programs will provide the appropriate Solution and Project files for Visual C++ Version 6 or Visual Studio.NET.[3] The GNU C++ compiler [4] has been successfully ported to Windows, but on Linux and UNIX systems it is the only real option. Apart from the OpenGL examples we will be developing, all the other programs are essentially Windows programs. The DirectInput- and DirectShow-based application programs require Microsoft's DirectX Version 9 SDK, which is not available for other operating systems. However, OpenGL has a long history of platform independence and so we will comment on how the OpenGL examples can be constructed using the GLUT (see Section 13.2) to make them completely platform-independent.

On Macintosh OS X platforms, the X-Code [1] suite is freely available and totally comprehensive. By using the GLUT, an OpenGL application can be ported to OS X with good success, as the middleware and scene-graph applications have been. Unfortunately, we have not yet seen any Macintosh systems with graphics chipsets that support Version 2 of OpenGL, so this rather restricts its use in VR applications where a high realism is required. (Alas, it is a sad fact that while there are several competing 3D chipset suppliers and many graphics card manufacturers who supply the PC market, the same can not be said for the Apple product line. Thus, Mac systems are not the first choice for cutting-edge graphics work. We will give one example of developing an OpenGL application for 3D mesh visualization on Mac OS X in Chapter 13).[4]

12.3   A Framework for the Applications

One way in which application programs can be developed is to start with a basic framework and extend it. In Part II, we will be looking at a broad range

---

[3]The readme.txt file on the accompanying CD will give more details of the latest versions of the compilers.

[4]Macs based on Intel hardware are now available, so perhaps OS X will enjoy the benefits of a vast selection of different graphics adapters after all.

of applications which are basically targeted at the *Wintel*[5] platform; therefore, we have a choice of two types of framework on which to base our code:

- *Win32.* Message-based API function library for Windows.
- *MFC.* Object-orientated class library encapsulating the Windows API.

These identify two styles of programming: *functional* or *object-orientated.* If you prefer to program exclusively in C then you have no choice but to use the Win32 application framework. Applications written in C++ can use either style. There really are no hard and fast rules as to which framework to use. MFC applications give you a fantastically functional application (toolbars, neat frames and robust document handling functions) with very little programmer effort, but they are bloated in code size. If you want to do something slightly unusual, MFC code can look quite inelegant as you break up the structure of the tightly integrated object-orientated classes.

Since all our VR applications need high-speed graphical display, for movie presentation or 3D visualization, we will need to use either, or both:

- OpenGL
- Direct3D

Both of these API libraries can fit neatly within either of our application frameworks. We will consider template codes that use these libraries for 3D rendering programs in Chapter 13. A template for using DirectInput to access devices such as joysticks will be given in Chapter 17 and one for using the extremely obscure DirectShow interface in Chapter 15.

In Appendices D and E, we provide two frameworks that are built on by the examples of later chapters, but do please note: we do not intend to give a comprehensive tutorial on writing application programs for Windows. For that you should consult one of the large volumes with something like "...Programming Windows..." in the title. A couple of good examples are [10] and [14].

From these frameworks, we are almost ready to proceed to build some working programs. However there is one small complication: COM. Microsoft has chosen to implement its strategy for directly accessing the computer's hardware, its HIDs and GPUs, through DirectX, which uses COM. Whilst we could avoid COM for 3D graphical work by using OpenGL, for any sort of video work (using DirectShow), we have no choice but to delve into COM.

---

[5]Wintel: a PC with an Intel x86 or compatible processor running the Microsoft Windows operating system.

12.4   The Component Object Model

If you haven't worked with COM before, it can be quite a shock to be confronted by some of its peculiarities. A lot of the difficulties are due to the sudden onslaught of a vast collection of new *jargon* and terminology. However, once you catch on to the key concepts and find, or are shown, some of the key ideas, it becomes much easier to navigate through the bewildering and vast reference documentation around COM. So, before jumping into the examples, we thought it important to highlight the concepts of COM programming and pull out a few key constructs and ideas so that you will see the structure. Hopefully, you will appreciate that it is not as overly complex as it might appear at first sight. Our discussion here will be limited to those features of COM needed for the parts of the DirectX system we intend to use.

12.4.1   What is COM?

COM grew out of the software technologies of *object linking and embedding (OLE)* and *ActiveX* [2]. It can be summed up in one sentence:

> COM is a specification for creating easily usable, distributable and upgradable software components (library routines by another name) and are independent of the high level language in which they were created, i.e., it is architecture-neutral.

The *jargon of COM* is well-thought-out but may be a little confusing at first. Whilst it is supposed to be language-independent, it is best explained in the context of the C++ language in which the majority of COM components are implemented anyway. Any COM functionality that one wishes to provide or use is delivered by a set of *objects*. These objects are instances of a *class* (called a CoClass) that describes them. The objects are not created, nor used, in the conventional C++ sense, i.e., by instantiating them directly with commands such as `new MyComObject1;`. Instead, each COM class provides a series of *interfaces* through which all external communication takes place. Every COM interface provides a number of *methods* (we would call them functions) which the application program can use to manipulate the COM object and use its functionality. Thus, the key terms when talking about COM are:

- *COM* object. An instance of a CoClass.

- *CoClass*. The class describing the COM object.

- *Interface.* A pointer by which to access all the methods and properties of a COM object.

- *Method.* The functions of a COM object's interface that do all the work of the COM object.

Earlier in this chapter, we hinted at the problems of versioning in DLLs. There are other problems, too, such as calling external functions from within a DLL, say, one written in C++, from BASIC language subroutines. COM seeks to overcome this by never upgrading any object or interface. If the authors of a COM library want to provide extra functionality, they must define a new interface but they must not remove the original one, either. This is one reason why one has interfaces such as *ICaptureGraphBuilder* and *ICaptureGraphBuilder2*. COM components can still be packaged in a DLL or as an executable. They can be located in different places or indeed moved if necessary. Application programs know where they are by looking in the Windows registry for their class identifier (CLSID) which has the form of a globally unique identifier (a GUID). Before COM objects can be used, therefore, they have to be installed into Windows. This effectively means writing the GUID into the registry and mapping their location on disk. In a typical Windows registry, there may be may hundreds of these COM classes.

> For example, the video codec Xvid (discussed in Section 5.5.2) can be used by any video player programs installed on the same computer because it is implemented as a COM component. On a typical system, the binary code for the codec is stored in the file C:\windows\system32\xvid.ax, but the key to making Xvid available is its registration through its GUID entry in the CLSID section of the registry (see Figure 12.1).

One of the biggest headaches for novice COM programmers looking for example code that they can reuse is the fact that there are often many ways to do exactly the same thing. If you read a book on COM and then turn to an example from the Direct3D SDK, it is likely that even getting hold of the first interface will break the rules so carefully explained by the COM purist in her book on COM. Sadly, that is one of the things those of us who only want to do 3D or video programming will just have to accept as the price to be paid for not having to write every line of an MPEG-4 DVD decoder ourselves. To try to find a path through the COM maze, all our example programs that need to use COM will stick to one or two standard ways to get the programs up and running.

Figure 12.1. A typical COM object has its unique class identifier stored in the Windows registry.

12.4.2   How COM Works

COM is a collection of components, each one of which can expose a number of interfaces. By convention, the names of these interfaces always begins with the letter I. They can be represented diagrammatically, as in Figure 12.2.

> At the top of Figure 12.2, we see an example of the interfaces that a COM object exposes. In this case, it is the COM object in DirectShow which controls the whole operation of a DVD/movie playing program. Like all COM objects, it exposes the interface IUnknown along with three other interfaces through which application programs control playback. At the bottom of Figure 12.2 is another example of an interface. This time, it is a DirectShow component (a filter) for a COM object that will display (render) a decoded DVD/movie to the screen. Unlike the GraphBuilder object (on the left), which is a major component in DirectShow, a renderer filter is often written for specific purposes. We shall be writing examples in Chapter 16. In the center of the figure is an illustration of the class hierarchy leading down to the CBaseRenderer class an application program can use to derive a class from in order to implement a rendering filter that fits the DirectShow specification and implements all the interfaces that a COM object requires. (It is the *CoClass* for the DirectShow renderer filter COM object.)

Like C++ classes, interfaces can inherit properties from their *parent interface*, and all COM objects must expose at least one interface, which is known as IUnknown. If IUnknown were the only interface that a COM object

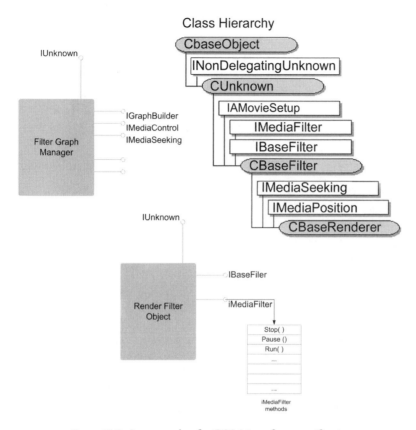

Figure 12.2. An example of a COM interface specification.

offered, it would not be particularly useful. However, IUnknown is essential because it is a COM object's way of giving access to the other interfaces that it supports, and it is the way memory usage by COM objects is managed. Instances of COM objects and their interfaces take up memory; therefore, when they are no longer required, that memory should be released (freed up). That sounds simple enough, but because the same COM object may be in use by the different interfaces that it offers, by different programs and even by different hosts, it is not easy to tell whether or not the memory should be released. For this reason, COM uses a reference-counting approach to count the number of programs using an interface. When a pointer to an interface is requested, $(+1)$ is added to its reference count, and (-1) is added when it is released. Once the reference count reaches zero, memory used by the interface may be freed, and when all the interfaces supported by a COM object are released, the memory used by the COM object itself may be freed. *But herein lies a major problem!* Get the reference counting wrong and your program could try to use an object that no longer exists, or it could go on creating new instances of the object. Not even exiting the program helps; COM objects exist outside specific applications. Now, you may be a good programmer who is very careful about freeing up the memory that you use, but it is not always easy to tell with the DirectX library how many COM objects are created when you call for a particular interface.

> You may be asking: what has this got to do with me? The answer is simple: you will be developing programs to make your virtual world a reality. Those programs will want to run in real time for a long time (days or weeks of continuous execution). You need to use COM if you want to use any component of DirectX (and you will), so if you get reference counting wrong, your program will either crash the computer or cause it to grind to a halt!

Luckily, a recent development called the C++ Active Template Library (ATL) has a special mechanism for automatically releasing COM objects when a program no longer needs them. We shall employ this strategy in many of our example programs. Of course, this adds yet another, and visually different, way of writing COM code to the list, confusing the programmer even more. To read more about COM, and there is a lot more to be read, the book by Templeman [13] shows how COM fits in with the MFC, and Grimes et al. [6] cover using COM in the context of the ATL.

Additional details of how to use the COM and a template for writing COM programs is given in Appendix F.

12.5   Summary

In this chapter, we have raced through a wide range of topics, topics that could easily occupy several volumes on their own. We have tried to keep the topics as brief as possible and referred to other sources where each of them is explored in much greater detail. We hope that you have gained an appreciation of the way the application programs in the rest of the book will be structured. It is not essential to have a comprehensive understanding of the detail here in order to appreciate the practical examples we are now turning to, but we hope this will make it easier to find your way through the code and thus focus on the more important parts, where the basic frameworks metamorphose into the specific applications.

You may have noticed that the Direct3D version mentioned here is Version 9. It may be that by the time you are reading, this it will be Version 10, 11, or even 97. The good news is because of COM, everything you read here should still work exactly as described. We have also noticed that as each Direct3D version changes, the differences become smaller and tend to focus on advanced features, such as the change from Shader Model 2 to Shader Model 3 or the changes that allow programs to take advantage of the new features of the GPU. Thus, even a new Direct3D version will almost certainly use initiation code very similar to that given here.

In the remainder of the book, we will get down to some specific examples of programs in 3D graphics, multimedia and custom interaction for VR.

Bibliography

[1] Apple Computer, Inc. *Learning Carbon.* Sebastopol, CA: O'Reilly & Associates, 2001.

[2] D. Chappell. *Understanding ActiveX and OLE.* Redmond, WA: Microsoft Press, 1996.

[3] R. Fernando (editor). *GPU Gems.* Boston MA: Addison-Wesley Professional, 2004.

[4] Free Software Foundation, Inc. "GCC, the GNU Compiler Collection". http://gcc.gnu.org/.

[5] K. Gray. *Microsoft DirectX 9 Programmable Graphics Pipeline.* Redmond, WA: Microsoft Press, 2003.

[6] R. Grimes et al. *Beginning ATL COM Programming.* Birmingham, UK: Wrox Press, 1998.

[7] Microsoft Corporation. "Visual C++ 2005 Express Edition". http://msdn.microsoft.com/vstudio/express/visualC/.

[8] M. Pesce. *Programming Microsoft DirectShow for Digital Video and Television.* Redmond, WA: Microsoft Press, 2003.

[9] M. Pharr (editor), *GPU Gems 2.* Boston, MA: Addison-Wesley Professional, 2005.

[10] B. E. Rector and J. M. Newcomer. *Win32 Programming.* Reading, MA: Addison-Wesley Professional, 1997.

[11] S. Stanfield. *Visual C++ 6 How-To*. Indianapolis, IN: Sams, 1997.

[12] J. Swanke. *Visual C++ MFC Programming by Example*. New York: McGraw Hill, 1998.

[13] J. Templeman. *Beginning MFC COM Programming*. Birmingham, UK: Wrox Press, 1997.

[14] C. Wright. *1001 Microsoft Visual C++ Programming Tips*. Roseville, CA: James Media Group, 2001.

13 Programming 3D Graphics in Real Time

Three-dimensional graphics have come a long way in the last 10 years or so. Computer games look fantastically realistic on even an off-the-shelf home PC, and there is no reason why an interactive virtual environment cannot do so, too. The interactivity achievable in any computer game is exactly what is required for a virtual environment even if it will be in a larger scale. Using the hardware acceleration of some of the graphics display devices mentioned in Chapter 12, this is easy to achieve. We will postpone examining the programmability of the display adapters until Chapter 14 and look now at how to use their *fixed functionality*, as it is called.

When working in C/C++, the two choices open to application programs for accessing the powerful 3D hardware are to use either the OpenGL or Direct3D[1] API libraries. To see how these are used, we shall look at an example program which renders a mesh model describing a virtual world. Our program will give its user control over his navigation around the model-world and the direction in which he is looking. We will assume that the mesh is made up from triangular polygons, but in trying not to get too bogged down by all the code needed for reading a mesh format, we shall assume that we have library functions to read the mesh from file and store it as lists of vertices, polygons and surface attributes in the computer's random access memory (RAM).[2]

---

[1] Direct3D is only available on Windows PCs.

[2] The mesh format that we will use is from our open source 3D animation and modeling software package called OpenFX. The software can be obtained from http://www.openfx.org. Source code to read OpenFX's mesh model .MFX files is included on the CD.

It always surprises us how richly detailed a 3D environment, modeled as a mesh, can appear with only a few tens or hundreds of polygons. The reason for this is simply that texture mapping (or image mapping) has been applied to the mesh surfaces. So our examples will show you how to use texture maps. If you need to remind yourself of the mathematical principles of texturing (also known as image mapping or shading), review Section 7.6. In the next chapter, we shall see how texturing and shading are taken to a whole new level of realism with the introduction of programmable GPUs.

13.1 Visualizing a Virtual World Using OpenGL

OpenGL had a long and successful track record before appearing on PCs for the first time with the release of Windows NT Version 3.51 in the mid 1990s. OpenGL was the brainchild of Silicon Graphics, Inc.(SGI), who were known as *the* real-time 3D graphics specialists with applications and their special workstation hardware. Compared to PCs, the SGI systems were just unaffordable. Today they are a rarity, although still around in some research labs and specialist applications. However, the same cannot be said of OpenGL; it goes from strength to strength. The massive leap in processing power, especially in graphics processing power in PCs, means that OpenGL can deliver almost anything that is asked of it. Many games use it as the basis for their rendering engines, and it is ideal for the visualization aspects of VR. It has, in our opinion, one major advantage: it is *stable*. Only minor extensions have been added, in terms of its API, and they all fit in very well with what was already there. The alternate 3D software technology, Direct3D, has required nine versions to match OpenGL. For a short time, Direct3D seemed to be ahead because of its advanced shader model, but now with the addition of just a few functions in Version 2, OpenGL is its equal. *And, we believe, it is much easier to work with.*

13.1.1 The Architecture of OpenGL

It is worth pausing for a moment or two to think about one of the key aspects of OpenGL (and Direct3D too). They are designed for *real-time* 3D graphics. VR application developers must keep this fact in the forefront of their minds at all times. One must always be thinking about how long it is going to take to perform some particular task. If your virtual world is described by more than 1 million polygons, are you going to be able to render it at 100 frames per second (fps) or even at 25 fps? There is a related question: how does one *do* real-time rendering under a multitasking operating system? This is not a trivial question, so we shall return to it in Section 13.1.6.

To find an answer to these questions and successfully craft our own programs, it helps if one has an understanding of how OpenGL takes advantage of 3D hardware acceleration and what components that hardware acceleration actually consists of. Then we can organize our program's data and its control logic to optimize its performance in real time. Look back at Section 5.1 to review how VR worlds are stored numerically. At a macroscopic level, you can think of OpenGL simply as a friendly and helpful interface between your VR application program and the display hardware.

Since the hardware is designed to accelerate 3D graphics, the elements which are going to be involved are:

- *Bitmaps*. These are blocks of memory storing images, movie frames or textures, organized as an uncompressed 2D array of pixels. It is normally assumed that each pixel is represented by 3 or 4 bytes, the dimensions of the image may sometimes be a power of two and/or the pitch[3] may be word- or quad-word-aligned. (This makes copying images faster and may be needed for some forms of texture interpolation, as in image mapping.)

- *Vertices*. Vertex data is one of the main inputs. It is essentially a floating-point vector with two, three or four elements. Whilst image data may only need to be loaded to the display memory infrequently, vertex data may need to change every time the frame is rendered. Therefore, moving vertex data from host computer to display adapter is one of the main sources of delay in real-time 3D. In OpenGL, the *vertex* is the only structural data element; polygons are implied from the vertex data stream.

- *Polygons*. Polygons are implied from the vertex data. They are rasterized[4] internally and broken up into *fragments*. Fragments are the basic visible unit; they are subject to lighting and shading models and image

---

[3]The pitch is the number of bytes of memory occupied by a row in the image. It may or may not be the same as the number of pixels in a row of the image.

[4]Rasterization is the process of finding what screen pixels are filled with a particular polygon after all necessary transformations are applied to its vertices and it has been clipped.

mapping. The final color of the visible fragments is what we see in the output frame buffer. For most situations, a fragment is what is visible in a pixel of the final rendered image in the frame buffer. *Essentially, one fragment equals one screen pixel.*

Like any microcomputer, a display adapter has three main components (Figure 13.1.1):

- *I/O.* The adapter receives its input via one of the system buses. This could be the AGP or the newer PCI Express interface[5]. From the program's viewpoint, the data transfers are blocks of byte memory (images and textures) or floating-point vectors (vertices or blocks of vertices), plus a few other state variables. The outputs are from the adapter's video memory. These are transferred to the display device through either the DVI interface or an A-to-D circuit to VGA. The most significant bottleneck in rendering big VR environments is often on the *input* side where adapter meets PC.

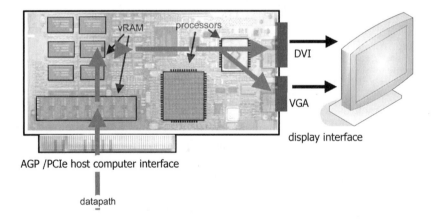

Figure 13.1. The main components in a graphics adapter are the video memory, parallel and pipeline vertex and fragment processors and video memory for textures and frame buffers.

---

[5]The PCI Express allows from much faster read-back from the GPU to CPU and main memory.

- *RAM.* More precisely video or vRAM. This stores the results of the rendering process in a frame buffer. Usually two or more buffers are available. Buffers do not just store color pixel data. The Z depth buffer is a good example of an essential buffer with its role in the hidden surface elimination algorithm. Others such as the stencil buffer have high utility when rendering shadows and lighting effects. vRAM may also be used for storing vertex vectors and particularly image/texture maps. The importance of texture map storage cannot be overemphasized. Realistic graphics and many other innovative ideas, such as doing high-speed mathematics, rely on big texture arrays.

- *Processor.* Before the processor inherited greater significance by becoming externally programmable (as discussed in Chapter 14), its primary function was to carry out floating-point vector arithmetic on the incoming vertex data ,e.g., multiplying the vertex coordinates with one or more matrices. Since vertex calculations and also fragment calculations are independent of other vertices/fragments, it is possible to build several processors into one chip and achieve a degree of parallelism. When one recognizes that there is a sequential element in the rendering algorithm—vertex data transformation, clipping, rasterization, lighting etc.—it is also possible to endow the adapter chipset with a pipeline structure and reap the benefits that pipelining has for computer processor architecture. (But without many of the drawbacks such as conditional branching and data hazards.) Hence the term often used to describe the process of rendering in real time, *passing through the graphics pipeline.*

So, OpenGL drives the graphics hardware. It does this by acting as a state machine. An application program uses functions from the OpenGL library to set the machine into an appropriate state. Then it uses other OpenGL library functions to feed vertex and bitmap data into the graphics pipeline. It can pause, send data to set the state machine into some other state (e.g., change the color of the vertex or draw lines rather than polygons) and then continue sending data. Application programs have a very richly featured library of functions with which to program the OpenGL state machine; they are very well documented in [6]. The official guide [6] does not cover programming with Windows very comprehensively; when using OpenGL with Windows, several sources of tutorial material are available in books such as [7].

Figure 13.2 illustrates how the two main data streams, polygon geometry and pixel image data, pass through the graphics hardware and interact with

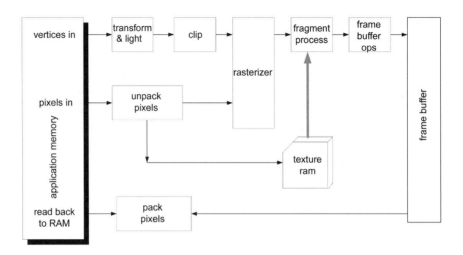

Figure 13.2. The OpenGL processing pipeline: geometric 3D data in the form of vertex data and image data in the form of pixels pass through a series of steps until they appear as the output rendered image in the frame buffer at the end of the pipeline.

each other. It shows that the output from the frame buffers can be fed back to the host computer's main memory if desired. There are two key places where the 3D geometric data and pixel image interact: in the rasterizer they are mixed together so that you could, for example, play a movie in an inset area of the window or make a Windows window appear to float around with the geometric elements. Pixel data can also be stored in the hardware's video RAM and called on by the fragment processor to texture the polygons. In OpenGL Version 1.x (and what is called OpenGL's fixed functionality), shading and lighting are done on a per-vertex basis using the Gouraud model. Texture coordinates and transformations are also calculated at this stage for each primitive. A *primitive* is OpenGL's term for the group of vertices that, taken together, represent a polygon in the 3D model. It could be three, four or a larger group of vertices forming a triangle strip or triangle fan (as detailed in Section 5.1). In fixed-functionality mode, the fragment processing step is fairly basic and relates mainly to the application of a texture. We shall see in Chapter 14 that with OpenGL Version 2, the fragment processing step becomes much more powerful, allowing lighting calculation to be done on a per-fragment basis. This makes Phong shading a possibility in real time. The full details of the processing which the geometry undergoes before rasteriza-

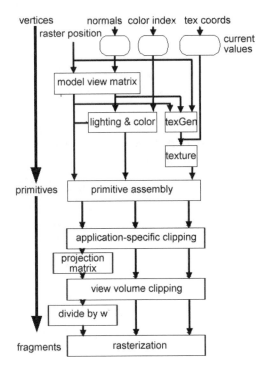

Figure 13.3. Geometry processing by OpenGL before the rasterization stage.

tion is given in Figure 13.3. It is worth also pointing out that the frame buffer is not as simple as it appears. In fact, it contains several buffers (Z depth, stencil etc.), and various logical operations and bit blits[6] can be performed within and among them.

The two blocks transform & light and clip in Figure 13.2 contain quite a bit of detail; this is expanded on in Figure 13.3. As well as simply passing vertex coordinates to the rendering hardware, an application program has to send color information, texture coordinates and vertex normals so that the lighting and texturing can be carried out. OpenGL's implementation as a state machine says that there are current values for color, surface normal and texture coordinate. These are only changed if the

---

[6]Bit blit is the term given to the operation of copying blocks of pixels from one part of the video memory to another.

application sends new data. Since the shape of the surface polygons is implied by a given number of vertices, these polygons must be assembled. If a polygon has more that three vertices, it must be broken up into primitives. This is primitive assembly. Once the primitives have been assembled, the coordinates of their vertices are transformed to the camera's coordinate system (*the viewpoint is at* $(0, 0, 0)$*; a right-handed coordinate system is used with* $+x$ *to the right,* $+y$ *up and* $-z$ *pointing away from the viewer in the direction the camera is looking*) and then clipped to the viewing volume before being rasterized.

To summarize the key point of this discussion: as application developers, we should think of vertices first, particularly how we order their presentation to the rendering pipeline. Then we should consider how we use the numerous frame buffers and how to load and efficiently use and reuse images and textures without reloading. *The importance of using texture memory efficiently cannot be overemphasized.*

In Figure 13.4, we see the relationship between the software components that make up an OpenGL system. The Windows DLL (and its stub library file) provides an interface between the application program and the adapter's

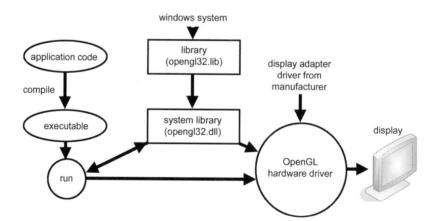

Figure 13.4. A program source uses a stub library to acquire access to the OpenGL API functions. The executable calls the OpenGL functions in the system DLL, and this in turn uses the drivers provided by the hardware manufacturer to implement the OpenGL functionality.

device driver, which actually implements the OpenGL functionality. With this component architecture, taking advantage of any new functionality provided by graphics card vendors is just a matter of replacing the driver with another version.

Now that we have a fair understanding of how OpenGL works, we can get down to some detail of how to use it to *visualize a virtual environment.*

13.1.2   Rendering OpenGL in a Windows Window

The template code in Appendix D shows that to render within a Windows window, an application program must provide a handler for the WM_PAINT message, obtain a drawing device context for the client area and use it to render the lines, bitmaps or whatever. In platform-independent OpenGL, there is no such thing as a device context. To draw something, one simply issues calls to the required OpenGL functions, as in:

```
glClear(GL_BUFFER);        // clear the screen
glBegin(GL_TRIANGLES);     // Draw a single triangle - just as an example,
glVertex3f(0.0.0,-1.0);    // it may not be visible until we describe
glVertex3f(0.0.0,-1.0);    // the view point correctly.
glVertex3f(0.0.0,-1.0);    // 3 calls to glVertex... make up a triangle
glEnd();                   // match the call to glBegin()
glFlush();                 // flush the OpenGL buffers
glFinish();                // finish drawing and make it visible
```

Drawing directly to the screen like this can give an unsatisfactory result. Each time we draw, we start by clearing the screen and then render one polygon after another until the picture is complete. This will give the screen the appearance of flickering, and you will see the polygons appearing one by one. This is not what we want; a 3D solid object does not materialize bit by bit. To solve the problem, we use the technique of double buffering. Two copies of the frame buffer exist; one is visible (appears on the monitor) and one is hidden. When we render with OpenGL, we render into the hidden buffer, and once the drawing is complete, we swap them over and start drawing again in the hidden buffer. The swap can take place imperceptibly to the viewer because it is just a matter of swapping a pointer during the display's vertical blanking interval. All graphics hardware is capable of this.

Under Windows, there is a very small subset of platform-specific OpenGL functions that have names beginning wgl. These allow us to make an implicit connection between the Windows window handle hWnd and OpenGL's drawing functions. All one has to do to get OpenGL working on Windows is run a small piece of initialization code during window creation (by handling the WM_CREATE message) and then act on the WM_PAINT message by rendering the 3D scene with calls to the appropriate OpenGL drawing functions. Listing 13.1 shows the key steps that enable us to use OpenGL to render into a

```
// the window message handler function is passed a
// handle to the window as its first argument hWnd
 HGLRC hRC; HDC hDC;              // local variables - HGLRC is a display
 ....                            // context for OpenGL windows
 case WM_CREATE:
   hDC = GetDC(hWnd);
   if(!bSetupScreenFormat(hDC))  // specify the 3D window's properties
     PostQuitMessage(0);         // NO OpenGL available !
   hRC = wglCreateContext( hDC );
   wglMakeCurrent( hDC, hRC );
   glClearColor(0.0,0.0,0.0,1.0);
   glClear(GL_COLOR_BUFFER_BIT | GL_DEPTH_BUFFER_BIT);
   break;
 case WM_PAINT:
   hDC = BeginPaint(hWnd, &ps);
   glBegin(GL_TRIANGLES);  // draw a single triangle - just as an example
   glVertex3f(0.0.0,-1.0); // it may not be visible until we describe
   glVertex3f(0.0.0,-1.0); // the view point correctly
   glVertex3f(0.0.0,-1.0); // 3 call to glVertex... make up a trianble
   glEnd();
   glFlush();     // flush the OpenGL buffers
   glFinish();    // OpenGL function to finish drawing and make it visible
   // these are the Windows specific functions
   hDC = wglGetCurrentDC();
   SwapBuffers(hDC);
   EndPaint(hWnd, &ps);
   break;
 case WM_DESTROY:
   hRC = wglGetCurrentContext(); // get and release the OpenGL
   hDC = wglGetCurrentDC();      // system
   wglMakeCurrent(NULL, NULL);
   if (hRC != NULL)wglDeleteContext(hRC);
   if (hDC != NULL)ReleaseDC(hWnd, hDC);
   break;
 ....
```

Listing 13.1. To use OpenGL in Windows, it is necessary to handle four messages: one to initialize the OpenGL drawing system WM_CREATE, one to perform all the necessary rendering WM_PAINT, one to release OpenGL WM_DESTROY and WM_RESIZE (not shown) to tell OpenGL to change its viewport and aspect ratio.

```
BOOL bSetupScreenFormat(HDC hDC){
  PIXELFORMATDESCRIPTOR pfd = {
    sizeof(PIXELFORMATDESCRIPTOR),     // size of this pfd
      1,                               // version number
    PFD_DRAW_TO_WINDOW |               // support window
    PFD_SUPPORT_OPENGL |               // support OpenGL
    PFD_DOUBLEBUFFER,                  // double buffered
    PFD_TYPE_RGBA,                     // RGBA type
      24,                              // 24-bit color depth
      0, 0, 0, 0, 0, 0,                // color bits ignored
      0,                               // no alpha buffer
      0,                               // shift bit ignored
      0,                               // no accumulation buffer
      0, 0, 0, 0,                      // accum bits ignored
      32,                              // 32-bit z-buffer
      0,                               // no stencil buffer
      0,                               // no auxiliary buffer
    PFD_MAIN_PLANE,                    // main layer
      0,                               // reserved
      0, 0, 0                          // layer masks ignored
      };
  int pixelformat;
  if ( (pixelformat = ChoosePixelFormat(hDC, &pfd)) == 0 ) {
    return FALSE;
  }
  if (SetPixelFormat(hDC, pixelformat, &pfd) == FALSE) {
    return FALSE;
  }
  return TRUE;
}
```

Listing 13.2. A pixel format tells OpenGL whether it is to render using double buffering, 24-bit color and many other attributes.

window. Listing 13.2 describes the key function to give the window certain attributes, such as double buffering and stereoscopic display.

13.1.3 Designing the Software

There are a vast number of ways one could write our visualization program. Design questions might include how the user navigates through the scene, whether it is a windowed or full screen program etc. Our desire is to try to keep the code as focused as possible on the 3D programming issues. Therefore, we won't include a fancy toolbar or a large collection of user dialog boxes—you can add these if desired. We will just give our program a simple menu and a few hot keys for frequently used commands. Neither will we present every aspect of the program here in print. All the programs are included on the CD; you can explore that using the book as a guide. In the text, we will stress the structure of the program and its most important features.

The options the program will offer its user are:

- Render the object in windowed or full screen mode. You may remember that to render a full screen using OpenGL, you first make a window that has the same dimensions as the desktop. A *child* window will be used to receive the OpenGL-rendered output, and when it is created, the decision can be made as to whether it will be full-screen or windowed.

- Change the color of the background. This is simply done by changing the background color during initialization.

- Slowly rotate the model. A Windows timer going off every 10 ms or so will change the angular orientation of the model.

- Allow the user to rotate the model in front of her by using the mouse.

- If the mesh makes use of image maps, display them. Try to use the map in a manner close to the way it would appear when rendered by the CAD or animation package which originated the mesh. (The fixed-functionality pipeline limits the way an image may be repeated; mosaic tiling is not normally available, for example.)

- Offer the best possible rendering acceleration by using display lists and where possible *polygon stripping* (described in Section 13.1.5).

In the following sections, we will look at the strategy used to implement the application by concentrating on the most important statements in the code.

13.1.4   Setting Up

A main window, with menu bar, is created in the standard manner; that is, in the manner described in Listings D.2 and D.3. Global variables (see Listing 13.3) are declared and initialized. The application-defined structures for the vertices, polygons and image maps which describe the mesh models (see Section 5.1) are shown in Listing 13.4. Code for the functions that parse the file for vertex, polygon and image map data and build lists of these key data in global memory buffers will not be discussed further in the text, because

```
VERTEX   *MainVp=NULL;              // Pointers to RAM buffers
FACE     *MainFp=NULL;              // for the 3D geometry description
MAP      *MapsFp=NULL;              // of the mesh models
long     Nvert=0,Nface=0,           // Numbers of vertices and polygons
         NvertGlue=0,Nmap=0;        // Number of maps and vwitices with
                                    // mapping coordinates
int      listID=0;                  // OpenGL display list
// User commands to  change the appearance of the models and rendering
BOOL     Mapped=FALSE,smooth_shading=TRUE,
         ffastmapping=TRUE,accelerated=TRUE,LightBackground=FALSE;
// User interactively changes these to alter their viewpoint
GLfloat ObjScale=0.5,view_scale=1.0;
GLfloat up_angle=0.0,round_angle=0.0;
GLfloat xpos_offset=0.0,ypos_offset=0.0;
```

Listing 13.3. The application's most significant global variables for the mesh data and viewpoint.

```
typedef GLfloat vector[3];       // application defined
typedef long    point[3];        // datatypes for 3D vectors
typedef GLfloat normal[3];       // of different kind

typedef struct tagVERTEX {       // the Vertex data structure
 point p;                        // 3D location
 GLfloat x,y;                    // mapping coordinates, where appropriate
} VERTEX;

typedef struct tagFACE {         // the polygon data structure
 long V[3],A[3],next;            // attached to vertices, faces adjacent
                                 // Next is used by the triangle stripping
                                 // optimzation algorithm
 unsigned char color[3],matcol[3], // surface attributes
               texture,map,axis;
} FACE;

typedef struct tagMAP {          // the image map data structure
       vector n,p,x,y;           // for maps that do no use mapping
                                 // coordinates
       char filename[128],       // the filenames with the images for the
           frefl[128],fbump[128];// 3 types of surface for each map
       ...                       // other minor elements
} MAP;
```

Listing 13.4. Data structures for the 3D entities to be rendered. The functions which read the OpenFX and 3ds Max 3DS data files convert the information into arrays of vertices, triangular polygons and image maps in this format.

it would take up too much space. If you examine the project on the CD, you will see that the code can handle not only OpenFX mesh models but also the commercial and commonly used 3DS mesh format. This brings up the interesting point of how different applications handle 3D data and how compatible they are with the OpenGL model.

The OpenFX and 3DS file formats are typical of most 3D computer animation application programs: they use planar polygons, with three or four

```
case WM_TIMER:                                       // handle timer messages
  if(IsWindow(hWndDisplay)){                          // if the Display window
                                                      // is visible
    round_angle += 0.5;                               // increment rotation
                                                      // by 1/2 degree
    if(round_angle > 180.0)round_angle -= 360;
    InvalidateRect(hWndDisplay, NULL, FALSE);         // update the display
  }
  break;
  .....
case WM_COMMAND:                                      // handle
  if (wParam == ID_FILE_QUIT)PostQuitMessage(0);      // tell application to exit
  else if(wParam == ID_CONTROL_STARTDISPLAY){         // open the display window
    if(hWndDisplay == NULL){
      hWndDisplay=DisplayEntry(hInstance,hWnd);       // create the OpenGL window
      if(IsWindow(hWndDisplay))Make3dListS();         // Make the OpenGL
                                                      // display list.
    }
    else{
      DisplayClose(hInstance,hWndDisplay);            // close the display window
      hWndDisplay=NULL;
    }
  }
  else if (wParam == ID_LOAD_MODEL){                  // load a mesh model
    .. // Use a file open dialog to select the model filename
    int type=WhatIsFileType(filename);                // MFX or 3DS file type
    LoadModel(filename,type);                         // read the mesh data into
    if(IsWindow(hWndDisplay)){                         // internal RAM structures
      Make3dListS();                                  // make teh OpenGL
                                                      // display list.
      InvalidateRect(hWndDisplay, NULL, TRUE);        // update the display
    }
  }
  ... // other menu commands
  break;
case ...    // other windows messages
```

Listing 13.5. Message processing of commands from the main window's menu.

vertices at most. Many 3D CAD software packages, however, use some form of parametric surface description, NURBS or other type of surface patch. OpenGL provides a few NURBS rendering functions, but these are part of the auxiliary library and are unlikely to be implemented in the GPU hardware, so they tend to respond very slowly.

Even within applications which concentrate on simple triangular or quadrilateral polygons, there can be quite a bit of difference in their data structures. OpenFX, for example, assigns attributes on a per-polygon basis, whereas 3DS uses a *materials* approach, only assigning a material identifier to each polygon, and thus whole objects or groups of polygons have the same attributes. OpenGL assumes that vertices carry the attributes of color etc., and so in the case of rendering polygons from an OpenFX mesh, many calls to the functions that change the color state variable are required. Another difference between OpenFX and 3DS concerns mapping. Fortunately, both packages support mapping coordinates, and this is compatible with the way OpenGL handles image mapping.

In the basic fixed-functionality pipeline of OpenGL, the lighting and shading models are somewhat limited when compared to the near-photorealistic rendering capability of packages such as 3ds Max, and some compromises may have to be accepted.

Often, hardware acceleration poses some constraints on the way in which the data is delivered to the hardware. With OpenGL, it is much more efficient to render triangle strips than individual triangles. However, often triangle meshes, especially those that arise by scanning physical models, are not recorded in any sort of order, and so it can be well worth preprocessing or reprocessing the mesh to try to build it up from as many triangle strips as possible.

Most 3D modeling packages use a right-handed coordinate system in which the z-axis is vertical. OpenGL's right-handed coordinate system has its y-axis vertical, and the default direction of view is from $(0, 0, 0)$ looking along the $-z$-direction. With any mesh model, we must be careful to remember this and accommodate the scales involved, too. In our 3D application examples, we will always scale and change the data units so that the model is centered at $(0, 0, 0)$, the vertex coordinates conform to OpenGL's axes and the largest dimension falls in the range $[-1, +1]$.

The main window's message-handling function processes a number messages. Listing 13.5 singles out the WM_TIMER message, which is used to increment the angular rotation of the model every few milliseconds and WM_COMMAND to handle commands from the user menu.

In response to the menu command to display the rendered model, a child window is created, as shown in Listing 13.6. At this stage, the user can choose to make a window which has a border and can be moved around the desktop or one that has no border and fills the screen. Messages for the child window are processed by its own message handling function, the key steps of which are featured in Listing 13.7. The handler for the WM_PAINT message is where most of the work for the application is done.

To allow the user to appear to *pick up and look the object from all sides*, mouse messages sent to the window's handler function are used to change the direction of the orientation variables. Listing 13.8 covers the main points of how this is accomplished: on a press of the left mouse button, its input

```
HWND GLWindowPr(HINSTANCE hInstance, HWND hWndParent){
 HWND hwnd;
 if(!FullScreen){                                      // global variable
   hwnd = CreateWindow(                                // a basic window
     CLASSNAME1,CLASSNAME1,                            // macro with text
                                                       // class name
     WS_POPUP | WS_CAPTION | WS_SYSMENU | WS_THICKFRAME,// standard decoration
     64,64, 64+512, 64+512,                            // screen position
                                                       // and size
     hWndParent,NULL,hInstance,NULL);
 }
 else{
   hwnd=CreateWindowEx(                                // a full screen window
     0,
     "FSCREEN",                                        // The class name for
     "FSCREEN",                                        // the window
     WS_POPUP,                                         // NO border menu etc.
     0,
     0,
     GetSystemMetrics(SM_CXSCREEN),                    // the size of
                                                       // the display
     GetSystemMetrics(SM_CYSCREEN),                    // screen
     hWndParent, NULL, hInstance, NULL );
 }
 ShowWindow(hwnd,SW_SHOW);
 InvalidateRect(hwnd,NULL,FALSE);                      // make sure contents
                                                       // are drawn

 return hwnd;
}
```

Listing 13.6. Creating the OpenGL display window in either full-screen or windowed mode.

```
LRESULT WINAPI GLProcessWndProc( HWND hWnd, UINT msg,
WPARAM wParam, LPARAM lParam ){
// these variables are used by the mouse button and move messages to
// allow the user to change the viewpoint from which they view the model
    static int mx,my,ms;
    static RECT rc;
    static BOOL bCapturedl=FALSE;
    static BOOL bCapturedr=FALSE;
    switch( msg )      {
      case WM_CREATE:{
            // OpenGL creation code with additional application specifics
            // in a separate function
            initializeGL(HWND hWnd);
        }
        break;
// if the uses changes the size of the window (which can't happen in full
// screen mode) the OpenGL viewport must be adjusted to still fill the
// window's client area
      case WM_SIZE:
        if(!FullScreen){
          GetClientRect(hWnd, &rc);
          glViewport(0, 0, rc.right,rc.bottom);
          InvalidateRect(hWnd,NULL,FALSE);
        }
        break;
      case WM_DESTROY:{
        // standard OpenGL window closure
        return 0;
      }
      case WM_PAINT:
        DrawScene(hWnd);                              // draw the scene !!!!!!!!!!!
        break;
..
// mouse handling messages
..        // watch for the "ESC" key pressed in full
      case WM_CHAR:
        // screen mode it is the only way to close the window
        if (wParam == VK_ESCAPE)
            PostMessage((HWND)GetWindowLong(hWnd,GWL_USERDATA),
                        WM_COMMAND,ID_CONTROL_CLOSE_WINDOW, 0);
        // otherwise pass on other keys to main window
        else PostMessage((HWND)GetWindowLong(hWnd,GWL_USERDATA),
                        WM_CHAR,wParam,lParam);
        break;
      default:
        break;
    }
    return DefWindowProc( hWnd, msg, wParam, lParam );
}
```

Listing 13.7. The OpenGL window's message handling function. The WM_CREATE message handles the standard stuff from Listing 13.1 and calls the function initializeGL(..) to handle application-specific startup code.

```
// mouse messages - local STATIC variables mx,my,bCaptuered etc.
// declared earlier
..
case WM_LBUTTONDOWN:
   if(bCapturedr)break;
   SetCapture(hWnd);            // all mouse messages are sent to this window
   mx=(int)LOWORD(lParam);      //
   my=(int)HIWORD(lParam);      // Get position of mouse in client area
                                // and dimensions
   GetWindowRect(hWnd,&rc);     //
   bCapturedl=TRUE;             // flag to indicated mouse is in capture mode
   break;
case WM_RBUTTONDOWN:            // similar to left button down message handling
   ...                          // code to handle right button
   break;
case WM_MOUSEMOVE:{
   double dx,dy;
   if(bCapturedl){              // the left must button is DOWN so change view
     mx=(short)LOWORD(lParam)-mx; my=(short)HIWORD(lParam)-my;
     dx=(double)mx/(double)(rc.right-rc.left);
     dy=(double)my/(double)(rc.bottom-rc.top);
     if(GetAsyncKeyState(VK_CONTROL) & 0x8000){ // Zoom in or out if the
       xpos_offset += (GLfloat)mx * 0.01;       // Control key held down
       ypos_offset -= (GLfloat)my * 0.01;
     }
     else{                      // change the direction of view
       round_angle += dx * 360;
       up_angle += dy * 180;
     }
     InvalidateRect(hWnd,NULL,FALSE);  // redraw
     mx=(short)LOWORD(lParam);  my=(short)HIWORD(lParam);
   }
   else if(bCapturedr){
   .. // do much the same thing for the right mouse button
   }
   break;
case WM_LBUTTONUP:
   if(bCapturedr)break;         // don't do anything if not captured
   bCapturedl=FALSE;
   ReleaseCapture();            // release the capture of mouse messages
   break;
case WM_RBUTTONUP:              // similar to left button down message handling
   .. // code to handle up on right button
   break;
..// other messages
```

Listing 13.8. Handling mouse messages allows users to change their viewpoint of the mesh model

```
GLvoid initializeGL(HWND hWnd){ // perform application specific GL startup
 makeCheckeredImageMap();        // make a small image map for use in case of
                                 // mapping error
 initializeRender(hWnd);         // set up the lighting and general
                                 // viewing conditions
 return;
}


#define checkImageWidth  64
#define checkImageHeight 64

#define CHECK_TEXTURE    1      // OpenGL texture identifier

GLubyte checkImage[checkImageWidth][checkImageHeight][3];

void makeCheckeredImageMap(void){
 // This function uses the code from the OpenGL reference
 // manual to make a small image map 64 x 64 with a checkerboard
 // pattern. The map will be used as a drop in replacement
 // for any maps that the mesh model uses but that can't be created
 // beause, for example, the image file is missing.
 ...
 // The KEY OpenGL functions that are used to make a texture map follow
 glBindTexture(GL_TEXTURE_2D,CHECK_TEXTURE);  // define a 2D (surface) map
 // this loads the texture into the hardware's texture RAM
 glTexImage2D(GL_TEXTURE_2D, 0, 3, checkImageWidth,
             checkImageHeight, 0, GL_RGB, GL_UNSIGNED_BYTE,
           &checkImage[0][0][0]);
 // very many parameters can be applied to image maps e.g.
 glTexParameterf(GL_TEXTURE_2D, GL_TEXTURE_WRAP_S, GL_REPEAT);
 ..  // consult the reference documentation for specific detail
 glHint(GL_PERSPECTIVE_CORRECTION_HINT,GL_NICEST);
 ... //
}
```

Listing 13.9. Additional OpenGL configuration for lighting, a default image map, viewport and those camera properties (including its location at the origin, looking along the $-z$-axis) that don't change unless the user resizes the window.

is captured for exclusive use by this window (even if it moves outside the window) and its screen position recorded. As the mouse moves, the difference in its current position is used to adjust the orientation of the mesh by rotating it about a horizontal and/or vertical axis. The right mouse button uses a similar procedure to change the horizontal and vertical position of the model relative to the viewpoint.

OpenGL initialization which uses the standard set-up shown in Listings 13.1 and 13.2 is extended by the application-specific function,

```
void initializeRender(HWND hWnd){
  GLfloat d[]={1.0,1.0,1.0,1.0};        // diffuse light color
  GLfloat a[]={0.1,0.1,0.1,1.0};        // ambient light color
  GLfloat s[]={1.0,1.0,1.0,1.0};        // specular light color
  GLfloat p[]={0.0,0.0,1.0,1.0};        // (x,y,z,w  w=1 => positional light
  glLightModeli(GL_LIGHT_MODEL_TWO_SIDE,GL_TRUE);
  glEnable(GL_LIGHT0);                  // enable the first (OGL has limited
                                        // number of lights)
  glLightfv(GL_LIGHT0,GL_DIFFUSE,d);// set the light parameters
  glLightfv(GL_LIGHT0,GL_AMBIENT,a);
  glLightfv(GL_LIGHT0,GL_SPECULAR,s);
  glLightfv(GL_LIGHT0,GL_POSITION,p);
  glLightf(GL_LIGHT0,GL_LINEAR_ATTENUATION,0.15);
  glLightf(GL_LIGHT0,GL_QUADRATIC_ATTENUATION,0.05);
  glClearDepth(1.0);                    // setup a Z buffer for hidden
                                        // surface drawing
  glEnable(GL_DEPTH_TEST);              // a depth of 1.0 is fathest away
  glClearColor(gbRed,gbGreen,gbBlue,1.0);
  // The following code sets up the view point. It uses some UTILITY
                                        // library functions
  RECT oldrect;
  GLfloat aspect;
  glMatrixMode(GL_PROJECTION);          // Going to define the viewing
                                        // transformation
  glLoadIdentity();
  GetClientRect(hWnd, &oldrect);
  aspect = (GLfloat) oldrect.right / (GLfloat) oldrect.bottom;
  // A utility library function makes it easier to define a field of
  // view 45 degrees,
  // an aspect ratio that matches the window dimensions and two clipping planes.
  // The clipping planes allow us to restrict which polygons are
  // rendered to a volume
  // space lying beyond the near_plane and on the viewpoint side of
  // the far_plane.
  gluPerspective( 45.0, aspect, near_plane, far_plane);
  glMatrixMode( GL_MODELVIEW );         // Going to define the object transformation
  glViewport(0, 0, oldrect.right,oldrect.bottom); // set to match the window
}
```

Listing 13.9. (continued).

initializeGL(..), that prepares the environment in which the model will be viewed. It enables and specifies lighting. It defines the field of view and a perspective projection with the viewpoint located at $(0, 0, 0)$ and the viewer looking along the $-z$-axis; the y-axis is vertical by default. OpenGL uses a right-handed coordinate system as illustrated in Figure 13.10.

All 3D coordinates are subject to a transformation, which is called the GL_MODELVIEW transformation. This allows hierarchical models to be built

```
void FreeObject(void){
 ///// Free up all the memory buffers used to hold
 ...// the mesh detail and reset pointers
 MainVp=NULL; MainFp=NULL; MapsFp=NULL;
 Nvert=0; Nface=0; NvertGlue=0; Nmap=0;
}

void LoadModel(char *CurrentDirFile, int fid){    // load the mesh
 FreeObject();  // make sure existing mesh is gone
 if(fid == FI_3DS)Load3dsObject(CurrentDirFile); // select loading function
 else            LoadOfxObject(CurrentDirFile); // by filename extension
 ListAdjFaces();            // find a list of all polygons which are adjaent
                           // to each other
 FixUpOrientation();   // make sure all the faces have consistent normals
 StripFaces();            // order the polygons into strips if possible
}

int LoadOfxObject(char *FileName){ // load the OpenFX model - the file
                                  // is segmented
 int fail=1;                      // into chunks
 long CHUNKsize,FORMsize;
 if( (fp=fopen(FileName,"rb")) == NULL)return 0;
 // check headers to see if this is an OpenFX file
 if (fread(str,1,4,fp) != 4 || strcmp(str,"FORM") != 0){ fail=0; goto end; }
 if (getlon(&FORMsize) != 1) goto end;
 if (fread(str,1,4,fp) != 4 || strcmp(str,"OFXM") != 0){ fail=0; goto end; }
loop:  // we have an OpenFX file so go get the chunks
 if (fread(str,1,4,fp) != 4 ) goto end;
 if (getlon(&CHUNKsize) != 1) goto end;
 if     (strcmp(str,"VERT") == 0){ReadVertices(CHUNKsize);}
 // read the vertices
 else if(strcmp(str,"FACE") == 0)ReadOldFaces(CHUNKsize,0);
 // read the polygons
 else if ... // read other required chunks
 else getchunk(CHUNKsize); // read any chunks not required
 goto loop;
end:
 fclose(fp);
 return fail;  // 1 = success 0 = failure
}
```

Listing 13.10. Loading the OpenFX and 3ds Max mesh models involves parsing the data file. The files are organized into chunks. After reading the data, it is processed to make sure that all the polygons have a consistent surface normal. The function StripFaces() provides a rudimentary attempt to optimize the order of the triangular polygons so that they are amenable to OpenGL's acceleration.

and lights and viewpoint positioned. Transformations are assembled using a matrix stack so that previous transformations can be undone easily (simply by popping the stack). A viewing transformation, the GL_PROJECTION transformation, is necessary to specify the viewing conditions such as the field of view. When rendering, these two transformations are combined to form the image of a 3D point on the 2D image plane. The image plane itself is scaled so as to fit into the *viewport*, which is dimensioned so that it fits into the encapsulating window's client area.

The initialization phase also creates a small image map (checker board pattern) to act as a substitute in the event that a map from the mesh file cannot be created during rendering. Making texture maps in OpenGL involves two main steps: preparing the image data, including assigning it an identifier, and writing that image data into the display adapter's texture memory. There are some restrictions on the format of the data that may be used to define pixel textures in OpenGL, such as the dimensions must be a power of two.

To perform this additional initialization, the code takes the form shown in Listing 13.9.

13.1.5   Loading the Model

If you want to see all the detail of how the vertex, polygon and image map information is extracted from the 3D data files, consult the actual source code. Listing 13.10 gives a brief summary. Both the OpenFX and 3DS mesh files

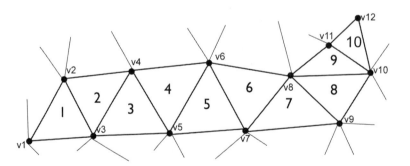

Figure 13.5. OpenGL provides a mechanism for sending strips of triangles to the rendering pipeline. In this example, we see that triangular polygons 1 to 8 (vertices 1 to 10) can be combined as a single triangle strip. Triangles 9 and 10 cannot be included in the strip because they violate the vertex/edge ordering. OpenGL has a structure called a *triangle fan* to deal with such cases.

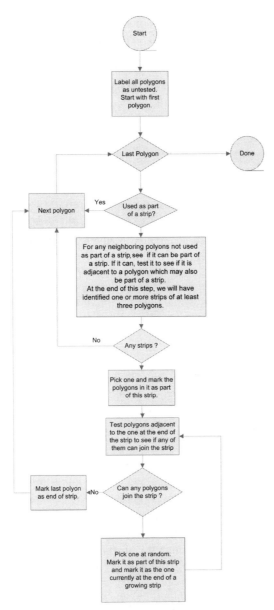

Figure 13.6. The stripping algorithm tries to build strips of adjacent triangular polygons. This helps to reduce the number of vertices that have to be rendered by OpenGL. For example, if you can make a strip from 100 adjacent polygons, you reduce the number of `glVertex..()` calls from 300 to 201.

use a storage scheme based on *chunks*. There is a chunk to store the vertex data, one to store the triangle face/polygon data and numerous others. Each chunk comes with a header (OpenFX has a four-character ID and 3DS a two-byte magic number) and a four-byte length. Any program parsing this file can jump over chunks looking for those it needs by moving the file pointer forward by the size of a chunk.

A fairly simplistic algorithm which attempts to locate strips of triangles such as those in the mesh illustrated in Figure 13.5 is provided by function StripFaces(). The strategy of the algorithm is summarized in Figure 13.6. The algorithm locates groups of polygons that can form strips and sets the member variable next in each polygon's data structure so that it behaves like a linked list to vector a route through the array of polygons stored in RAM at an address pointed to by MainFp. By following this trail, the polygons can be rendered as a triangle strip, thus taking advantage of OpenGL's rendering acceleration for such structures.

13.1.6   Rendering the 3D Data

We shall discuss rendering with the use of five listings so as to highlight the differences which occur when one uses OpenGL to render simply-shaded polygons, polygons that make up the whole or part of a smooth, rather than faceted, object. We shall also introduce the *display list*, which allows further optimization of the rendering process.

Display lists are very useful because they allow us to prepare a kind of ready-made scene, one that can be quickly configured and rendered without having to load all the vertices and images from the host computer's RAM. For a VR application, a display list can be used to store all the OpenGL commands to render a mesh model, independent of taking any particular view. The program's user can then view the object from any angle by calling the display list after position and rotation transformations are set up. *Using a display list works in this case because the mesh does not change over time. If the mesh were deforming or moving, the list would have to be regenerated after each change to the mesh.*

The function which is called to render a particular view or in response to a WM_PAINT message appears in Listing 13.11. Note that a transformation of -5.0 is applied to the negative z-axis; this moves the model away from the camera so that it all is visible, i.e., we see the outside. If the model were of a virtual environment, we would wish to keep it centered at $(0, 0, 0)$ and

```
GLvoid Draw3dScene(HWND hWnd,BOOL shaded){      // called when window is drawn
  glClear(GL_COLOR_BUFFER_BIT |
          GL_DEPTH_BUFFER_BIT);                 // clear frame buffer
                                                // and Z-buffer
  glShadeModel(GL_SMOOTH);                      // draw using GOURAUD shading
  glPushMatrix();                              // store existing
                                                // transformations
  glLoadIdentity();                            // start with NO transformation
  glTranslatef(xpos_offset, ypos_offset, -5.0); // move mesh model
  glRotatef(up_angle,1.0,0.0,0.0);             // rotate up or down
  glRotatef(round_angle+180.0,0.0,1.0,0.0);    // rotate to get side view
  glScalef(view_scale, view_scale, view_scale); // scale to fit into view
  if(listID > 0){                              // If a display list of the
    glEnable(GL_LIGHTING);                     // mesh with maps etc is ready
    glCallList(listID);                        // draw it with lighting
                                                // enabled.
    glDisable(GL_LIGHTING);
  }
  glPopMatrix();                               // restore the previous
                                                // transformation
  glFlush();                                   // make sure OGL pipeline is
                                                // all done
}
```

Listing 13.11. Rendering the scene using a display list.

expand it so that we can move about inside it. We should, however, still make sure that it all lies on the camera side of the rear clipping plane.

Compiling a display list is not a complex process because the normal rendering commands are issued in exactly the same way that they would be when rending directly to the frame buffer:

```
  ..
  glNewList(listID,GL_COMPILE);
  // start to build a Display List
  glBegin(GL_TRIANGLES);
  // render triangular polygons by calling glVertex() three times
  glEnd();
  glEndList();
  // end the display list
  ..
```

To build a display list for the mesh, we must loop over every face/polygon and call glVertex..() for each of its three vertices. To achieve the correct lighting, a surface normal will also have to be defined at each vertex. When a

```
void Make3dListS(void){ // make display lists
 GLfloat x,y,z,n[3],color[4],spec_color[]=(1.0,1.0,1.0};
 FACE *fp;
 VERTEX *v,*v1,*v2,*v0;
 normal *nv;
 int i,j,k,Vi,V[3],jf,jl,count,strip;
 if(Nface == 0)return;

 scale=GetScaleValue(c)*ObjScale;

 if((nv=(normal *)malloc(sizeof(normal)*Nvert)) == NULL)return;
 ..
 // make vertex normals from adjacent vertices
 ..

 if(listID > 0){glDeleteLists(listID,listID);listID=0;}
 listID=1;  glNewList(listID,GL_COMPILE);
 count=0;

 glBegin(GL_TRIANGLES);
 fp=MainFp; for(i=0;i<Nface;i++,fp++){  // loop over all polygons
   if(fp->mapped)continue;
 // skip any polygons with image maps applied
   v0=(MainVp+fp->V[0]); v1=(MainVp+fp->V[1]); v2=(MainVp+fp->V[2]);
   if(!same_color(fp->color)){
     glMaterialfv(GL_FRONT_AND_BACK,GL_AMBIENT_AND_DIFFUSE,fp->color);
     glMaterialfv(GL_FRONT_AND_BACK,GL_SPECULAR,spec_color);
   }
   if(Normalize(v0->p,v1->p,v2->p,n)){
     for(j=0;j<3;j++){
         Vi=fp->V[j]; v=(MainVp+Vi);
         x=((GLfloat)(v->p[0]-c[0]))*scale;
 // Scale and center the model in the
         y=((GLfloat)(v->p[1]-c[1]))*scale;  // view volume.
         z=((GLfloat)(v->p[2]-c[2]))*scale;
         if(smooth_shading && ((fp->texture >> 7) != 0)){ // smoothed
           glNormal3f( nv[Vi][0], nv[Vi][1], nv[Vi][2]);
         }
         else{                                        // not smoothed
           glNormal3f( n[0], n[1], n[2]);
         }
         glVertex3f(x,y,z);
     }
   }
 }
}
```

Listing 13.12. A display list holds all the commands to render the polygons in a mesh model. Changes of color apply to all subsequent vertices added to the list, and so each time a change of color is detected, a new set of material property specifications have to be added to the display list.

```
glEnd();
if(Mapped){
..... // Render any polyongs with image maps - see next listing
}
glEndList();
free(nv);
}
```

Listing 13.12. (continued).

smooth shading model is required, a precalculated average normal (obtained from all the polygons attached to a particular vertex) is sent to the rendering pipeline along with the vertex coordinates. Listing 13.12 brings all the required elements together to compile the display list.

13.1.7   Image Mapping

One of OpenGL's major strengths is its image-mapping[7] capability which has improved considerably in the incremental Versions 1.1 through 1.5 and, as we shall see in the next chapter, is now (in Version 2.0) nearly as powerful

```
if(Mapped){     // Render any polygon in the model that uses an image map
   ..           // declare any local variables here
   if(Nmap > 0)for(k=0;k<Nmap;k++){       // loop over all the maps
     if(MapsFp[k].pp < 0){                // if the map is a
                                          // reflection map
       Image=LoadMAP(MapsFp[k].frefl,&xm,&ym); // load the map & uncompress it
                                          // into pixel array
       glEnable(GL_TEXTURE_GEN_S);        // For reflection maps the
                                          // texture coordinates
       glEnable(GL_TEXTURE_GEN_T);        // are generated automatically.
       refmap=TRUE;
     }
```

Listing 13.13. Rendering image mapped polygons by looping over the image maps and polygons. When a polygon is textured with a map, the appropriate texture coordinates are calculated and the vertices passed to OpenGL.

---

[7]The terms *image map, texture, shader* are often loosely used to refer to the same thing because they all result in adding the appearance of detail to a polygonal surface without that detail being present in the underlying mesh and geometry. Whilst we would prefer to reserve the words shader and texture for algorithmically generated surface detail, we will, when talking generally, use them interchangeably.

```
           else {                              // and ordinary map painted onto surface
           Image=LoadMAP(MapsFp[k].filename,&xm,&ym);
           refmap=FALSE;
       }
   if(Image != NULL)makeNextMap(Image,xm,ym); // If map pixels were created
                                              // make the OGL map in
                                              // texture RAM.
       glEnable(GL_TEXTURE_2D);               // render from a map instead of
                                              // a color
       // Tell OGL which type of texture to use and which image map to use.
       // If we could
       // not load the map we tell OGL to use the simple checker board texture
       if(Image != NULL)glBindTexture(GL_TEXTURE_2D,IMAGE_TEXTURE);
       else             glBindTexture(GL_TEXTURE_2D,CHECK_TEXTURE);
       // For OpenFX models we need to get its map parameters so that
       // we can generate
       // texture coordinates.
       GetMapNormal(MapsFp[k].map,MapsFp[k].p,MapsFp[k].x,MapsFp[k].y,mP,mX,mY,mN);
       glBegin(GL_TRIANGLES);                 // now render the polygons
       fp=MainFp; for(i=0;i<Nface;i++,fp++){
         if((fp->texture & 0x40)!=0x40)continue; // skip if polygon not mapped
         brushID=(fp->brush & 0x1f);          // which image ?
         if(brushID != k)continue;            // skip if not this image map
         v0=(MainVp+fp->V[0]); v1=(MainVp+fp->V[1]); v2=(MainVp+fp->V[2]);
         if(Normalize(v0->p,v1->p,v2->p,n)){   // get polygon normal
           for(j=0;j<3;j++){
             Vi=fp->V[j]; v=(MainVp+Vi);
             x=((GLfloat)(v->p[0]-c[0]))*scale;  // scale and center model
             ..//  same for y and z ccords   ;  // in view volume
             if(smooth_shading && ((fp->texture >> 7) != 0)){ // smoothed
               glNormal3f(nv[Vi][0], nv[Vi][1],nv[Vi][2]);
             } else glNormal3f( n[0], n[1], n[2]);         // not smoothed
             if(MapsFp[k].map == MAP_BY_VERTEX){
               // texture coords exist - use them!!
                 alpha=(GLfloat)v->x; beta=(GLfloat)v->y;
             }
             else if(MapsFp[k].map == CYLINDER || // OpenFX model - must
                                             // calculate texture coords
                     MapsFp[k].map == CYLINDER_MOZIAC)
                GetMappingCoordC(mN,mX,mY,mP,MapsFp[k].angle,
                                            v->p,&alpha,&beta);
             else GetMappingCoordP(mN,mX,mY,mP,v->p,&alpha,&beta);

   if(!refmap)glTexCoord2f(alpha,beta);
             glVertex3f(x,z,-y);
           } } }
     glEnd();                                 // all polygons done with this map
     glDisable(GL_TEXTURE_2D);                // done texturing with this map
     if(refmap){
       glDisable(GL_TEXTURE_GEN_S);           // We were rendering reflection
       glDisable(GL_TEXTURE_GEN_T);           // mapping.
     }
   }                                          // end of map loop
 } //  end of rendering image maps
```

Listing 13.13. (continued).

as the non-real-time rendering engines. There are few effects that cannot be obtained in real-time using image mapping.

> In the example we are working on, we shall not deploy the full spectrum of compound mapping where several maps can be mixed and applied at the same time to the same polygons. Neither will we use some of the tricks of acceleration, such as loading all the pixels for several maps as a single image. In many cases, these tricks are useful, because the fixed-functionality pipeline imposes some constraints on the way image maps are used in the rendering process.

Listing 13.13 shows how the parts of the mesh model which have textured polygons are passed to OpenGL. The strategy adopted is to run through all the image maps used by the mesh model (there could be a very large number of these). For each map (map i), its image is loaded and the pixels are passed to OpenGL and loaded into texture RAM. After that, all the polygons are checked to see whether they use map i. If they do then the texture coordinates are obtained. OpenGL requires per-vertex texture coordinates, and, therefore, if we are loading an OpenFX model, we must generate the vertex texture coordinates. 3ds Max already includes texture coordinates for each vertex, and so these meshes they can be passed directly to OpenGL.

The code that is necessary to process the image by loading it, decoding it and making the OpenGL texture is shown in Listing 13.14. To round off the discussion on image mapping, if you look back at Listing 13.13, you will notice the calls to the functions `glEnable(GL_TEXTURE_GEN_S);` and `glEnable(GL_TEXTURE_GEN_T);`. If these parameters are enabled, OpenGL will ignore any texture coordinates specified for vertices in the model and instead generate its own based on the assumption that what one sees on the mesh surface is a reflection, as if the image were a photograph of the surrounding environment. This is the way OpenGL simulates reflective/mirrored surfaces and the extremely useful application of environment maps as discussed in Section 7.6.

This takes us about as far as we need to go with OpenGL for the moment. We have seen how to render image-mapped polygons from any viewpoint and either move them or move the viewpoint. We can use a good lighting model and can draw objects or virtual worlds constructed from many thousands of planar triangular or other polygonal primitives, in real time. It is possible to extend and use this code, so we shall call on it as the basis for some of the

```
    // load the Image map into pixel array
unsigned char *LoadMAP(char *CurrentFile,int *xx,int *yy){
    // JPEG type image
    if(strstr(CurrentFile,".JPG")||strstr(CurrentFile,".jpg"))
    // read, decompress and return pointer
    return LoadJpeg(CurrentFile,xx,yy);
    // other maps
    else if(....
    return NULL;
}

    // This function makes the OpenGL image map from a block of 24 bit pixels
in RGB order void makeNextMap(unsigned char *pixels24, int x, int y){
    int xi,yi;
    // OpenGL ony accepts image maps whose height and width
    // are a power of 2. In order to make that map we
    // must find a suitable power of 2 and use the utility
    // library to interpolate from the original dimensions.
    // Because this must be done in the CPU it will be very slow.
    // However because we are building a display list this
    // only has to be done ONCE and not when we are running the list.
    unsigned char *px;
    if       (x <  96)xi=32;
    else if(x < 320)xi=64;
    else xi=128;
    if       (y <  96)yi=32;
    else if(y < 320)yi=64;
    else yi=128;
    if((px=(unsigned char *)malloc(xi*yi*3)) == NULL){
      free((char *)pixels24);
      return;
    }
// scale the image to chosen power of two dimension
    gluScaleImage(GL_RGB,x,y,GL_UNSIGNED_BYTE,pixels24,xi,yi,GL_UNSIGNED_BYTE,px);
    free((char *)pixels24);     // we don't need the original image any more
                                // Next make the texture in GPU RAM
    glBindTexture(GL_TEXTURE_2D,DUMMY_TEXTURE);
    glTexImage2D(GL_TEXTURE_2D,0,3,xi,yi,0,GL_RGB,GL_UNSIGNED_BYTE,(GLvoid *)px);
    free((char *)px);           // we don't need any CPU memory it's all in GPU
}
```

Listing 13.14. Creating the image map requires three steps: it must be loaded and decompressed, then it has to be converted into a form that OpenGL can use (i.e., its height and width must be a power of two) then it can be passed to OpenGL and loaded into the texture RAM on the display adapter.

```
//  The following code illustrates the format of the loading code
//  - an external JPEG image manipulation library written in
//  C is used from our C++ application
extern "C" {
  extern long ReadJPG(char *filename, short info_only);
  long x_sizeJ, y_sizeJ; // used by Library code
  unsigned char *lpImage=NULL;
};


static unsigned char *LoadJpeg(char *filename, int *x, int *y){
 unsigned char *screen;
 long xs,ys,imagesize;
 if(ReadJPG(filename,1) == 0)return NULL;
// test image is OK - if not return no pixels
 xs=x_sizeJ; ys=y_sizeJ;
// if OK get width and height
 imagesize=xs*ys*3*sizeof(unsigned char);
// allocate space for image pixel array
 if((screen=(unsigned char *)malloc(imagesize)) == NULL)return NULL;
 x_sizeJ=xs; y_sizeJ=ys;
 // We need an image with (0,0) at bottom left
 // JPEG uses (0,0) at top left so must read in
 // opposite order.
 lpImage=(screen+(ys-1)*xs*3);
 if(ReadJPG(filename,0) == 0){
   free(screen);
   return NULL;
 }
 // also return image width and height
 *x = xs; *y = ys;
 return screen;
}
```

Listing 13.14. (continued).

projects in Chapter 18. We shall now turn to look briefly at how OpenGL
can be used in a system independent manner.

13.2 System Independence with the GLUT Libraries

The OpenGL Utility Toolkit [3] is a useful library of functions that removes
completely the system dependent element from OpenGL application programs. It is available in binary form or as source code for UNIX, Windows and Macintosh platforms. Using it, programs can be put together very

```
.. // standard headers
#include <GLUT/glut.h>
.. // other headers, program variables and function prototypes
   // (all standard C data types! )

void Render() {    // Called by GLUT   render the 3D scene
 glClear(GL_COLOR_BUFFER_BIT | GL_DEPTH_BUFFER_BIT);
 ..                  // set up view transformations
 glCallList(listID);

 ..
 glFinish();
 glutSwapBuffers();    // GLUT function
}
```

Listing 13.15. Using the GLUT makes an application program platform-independent. The outline program in this listing shows a version of the mesh viewer developed in Section 13.1 which uses the GLUT. It was developed for the Mac OS X using the Xcode tools.

quickly. If you are writing a program and do not need to achieve a specific look or feel or tie into some system-dependent functionality, the GLUT is the obvious way to use OpenGL.

The GLUT has its own programming model which is not that dissimilar from the Windows approach. It uses a window of some kind to display the contents of the frame buffer. Initialization functions create this window and pass to the GLUT library a number of application-specific callbacks[8] to support mouse and keyboard input. The application programmer must supply the appropriate OpenGL drawing and other commands in these callback functions. Listing 13.15 shows the key elements of a version of our mesh model viewer using the GLUT and running under the Macintosh OS X operating system. The application would be equally at home on Windows or an SGI UNIX system. It is a good illustration of just how platform-independent OpenGL can be.

The example given in this section only demonstrates a few key features of the utility library. Menu commands, mouse clicks and multiple windows can also be used. Unless one really must have some system-specific functions in a 3D application, it is a good idea to use the GLUT or some other multi-platform framework so as to maintain as much host flexibility as possible.

---

[8]A callback is a function like a window message handler that is part of the program, but it is not called explicitly by the program itself but by another program or system or even the operating system. It is called indirectly, or put it another way, it is *called back*, hence the name.

```
void Reshape(int w,int h){ // called by GLUT when window changes shape
  glViewport(0, 0, w, h);
}

void Init(){ // initialize openGL - called after window is created
  ..
  glClearColor(gbRed,gbGreen,gbBlue,1.0);
  ..
  glEnable(GL_LIGHT0);
}

void KeyHandler(unsigned char key,int x,int y){ // called by GLUT
  switch(key){
    case 'l':                          // user pressed the 'l' key to
                                       //          load new mesh
      LoadOFXmesh(GetFileName());      // or 3DS mesh
      Make3dListS();                   // make the openGL display list
      break;
  };
}

void TimerFunc(int value){ // called by GLUT
  round_angle += 5.0; up_angle += 1.0; Render();  // change view and render
  glutTimerFunc(20,TimerFunc,0);       // run again in 20 ms
}

int main(int argc,char **argv){        // main program entry point
  glutInit(&argc,argv);
  glutInitWindowSize(640,480);         // set up window (size given)
  glutInitWindowPosition(100,100);     // postion on screen
  glutInitDisplayMode(GLUT_RGBA|GLUT_DOUBLE|GLUT_DEPTH);
  glutCreateWindow("Mesh Viewer for Mac OS X");
  glutDisplayFunc(Render);             // Tell GLUT the function to use
                                       //             to render the scene
  glutReshapeFunc(Reshape);            // Tell GLUT the function to use
                                       //             for window size change
  glutKeyboardFunc(Keyhandler);        // Tell GLUT the function to handle
                                       //             keyboard keys
  glutTimerFunc(100,TimerFunc,0);      // Tell GLUT to run this function
                                       //             in 100ms time
  Init();                              // initialize our OpenGL requirements
  glutMainLoop();                      // Wait in a loop and obey commands
                                       //             till window closed

  return 0;
}
```

Listing 13.15. (continued).

13.3   Visualizing a Virtual World Using Direct3D

Direct3D sets out to achieve the same result as OpenGL, but it does not have such a long history and it is evolving at a much faster rate. For this reason, we will not use it as our main framework for developing the visualization examples. We will use the other components of DirectX, i.e., DirectInput and DirectShow for all other software/hardware interfacing because it works well and there are very few alternatives for Windows applications.

13.3.1   A Bit of History

DirectX [4] first appeared under the guise of the *Games SDK for Windows 95* in late 1995, and it has been refined and updated many times. At first, it only gave the ability to draw directly to the display hardware, and for the graphics programmer that was a breath of fresh air. This is the DirectDraw component of DirectX. It offered drawing speeds as fast as could be obtained if one accessed the display adapter without using operating system mediation, say by using assembler code. DirectDraw was designed with 2D computer games in mind, so it had excellent support for *blits*, bit-block memory transfers, including transparent and "key color" blits. Many operations are performed asynchronously with page flipping and color fills done in the display adapter's hardware.

All that the DirectDraw API provides is fast access to a drawing canvas, the equivalent of the output frame buffer. To render virtual worlds or anything else in 3D, one still has to render into the frame buffer, and this is the function of Direct3D. The first few versions of Direct3D appeared in the 1990s. Back then, hardware support for 3D functions was still very limited and so Direct3D had two modes of operation, called retained mode and immediate mode. Retained mode [2] implemented most of the 3D functionality in library code running in the host's own processor, not on the GPU. Immediate mode was limited to a few simple functions and only managed to provide some very basic lighting models, for example. Its functions matched the capabilities of the hardware rather than the desires of the application programs. Now everything has changed. Retained mode has disappeared,[9] and immediate mode has evolved to map onto the powerful features provided in hard-

---

[9]Not quite, because it uses COM architecture and so is still in there, inside the Direct3D system. It is just not documented.

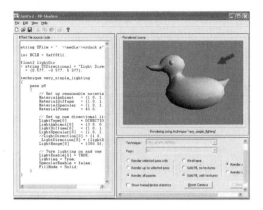

Figure 13.7. Two examples of shader development systems. On the left, a basic shader development program is illustrated. It presents the C-like shader code that Microsoft calls the *High Level Shader Language*; NVIDIA's similar version is *Cg*. On the right is a screen from ATI's RenderMonkey program, which can produce fabulously exciting shader programs.

ware. Indeed, it has driven the implementation of many hardware features. The immediate mode's *shader model* (as it is called) which allows all sorts of interesting lighting and mapping effects is developing fast and will probably continue to evolve and change extensively. It is now possible to write *Shader programs* for Direct3D that give the impression of chrome, wood, marble etc. For example, Figure 13.7 illustrates two programs that can be used to write and test shader programs. Much more elaborate shader design tools are available. NVIDIA's Cg [5] is a C like language dedicated to GPU programming. By using it or ATI's RenderMonkey [1] fascinating surface textures, lighting models and other effects can be formulated for use in real time.

13.3.2 The Architecture of Direct3D

Unsurprisingly, since it runs on the same hardware, Direct3D's architecture is not too dissimilar to that of OpenGL. Vertices still play a major part; the only differences are that in Direct3D, the term *fragment* is replaced with *pixel* and primitives are obtained from vertices by tessellation. Figure 13.8 shows the Direct3D rendering pipeline. In this diagram, a programmable pipeline is illustrated. The latest version of OpenGL also has a programmable pipeline; this will be discussed in the next chapter.

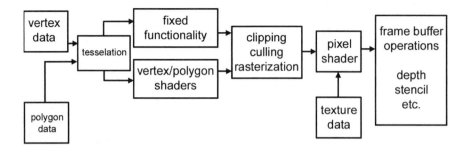

Figure 13.8. Direct 3D system overview. Note the programmable branch of the graphics pipeline. Primitive data refers to bitmaps and any other non-vertex input.

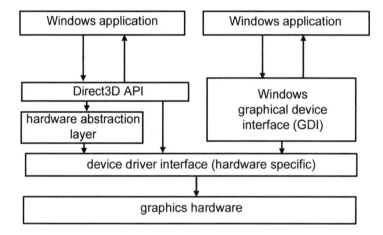

Figure 13.9. Direct3D/Windows system communication.

Direct3D's relationship with an application program is mediated by the device driver. A diagrammatic indication of this is presented in Figure 13.9. You will notice a special block called the HAL (the hardware abstraction layer). This is a device-specific interface provided by the hardware manufacturer. The HAL can be part of the display driver or contained in a separate dynamic link library (DLL) that communicates with the display driver through a private interface which the driver's creator defines. The Direct3D HAL is implemented by the GPU chip manufacturer or display adapter vendor. The HAL implements only device-dependent code and performs no emulation. If a function cannot be performed by the hardware, the HAL does not report it as a hardware capability. In DirectX9, the HAL can take advantage of the hardware's ability to do all the vertex processing or it can fall back to software processing. It's up to the display manufacturer to decide.

It is difficult to say whether Direct3D is a better environment for doing 3D graphics than OpenGL. Certainly, the architects of OpenGL had the luxury of designing their system for top-of-the-range workstations whilst those putting Direct3D together had to be content with the vagaries of an operating system that did not have a good reputation for any sort of graphics at all. Since both systems rely so heavily on the hardware of the graphics cards, one should comment to developers of 3D chipsets and their drivers if either Direct3D or OpenGL appear to differ markedly in their speed or performance.

13.3.3   Comparing OpenGL (OGL) and Direct3D (D3D)

The similarities between a 3D application written in OGL and one written in D3D are striking. The same concept of initialization is required, the window has to be attached to the 3D rendering engine and the same concept of vertex-defined polygons applies, as does the idea of image mapping. However, it is more interesting to investigate the differences. First, the coordinate systems differ. If the application uses the wrong one, the scene will appear to be upside-down, a mirror image of what you intended or may even look like it is inside out. OGL uses a *right-handed* coordinate frame of reference with the y-axis vertical. D3D uses a *left-handed* coordinate system with the y-axis vertical. This means that both libraries use depth information in the z direction, but for OGL, the further away an object is, the more negative its z-coordinate. For D3D, the further away an object is, the more positive is its z-coordinate. *Confusing? Yes! And it gets even more confusing when calculating surface normals and trying to determine whether we are looking at the inside or outside of an object.* We have been using the traditional right handed system, z vertical, in Part I and so to use algorithms described there in our code, we must swap the y- and z-coordinates. Figure 13.10 summarizes the coordinate system, surface normal determination and definition of inside/outside using the vertex ordering in OpenGL and Direct3D. The surface normal vector **n** points away from the surface to indicate the outside of a solid object, as indicated in the cubes. If the vertices labeled 0, 1 and 2, with vector coordinates $\mathbf{p}_0$, $\mathbf{p}_1$ and $\mathbf{p}_2$, define a polygon's position in the appropriate frames of reference then for OpenGL $\mathbf{n} = \mathbf{a} \times \mathbf{b}$ and Direct3D $\mathbf{n} = \mathbf{b} \times \mathbf{a}$. (Suppose that the cube shown is centered at the origin, axis-aligned, and that each side is of

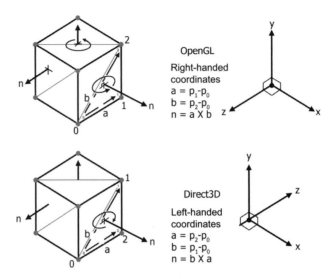

Figure 13.10. The geometry, coordinate systems and ordering of polygon vertices for surface normal calculations in OpenGL and Direct3D.

length 2. In the OGL coordinate system, the point 0 is located at $(1, -1, 1)$. In the D3D system, it is at $(1, -1, -1)$).

Another difference is the approach taken when rendering in a full-screen mode. Since D3D is a Microsoft creation, it can hook directly into the operating system and render directly to the screen without using the normal niceties of *giving way* to other applications. As a result, trying to mix D3D full-screen output with other windows does not work awfully well. In fact, for some games that use D3D, the display may be switched into an entirely different screen resolution and color depth.[10] On the other hand, OGL obtains its full-screen performance by drawing into a conventional window devoid of decoration (border menu etc.) and sized so that its client area exactly matches the size of the desktop.

If one looks at the sample programs that ship with the D3D SDK, at first sight it can be quite frightening because there seems to be a massive quantity of code dedicated to building even the simplest of programs. However, things are not quite as complex as they seem, because most of this code is

---

[10]You can sometimes detect this on CRT display by hearing the line/frame scan oscillator changing frequency.

due to D3D's desire to optimize the use of display adapter's capabilities. So for example, it might decide that a 64,000 color mode in a resolution of 640 × 480 is best suited for full-screen rendering and switch into that mode. If you are prepared to accept drawing into a desktop window, the complexity of the code required to set up the display is no more daunting than it is with OpenGL.

The final small difference we will mention now is in the way D3D applications typically carry out their rendering. Instead of responding to WM_PAINT messages, the rendering functions are called continuously from within the WinMain() function's message loop. This means that rendering takes place continuously (as fast as the processor can go, at frame rates greater than 100 Hz, for example). So typically the message loop will take the form:

```
MSG msg;
ZeroMemory( &msg, sizeof(msg) );
// infinite loop
while( msg.message!=WM_QUIT ){
// are there any messages ?
  if( PeekMessage( &msg, NULL, 0U, 0U, PM_REMOVE ) ){
    TranslateMessage( &msg );
    DispatchMessage( &msg );
  }
// Do D3D rendering
  else  RenderScene();
}
```

13.3.4 Rendering with Direct3D into a Windows Window

To configure an application program to render a 3D world into a window client area, the steps taken are remarkably similar to those we used earlier to do the same thing with OpenGL (see Listing 13.16). To render the scene, the code in Listing 13.17 is used. D3D offers a broad selection of options for describing the scene's geometry and its storage format. This code fragment offers an example of only one of them: a list of triangles defined by three vertices each. Each vertex carries two vectors with it, a position vector and a surface normal vector. In Direct3D, it is possible to customize the format in which the vertex data is stored, hence the use of the keyword CUSTOMVERTEX.

```
#include <d3dx9.h>                          // Header file from SDK
// these are local poiters to the D3D opbjects
LPDIRECT3D9            g_pD3D       = NULL; // Used to create the D3DDevice
LPDIRECT3DDEVICE9      g_pd3dDevice = NULL; // Our rendering device
...
// Set up D3D when window is created
  case WM_CREATE:
    // Create the D3D object.
    g_pD3D = Direct3DCreate9( D3D_SDK_VERSION );
    // Set up the structure used to create the D3DDevice.
    D3DPRESENT_PARAMETERS d3dpp;
    ZeroMemory( &d3dpp, sizeof(d3dpp) );
    d3dpp.Windowed = TRUE;
    d3dpp.SwapEffect = D3DSWAPEFFECT_DISCARD;
    d3dpp.BackBufferFormat = D3DFMT_UNKNOWN;
    d3dpp.EnableAutoDepthStencil = TRUE;
    d3dpp.AutoDepthStencilFormat = D3DFMT_D16;
    // Create the D3DDevice
     g_pD3D->CreateDevice( D3DADAPTER_DEFAULT, D3DDEVTYPE_HAL, hWnd,
                           D3DCREATE_SOFTWARE_VERTEXPROCESSING,
                           &d3dpp, &g_pd3dDevice ) ) );
    // Z buffer
    g_pd3dDevice->SetRenderState( D3DRS_ZENABLE, TRUE );
    return 0;
  case WM_DESTROY:                          // Relese the D3D object
    if( g_pd3dDevice != NULL )g_pd3dDevice->Release();
    if( g_pD3D != NULL )g_pD3D->Release();
    return 0;
```

Listing 13.16. Message handlers to create and destroy a window give convenient places to initialize and release the Direct3D rendering system. Many applications also put these steps in the `WinMain()` function on either side of the message loop.

To render the 3D scene with its geometry and image maps, we use a function called from within the message loop.

Despite the significant difference in program style and implementation detail between Direct3D and OpenGL, they both start with the same 3D data and deliver the same output. Along the way, they both do more or less the same thing: load a mesh, move it into position, set up a viewpoint, apply a lighting and shading model and enhance the visual appearance of the surface using image maps. Consequently, we can relatively easily rewrite any 3D virtual world visualizer using the Direct3D API instead of the OpenGL one. Quite a considerable fraction of the code can be reused without modification. Loading the mesh models, establishing the Windows window and handling

```
VOID RenderScene(void) {
    // Clear the backbuffer and the zbuffer
    g_pd3dDevice->Clear( 0, NULL, D3DCLEAR_TARGET|D3DCLEAR_ZBUFFER,
                         D3DCOLOR_XRGB(0,0,255), 1.0f, 0 );
    // Begin the scene
    if( SUCCEEDED( g_pd3dDevice->BeginScene() ) )      {
        // Setup the lights and materials
        // and viewing matrices to describe the scene
        SetupLightsMatricesMaterials();    // USER FUNCTION
        // Render the vertex buffer contents NOTE: g_pVB is a pointer to
        // an object describing the scene geometry which must be loaded
        // before rendering begins
        g_pd3dDevice->SetStreamSource( 0, g_pVB, 0, sizeof(CUSTOMVERTEX) );
        g_pd3dDevice->SetFVF( D3DFVF_XYZ|D3DFVF_NORMAL );
        g_pd3dDevice->DrawPrimitive( D3DPT_TRIANGLELIST, 0, 12);
        // End the scene
        g_pd3dDevice->EndScene();
    }
    // Carry out double buffering swap
    g_pd3dDevice->Present( NULL, NULL, NULL, NULL );
}
```

Listing 13.17. Render the scene by attaching the vertex objects to the D3D device and telling D3D to render them as a list of triangular polygons.

the user commands requires very little change. We must remember to switch to the left-handed coordinate system that Direct3D uses. Direct3D has no equivalent of the *display list*, so rendering will have be done explicitly, every time the window has to be refreshed.

We have already seen that in OpenGL, if we allow the color or map to differ from one polygon to the next, rendering will slow down and it will interfere with hardware optimization. In Direct3D, it is even more inefficient. Consequently, we will make the assumption that our mesh model contains only a few colors and image maps. This assumption requires us to carry out a post-load process where we break up the model into sets of polygons. There will be a set for every image map and a set for every color used. During the post-load process, we will also generate texture coordinates and load the image maps. When it comes time to render the mesh, each set of polygons will be considered in turn and appropriate attributes or maps applied as necessary.

Now we turn to look at some of the key implementation details. However, we will omit code that performs similar tasks to that discussed in the OpenGL examples. Using the listings given here, you should be able to follow the logic of the full sources and modify/extend them for your own needs.

```
struct CUSTOMVERTEX_0 {    // used for polygons without image maps
    D3DXVECTOR3 position; // The position
    D3DXVECTOR3 normal;   // The surface normal for the vertex
};
struct CUSTOMVERTEX_1 {    // used for polygons with image maps
    D3DXVECTOR3 position; // The position
    D3DXVECTOR3 normal;   // The surface normal for the vertex
    FLOAT       u, v;     // Mapping coordinates
};

// These tells Direct3D what our custom vertex structures will contain
#define D3DFVF_CUSTOMVERTEX_0 (D3DFVF_XYZ|D3DFVF_NORMAL)
#define D3DFVF_CUSTOMVERTEX_1 (D3DFVF_XYZ|D3DFVF_NORMAL|D3DFVF_TEX1)

struct COLOR {unsigned char r,g,b};

LPDIRECT3D9             g_pD3D        = NULL; // Used to create the D3DDevice
LPDIRECT3DDEVICE9       g_pd3dDevice = NULL; // The rendering device
LPDIRECT3DVERTEXBUFFER9 g_pMesh[32];         // Buffers to hold mesh
                                             // objects (16)
                                             // with & without textures
LPDIRECT3DTEXTURE9      g_pMap{16};          // Lists of up to 16 image maps
long g_MeshVertexCount[32];                  // Number of vertices in each mesh
COLOR g_color[16];                           // globals to store color
long g_NumberOfMeshes=0,g_NumberOfTextures=0;// count number of textures
```

Listing 13.18. The application uses two types of custom vertex, depending on whether it is rendering part of a polygon mesh with a texture or a basic color. In this code, the structures are defined and two global arrays are declared that will hold pointers to the mesh objects and the texture maps.

13.3.5   Preparing the Data for Use by Direct3D

The program begins by creating a Direct3D device (as in Listing 13.17) and child window to display the rendered view. In response to a user command to load the model, all the same procedures applied in the OpenGL program are repeated. In addition, vertex coordinates are switched to conform to Direct3D's requirements. Once this is done, using the structures defined in Listing 13.18, the post-load function of Listing 13.19 prepares an array of *vertex buffers* and *textures*.[11] We will limit the number of vertex buffers (holding triangles with the same color and smoothing state) and textures to 16 each. Separate vertex buffers are built from all the vertices associated with polygons having the same properties, including the same color, and they are similarly

---

[11]Direct 3D always describes image maps as textures.

```
void CreateMeshes(void){                        // create the Direc3D meshes
 .. // declare local variables
 .. // make list of average surface normals as in OpenGL example
 fp=MainFp; for(i=0;i<Nface;i++,fp++){  // count number of different colors
   if(fp->mapped)continue;                    // skip any polygons with image
                                              // maps applied
   if(!existing_color(fp->color)){           // check to see if color exists
    g_color[g_NumberOfMeshes]=fp->color;// keep record of color
    g_NumberOfMeshes++;                       // do no let exceed 16
   }
   g_MeshVertexCount[g_NumberOfMeshes]+=3; // 3 more vertices
 }
 // using color as a guide make up lists of vertices
 // that will all have same color
 for(k=0;k<g_NumberOfMeshes;k++){
   // make a vertex buffer for this little mesh
   if( FAILED( g_pd3dDevice->CreateVertexBuffer(
       g_MeshVertexCount[g_NumberOfMeshes]*sizeof(CUSTOMVERTEX_0),
       0, D3DFVF_CUSTOMVERTEX_0, D3DPOOL_DEFAULT, &g_pVB[k], NULL)))
       return E_FAIL;
   CUSTOMVERTEX_0* pVertices;
   g_pVB[k]->Lock(0, 0, (void**)&pVertices, 0 ); // get pointer to vertex buffer
   vtx=0;                                          // vertex index into buffer
   fp=MainFp; for(i=0;i<Nface;i++,fp++){  // loop over all polygons
     if(fp->mapped)continue;                    // skip any polygons with image
                                                // maps applied
     if(fp->color == g_color[k]){               // this face has the color add
                                                // a triangle to mesh
       v0=(MainVp+fp->V[0]); v1=(MainVp+fp->V[1]); v2=(MainVp+fp->V[2]);
       if(Normalize(v0->p,v1->p,v2->p,n)){// calculate surface normal
         for(j=0;j<3;j++){
           Vi=fp->V[j]; v=(MainVp+Vi);
           x=((GLfloat)(v->p[0]-c[0]))*scale;  // Scale and center the model in
           y=((GLfloat)(v->p[2]-c[2]))*scale;  // the view volume - correct for
           z=((GLfloat)(v->p[1]-c[1]))*scale;  // different coordinate system
           pVertices[vtx].position = D3DXVECTOR3(x,y,z);
           pVertices[vtx].normal   = D3DXVECTOR3(n[0],n[1],n[2]);
           if(smooth_shading && ((fp->texture >> 7) != 0)) // smoothed
            pVertices[vtx].normal = D3DXVECTOR3(nv[Vi][0],nv[Vi][1],nv[Vi][2]);
           else                                        // not smoothed
            pVertices[vtx].normal = D3DXVECTOR3(n[0],n[1],n[2]);
           vtx++;  // increment vertex count for this bit of mesh
         }
       }
     }
   }
   g_pVB[k]->Unlock(); // ready to move on to next patch of colored vertices
 }
 // Now do the same job for polygons that have maps applied (next listing)
 CreateMapMeshes();
}
```

Listing 13.19. Convert the vertex and polygon data extracted from the mesh model
file into a form that Direct3D can use.

```
void CreateMapMeshes(){                         // create vertex mesh for polygons
                                                // with maps
 .. //                                          // local variables
 if(Nmap > 0)for(k=0;k<Nmap;k++){               // loop over all the maps
   // Direct3D can make the map directly from a wide range of image file types
                                                // directly jpeg etc.
   D3DXCreateTextureFromFile( g_pd3dDevice, MapsFp[k].filename, &g_pMesh[k]);
   // For OpenFX models we need to get its map parameters so that we can
   // generate texture coordinates.
   GetMapNormal(MapsFp[k].map,MapsFp[k].p,MapsFp[k].x,MapsFp[k].y,
               mP,mX,mY,mN);
   // make vertex buffer
   if( FAILED( g_pd3dDevice->CreateVertexBuffer(
       g_MeshVertexCount[g_NumberOfMeshes]*sizeof(CUSTOMVERTEX_1),
       0, D3DFVF_CUSTOMVERTEX_1, D3DPOOL_DEFAULT, &g_pVB[k+16], NULL)))
       return E_FAIL;
   CUSTOMVERTEX_0* pVertices;
   // get pointer to vertex buffer
   g_pVB[k+16]->Lock(0, 0, (void**)&pVertices, 0 );
   vtx=0;                                       // vertex index into buffer
   fp=MainFp; for(i=0;i<Nface;i++,fp++){
     if((fp->texture & 0x40)!=0x40)continue; // skip if polygon not mapped
     brushID=(fp->brush & 0x1f);               // which image ?
     if(brushID != k)continue;                 // skip if not this image map
     v0=(MainVp+fp->V[0]); v1=(MainVp+fp->V[1]); v2=(MainVp+fp->V[2]);
     if(Normalize(v0->p,v1->p,v2->p,n))for(j=0;j<3;j++){
       Vi=fp->V[j]; v=(MainVp+Vi);
       x=((GLfloat)(v->p[0]-c[0]))*scale;  // scale and center model
       y=((GLfloat)(v->p[2]-c[2]))*scale;  // in view volume
       z=((GLfloat)(v->p[1]-c[1]))*scale;
       pVertices[vtx].position = D3DXVECTOR3(x,y,z);
       pVertices[vtx].normal    = D3DXVECTOR3(n[0],n[1],n[2]);
       if(smooth_shading && ((fp->texture >> 7) != 0)) // smoothed
         pVertices[vtx].normal    = D3DXVECTOR3(nv[Vi][0],nv[Vi][1],nv[Vi][2]);
       else                                      // not smoothed
         pVertices[vtx].normal    = D3DXVECTOR3(n[0],n[1],n[2]);
       if(MapsFp[k].map == MAP_BY_VERTEX){  // texture coords exist - use them!!
         alpha=(GLfloat)v->x;
         beta=(GLfloat)v->y;
       }
       else if(MapsFp[k].map == CYLINDER || // OpenFX model - must calculate
                                             // texture coords
       MapsFp[k].map == CYLINDER_MOZIAC)
       GetMappingCoordC(mN,mX,mY,mP,MapsFp[k].angle, v->p,&alpha,&beta);
       else GetMappingCoordP(mN,mX,mY,mP,v->p,&alpha,&beta);
     }
```

Listing 13.20. Load the image maps into Direct3D textures and create vertex buffers holding all the vertices associated with polygons carrying that texture.

```
        pVertices[2*i+0].u = alpha;        // add mapping coordinates to the
        pVertices[2*i+0].v = beta;         // custom vertex
        vtx++;
      }
    }
    g_pVB[k+16]->Unlock(); // ready to move on to next texture
  }
} // all done
```

Listing 13.20. (continued).

shaded. We need to do this because it is too inefficient to use Direct3D to render one polygon at a time. The limit of 32 different groups of vertices is not too restrictive, and a production version of the code could make these arrays dynamically sized.

Listing 13.20 performs a similar function for those parts of the model that have textured polygons. Function `CreateMapMeshes()` loops over every map in use. At the same time, it also loads the map from the file and creates a Direct3D texture with it. It finds all the polygons that use this map and builds up a vertex buffer by adding three vertices per polygon. Then, it determines the (u, v) mapping coordinates and vertex normals and puts everything in the vertex buffer using a custom vertex format. Once the maps are loaded and the mesh is formed in the collection of vertex buffers, rendering can begin.

13.3.6   Setting Up the View Transformations

There are three parts to a transformation applied to a mesh. In addition to the transformation that accounts for the position and scale of the mesh itself, there is the transformation that sets the location of the viewpoint and the camera projection transformation. To define a viewpoint, we will use the same variables as were used in the OpenGL example. The viewing transformation is defined by giving a location for the camera and a point for it to look at. An initial 45° field of view defines the perspective transformation (it can be changed in response to user commands), and the same Direct3D function sets front and back clipping planes, too. In this example, we will redefine the view every time the rendering routine is called. Listing 13.21 give the code for the function which fulfills this task.

```
VOID SetupView()
{
    // Set the model in the correct state of rotation given
    D3DXMATRIXA16 matWorld;
    D3DXMatrixIdentity( &matWorld );
    D3DXMatrixRotationX( &matWorld,up_angle );           // tilt up./down
    D3DXMatrixRotationY( &matWorld, round_angle );       // rotation
    D3DXMatrixTranslation(&matWorld,0.0f,0.0f,1.0f);     // move along z axis
    g_pd3dDevice->SetTransform( D3DTS_WORLD, &matWorld );

    // look along the Z axis from the coordinate origin
    D3DXVECTOR3 vEyePt( 0.0f, 0.0f,0.0f );
    D3DXVECTOR3 vLookatPt( 0.0f, 0.0f, 5.0f );
    D3DXVECTOR3 vUpVec( 0.0f, 1.0f, 0.0f );
    D3DXMATRIXA16 matView;
    D3DXMatrixLookAtLH( &matView, &vEyePt, &vLookatPt, &vUpVec );
    g_pd3dDevice->SetTransform( D3DTS_VIEW, &matView );
    // Set a conventional perspective view with a 45 degree field of view
    // and with clipping planes to match those used in the OpenGL example
    D3DXMATRIXA16 matProj;
    D3DXMatrixPerspectiveFovLH( &matProj, D3DX_PI/4, 0.5f, 0.5f, 10.0f );
    g_pd3dDevice->SetTransform( D3DTS_PROJECTION, &matProj );
}
```

Listing 13.21. In Direct3D, the viewing transformation is defined in three steps: mesh/object transformation, camera/viewer transformation and perspective transformation.

```
VOID SetupLights(){
    D3DXVECTOR3 vecDir;          // direction of light
    D3DLIGHT9   light;
    ZeroMemory( &light, sizeof(D3DLIGHT9) );
    light.Type       = D3DLIGHT_DIRECTIONAL;
    light.Diffuse.r  = 1.0f;  // white light - simple - no specular
    light.Diffuse.g  = 1.0f;
    light.Diffuse.b  = 1.0f;
    vecDir = D3DXVECTOR3(1.0f,1.0f,0.0f ); // over the viewers right shoulder
    D3DXVec3Normalize( (D3DXVECTOR3*)&light.Direction, &vecDir );
    light.Range      = 1000.0f;
    g_pd3dDevice->SetLight( 0, &light );
    g_pd3dDevice->LightEnable( 0, TRUE );
    g_pd3dDevice->SetRenderState( D3DRS_LIGHTING, TRUE );
    g_pd3dDevice->SetRenderState( D3DRS_AMBIENT, 0x00202020 );
}
```

Listing 13.22. Lighting is set up every time a frame is rendered. The light is set to a fixed position, but it could be adjusted interactively if desired.

13.3.7   Lighting

Lighting is achieved by defining the location and properties of a light, as in Listing 13.22.

13.3.8   Executing the Application

Listing 13.16 forms the template of a suitable `WinMain()` function with message processing loop. Rendering is done by placing a call to `Render()` inside the loop. The code for the `Render()` function is given in Listing 13.23. This code is quite short, because all the difficult work was done in the post-load stage, where a set of triangular meshes and textures were created from the data in the file. To render these meshes, the procedure is straightforward. We loop over all the meshes for the non-textured part of the model. A *material* is created with the color value for these triangles and the Direct3D device is instructed to use it. Once the non-textured triangles have been rendered, we turn our attention to each of the textures. Again, we loop through the list of textures, rendering the triangular primitives for each in turn. The code looks quite involved here, talking about *TextureStages*, but that is only because Direct3D has an extremely powerful texturing engine that can mix, blend and overlay images to deliver a wonderful array of texturing effect.

Finally, add a menu and some initialization code. Just before the program terminates, the COM objects are freed, along with any memory buffers or other assigned resources. Once running, the program should fulfill the same task as the OpenGL example in Section 13.1.

13.4   Summary

It has been the aim of this chapter to show how real-time rendering software APIs are used in practice. We have tried to show that to get the best out of the hardware, one should appreciate in general terms how it works and how the 3D APIs match the hardware to achieve optimal performance. We have omitted from the code listings much detail where it did not directly have a bearing on 3D real-time rendering, e.g., error checking. Error checking is of course vital, and it is included in all the full project listings on the accompanying CD. We have also structured our code to make it as adaptable and generic as possible. You should therefore be able to embed the 3D mesh model viewer in container applications that provide toolbars, dialog box controls and other

```
VOID Render(){ // render the scene
 .. // local variables
 g_pd3dDevice->Clear( 0, NULL, D3DCLEAR_TARGET|D3DCLEAR_ZBUFFER,
                      D3DCOLOR_XRGB(0,0,255), 1.0f, 0 );
 if( SUCCEEDED( g_pd3dDevice->BeginScene() ) ){  // begin the rendering process
   SetupLights();                                // set up the light
   SetupView();                                  // set up the viewpoint
   // render all the non textured meshes
   for(k=0;k<g_NumberOfMeshes;k++){
     D3DMATERIAL9 mtrl;
     ZeroMemory( &mtrl, sizeof(D3DMATERIAL9) );
     mtrl.Diffuse.r = mtrl.Ambient.r = (float)g_color[k].r/255.0f;
     mtrl.Diffuse.g = mtrl.Ambient.g = (float)g_color[k].g/255.0f;
     mtrl.Diffuse.b = mtrl.Ambient.b = (float)g_color[k].b/255.0f;
     mtrl.Diffuse.a = mtrl.Ambient.a = 1.0f;
     g_pd3dDevice->SetMaterial( &mtrl );         // set the colour of this mesh
     g_pd3dDevice->SetStreamSource( 0, g_pVB[k], 0, sizeof(CUSTOMVERTEX_0));
     g_pd3dDevice->SetFVF( D3DFVF_CUSTOMVERTEX_0 );
     g_pd3dDevice->DrawPrimitive( D3DPT_TRIANGLES,0,g_MeshVertexCount[k] );
   }
   // now do the mapped mesh elements
   for(k=0;k<Nmap;k++){                          // go through all maps
     g_pd3dDevice->SetTexture(0,g_pTexture[k]);  // this texture has the map
     g_pd3dDevice->SetTextureStageState( 0, D3DTSS_COLOROP,   D3DTOP_MODULATE );
     g_pd3dDevice->SetTextureStageState( 0, D3DTSS_COLORARG1, D3DTA_TEXTURE );
     g_pd3dDevice->SetTextureStageState( 0, D3DTSS_COLORARG2, D3DTA_DIFFUSE );
     g_pd3dDevice->SetTextureStageState( 0, D3DTSS_ALPHAOP,   D3DTOP_DISABLE );
     // this mesh has the mapping coordinates embedded in
     // its custom vertex structure
     g_pd3dDevice->SetStreamSource(0,g_pVB[k+16],0,sizeof(CUSTOMVERTEX_1));
     g_pd3dDevice->SetFVF( D3DFVF_CUSTOMVERTEX_1 );
     g_pd3dDevice->DrawPrimitive( D3DPT_TRIANGLES,0,g_MeshVertexCount[k+16] );
   }
   g_pd3dDevice->EndScene();                      // end the scene
   g_pd3dDevice->Present( NULL,NULL,NULL,NULL);   // all done
 }
```

Listing 13.23. Rendering the scene involved establishing a viewpoint and lighting conditions and then drawing the triangular primitives from one of two lists: a list for sets of colored triangles and a list for sets of textured triangles.

user interface decorations of your choice. Subsequent chapters will build on and use sections of the code developed here.

In the few pages of this chapter, we cannot hope to cover all the intricate and subtle points of the real-time APIs. Nor can we explore their power for realistic rendering with complex shaders, shadowing or exotic lighting which are possible with both the OpenGL and Direct3D libraries. This, however, is the subject of the next chapter.

Bibliography

[1] Advanced Micro Devices, Inc. "RenderMonkey Toolsuite". http://ati.amd.com/developer/rendermonkey/, 2006.

[2] R. S. Ferguson. *Practical Algorithms for 3D Computer Graphics*. Natick, MA: A K Peters, 2001.

[3] M. J. Kilgard. "The OpenGL Utility Toolkit". http://www.opengl.org/resources/libraries/glut/, 2006.

[4] Microsoft Corporation. "DirectX SDK". http://msdn.microsoft.com/directx/sdk/, 2006.

[5] NVIDIA Corporation. "Cg Toolkit 1.5". http://developer.nvidia.com/object/cg_toolkit.html, 2006.

[6] D. Shreiner, M. Woo, J. Neider and T. Davis. *OpenGL Programming Guide,* Fifth Edition. Reading, MA: Addison-Wesley Professional, 2005.

[7] R. Wright and M. Sweet. *OpenGL Superbible,* Second Edition. Indianapolis, IN: Waite Group Press, 1999.

14

High-Quality 3D with OpenGL 2

Chapter 13 describes how to use OpenGL to deliver realistic 3D graphics on any PC computer hardware. Even for scenes with several hundred thousand polygons, the specially designed GPU on the display adapter is able to render high-resolution images at more than 50 frames per second, i.e., in real time. OpenGL provides an API which hides from the application program any detail of whether the 3D pictures appear as a result of a traditional software renderer or a hard-wired version of the rendering pipeline. Despite the undoubted quality of the 3D that OpenGL, delivers it has limitations; it utilizes Gouraud shading, for example, instead of Phong shading.

In the advent of the GPU, with its increasing speed and power, the question arose as to how to exploit this technology to achieve the most advantageous leap forward in graphics quality and performance. And the simple answer was *make the GPU programmable*. Give to the application programs the power to tell the GPU exactly how to process the vertex, polygon and texture data. Of course there is a catch. The GPU is not simply a faster computer processor. It has characteristics that make it unique, and so a new programming language and programming paradigm needed to be created for these blisteringly fast little chips.

In this chapter, we shall look at how OpenGL has evolved to support the new programmable GPU hardware, which is vital for rendering with much greater realism in real time, and how its extra features allow application programs easy access to it. Specifically, we will show you how to write and use programs for the GPU written in the *OpenGL Shading Language* (GLSL).[1]

The definitive reference source for OpenGL Version 2 is the official guide [10], but it has few examples of shading-language programs. The original book on the shading language by Rost [9] explains how it works and gives numerous examples which illustrate its power. To get the most out of OpenGL, one should have both these books in one's library.

Of course, it is possible to use the Direct3D API to achieve similar results, but our opinion is that OpenGL 2 is a lot simpler to use and apply. Most importantly, it is completely compatible with stereoscopic rendering and cross-platform development, which is obviously very important in VR.

14.1   OpenGL Version 2

Turning back to Figures 13.2 and 13.3 in Chapter 13, you can see how the special graphics processing chips can turn a data stream consisting of 3D vertex and pixel information into rendered images within the frame buffer. In OpenGL, this is termed the *fixed functionality*. The fixed-functionality pipeline is also a bit restrictive because, for example, some application programs would like to make use of more realistic lighting models such as BRDF and subsurface reflection models that are so important in rendering human skin tones. Or, for that matter, more sophisticated shading algorithms, such as Phong shading. The list goes on. With such a long shopping list of complex algorithms vying to be in the next generation of GPUs, doing it all with a fixed-functionality processor was always going to be a challenge.

After some attempts to extend the hardware of the fixed functionality, all the graphics chip manufacturers realized that the only solution was to move away from a GPU based on fixed functionality and make the pipeline programmable. Now, one might be thinking that a programmable GPU would have an architecture not too dissimilar to the conventional CPU. However, the specialized nature of the graphics pipeline led to the concept of several

---

[1]The language is called a shading language because it tells the GPU how to shade the surfaces of the polygonal models in the rendering pipeline. This can include arbitrarily complex descriptions of light absorption, diffusion, specularity, texture/image mapping, reflection, refraction, surface displacement and transients.

forms of programmability. The things that one would typically want to do with the vertices (transform and lighting, for example) are different from what one would want to do with the fragments (texture and shading, for example) and different again from the logical operations on output images in the various frame buffers (masking for example).

Thus the concept of adding programmability to the fixed-functionality architecture gave rise to three interrelated, but distinct, processing subsystems:

1. Programmable vertex shaders perform calculations that relate to *per-vertex* operations, such as transformations, transients, lighting mapping coordinates.

2. Programmable fragment shaders do the calculations for *per-fragment* operations, such as coloring and texturing. Some lighting can be done in these processors as well.

3. Programmable pixel shaders perform logical and arithmetic operations on the system's various buffers.

The term *programmable shader* is used for all these, even though a vertex shader may modify the geometry (as in mesh skinning) and have nothing to do with shading at all. The fragment shader is usually combined with the pixel shader, and they both can be referred to as either *fragment* or *pixel* shaders.

Thus we can redraw the system diagram of the GPU (extending Figure 13.2) to show how these new programmable elements fit into the rendering pipeline (see Figure 14.1).

Having this hardware picture clear in our minds is very useful when designing programs to execute in the GPU processors. It makes it clear why there are restrictions regarding what can and can't be done in a shader program. It also illustrates why the flow of execution is as follows: run the vertex program first for every vertex and then, later on in the pipeline, run the fragment program. The fact that all vertices have the same function applied to each of them and that fragment processing can be done simultaneously allows for massive parallelism to be employed. At the time of writing, a good-quality programmable GPU, such as the 3Dlabs Wildcat Realizm, has 16 vertex processors and 48 fragment processors. Each processor operates concurrently, i.e., whilst the vertex processors may all run the same program, they do so in parallel. Similarly, all the fragment processors execute the same fragment-specific program simultaneously. In this specialized environment, there is a restriction

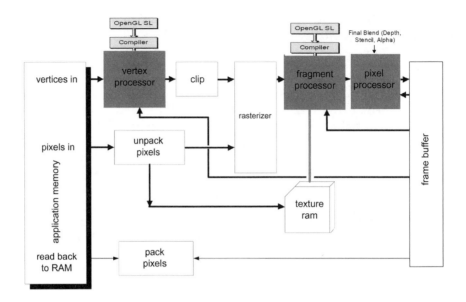

Figure 14.1. The OpenGL processing pipeline with its programmable processors. Programs are written in a high-level language, compiled and loaded into their respective processing units.

on the size of a program. Again, for example, the Realizm chip allows 1k instructions for a program in the vertex processor and 256k instructions in the fragment processor.

14.1.1 How It Works

In Version 2 of OpenGL, a small number of extra functions have been included to enable an application program to load the *shader program code* into the GPU's processor and switch the operation of the GPU from its fixed-functionality behavior to a mode where the actions being carried out on vertices and pixel fragments are dictated by their shader programs. The shader program code is written in a language modeled on C and called the *OpenGL Shading Language* (GLSL) [3]. Like C, the GLSL programs have to be compiled for their target processor, in this case the GPU. Different GPUs from different manufacturers such as NVIDIA, 3Dlabs and ATI will require different machine code, and so the GLSL compiler needs to be implemented as part of the graphics device driver which is supplied with a specific GPU.

After compilation, the device driver forwards the machine instructions to the hardware. Contrast this with Direct3D where its shader language is compiled by Direct3D, passed back to the application and then sent to the rendering hardware.

The GLSL syntax is rich, and we do not have space to go into it in any detail here. Rost et al. [3, 9] provide a comprehensive language definition and many examples. However, it should not be difficult to follow the logical statements in a GLSL code listing for anyone who is familiar with C. Nevertheless, it is important to appreciate the special features and restrictions which arise because of the close relationship with the language of 3D graphics and the need to prevent a shader code from doing anything that does not match well with the GPU hardware; for example, trying to access vertex data in a fragment shader. This is because at the fragment processing stage in the rendering pipeline, the vertex data has been interpolated by rasterization and the fragment does not know from which polygon it came. So it not possible to determine whether the fragment came from a triangular polygon or a quadrilateral polygon or what the coordinates of the polygon's vertices were. We shall now focus on what we believe to be the four key concepts in OpenGL Version 2:

1. The functions to compile, load and interact with the shader language programs.

2. The GLSL program data-type qualifiers: `const`, `attribute`, `uniform` and `varying`.

3. Built-in variables allow the shader programs to access and alter the vertex and fragment data as it is passed through the processors.

4. Samplers: these unique data types allow fragment shaders access to OpenGL's texture maps.

Perhaps the most important concept in shader programming is this: *the output from the vertex shader, which includes some of the built-in and all the user-defined varying variables that have been calculated in the vertex shader or set on a per-vertex basis, are interpolated across a polygon and passed to the fragment shader as a single value specifically for that fragment (or pixel) alone.*

14.1.2 Compiling, Installing and Using OpenGL Shaders

Putting this is terms of a simple recipe:

1. Read the shader source code from text files or store them in character memory buffers. The vertex and fragment programs are separate pieces of code; a typical shader will have one of each. The shader file or text buffer will contain text like this:

```
varying float LightIntensity;

void main(void){
    gl_FragColor = gl_Color * LightIntensity;
}
```

In a C text array, this might look like:

```
char gouraud_fragment_shader[] = "varying float LightIntensity;void main(void){"
                                 "gl_FragColor = gl_Color * LightIntensity;}";
```

2. Create a shader object using `glCreateShader(..)` which returns a handle to the program.

3. Tell the shader object what its source code is with `glShaderSource(..)`.

4. Compile the shader code with `glCompileShader(..)`.

5. Create the shader program with `glCreateProgram(..)`.

6. Attach the shader object to the program with `glAttachShader(..)`. A vertex and fragment shader can be attached to a single shader program.

7. Link the program with `glLinkProgram(..)`.

Listing 14.1 shows how this recipe would appear in an application program. The `progList` handle is used to identity the shader program. In this case, the shader program consists of both a vertex and fragment shader. It is possible to create shader programs that have either a vertex or fragment shader alone. Multiple shader programs can be compiled in this way, and when it

is time to use them, they are identified by their handle, e.g., the variable
progList.

```
GLuint vertList, fragList, progList;      // handles for shader codes and program
char VertexSource[]=" ....";
char fragmentSource[]=" ....";
vertList = glCreateShader(GL_VERTEX_SHADER);
fragList = glCreateShader(GL_FRAGMENT_SHADER);
glShaderSource(vertList, 1, &VertexSource,NULL);
glShaderSource(fragList, 1, &FragmentSource,NULL);
glCompileShader(vertList);
glCompileShader(fragList);
progList = glCreateProgram();
glAttachShader(progList,vertList);
glAttachShader(progList,fragList);
glLinkProgram(progList);
```

Listing 14.1. Installing a vertex and fragment shader pair. The OpenGL functions set
an internal error code and this should be checked.

To execute the shader, it is necessary to place a call to glUseProgram
(GLuint programHandle) in the drawing code before using any
glVertex...() commands. To switch to a different program, place a call to
glUseProgram(GLuint differentHandle) with a different program handle. To switch back to fixed-functionality rendering, call glUseProgram(0).
Note the use of an argument of 0. When the program and shader are no
longer required, they can be deleted with calls to glDeleteShader(GLuint
ShaderID) and glDeleteProgram(GLuint ProgramID).

Note: When a vertex or fragment program is in use, it disables fixed-
functionality processing. Thus, for example, a vertex shader must ensure any
geometric transformations you would have expected to occur under normal
circumstances are applied.

So far, we haven't said anything about passing information from application program to GPU shaders. This is done with a few additional functions
that write into the special variable types which are unique to GLSL. This is
the subject of the next section.

14.1.3   The Data-Type Qualifiers and Samplers

The OpenGL Shading Language supports the usual C-style data types of
float, int etc. To these it adds various types of vector and matrix decla-
rations such as vec3 (a 3×1 vector), vec2, and mat4 (a 4×4 array), for

example. Vectors and matrices are particularly pertinent for 3D work. Whilst these additional data types occur frequently in shader program, perhaps the most interesting element of GLSL variables are the *qualifiers*, e.g., `varying`. These are important because when applied to global variables (declared outside the `main` or other functions), they provide a mechanism for parameters to be passed to the shader code from the application. They also provide a mechanism for the vertex shader to pass information to the fragment shader and to interact with the *built-in variables* which provide such things as the transformation matrices calculated in the fixed-functionality pipeline. A good understanding of the meaning and actions of the qualifiers is essential if one wants to write almost any kind of shader. We will now consider a brief outline of each:

- `const`. As the name implies, these are constants and must be initialized when declared, e.g., `const vec3 Xaxis (1.0,0.0,0.0);`.

- `attribute`. This qualifier is used for variables that are passed to the vertex shader from the application program on a per-vertex basis. For example, in the shader they are declared in the statement `attribute float scale;`. A C/C++ application program using a shader with this global variable would set it in a call to the function `glVertexAttrib1f(glGetAttribLocation(Prog,"scale"),2.0)`. Variables declared with this qualifier are not accessible in a fragment shader. They are expected to change often, i.e., for every vertex.

- `uniform`. Variables declared with this qualifier are available in both vertex and fragment shaders because they are uniform across the whole primitive. They are expected to change infrequently. For example, a vertex and fragment shader might declare `uniform vec3 BaseColor;`. The application program would set this uniform variable with a call to `glUniform3f(getUniLoc(Prog, "BaseColor"), 0.75, 0.75, 0.75)`.

- `varying`. Variables qualified in this way are used to pass information from a vertex shader to a fragment shader. They are set on a per-vertex basis and are interpolated to give a value for the fragment being processed in the fragment shader.

- *Samplers.* Samplers are the way in which the OpenGL shading language uses image maps. There are 1D, 2D, 3D, cube, rectangular and shadow samplers.

14.1.4 The Built-in Variables

In both the vertex and fragment shader, the ability to *read from* and *write to* predefined (called built-in) qualified variables provides the interface to the fixed-functionality pipeline and constitutes the shader's input and output. For example, `attribute vec4 gl_Vertex;` gives the homogeneous coordinates of the vertex being processed. The built-in variable `vec4 gl_Position;` must be assigned a value at the end of the vertex shader so as to pass the vertex coordinates into the rest of the pipeline. A vertex shader must also apply any transformations that are in operation, but this can be done by a built-in shader function: `ftransform();`. Thus a minimal vertex shader can consist of the two lines:

```
void main(void){
  // apply the fixed functionality transform to the vertex
  gl_Position=ftransform();
  // copy the vertex color to the front side color of the fragment
  gl_FrontColor=gl_Color;
}
```

Qualifier	Type	Variable	R/W	Information
attribute	vec4	gl_Vertex	Read	Input vertex coordinates
attribute	vec4	gl_Color	Read	Input vertex color
attribute	vec4	gl_Normal	Read	Input vertex normal
attribute	vec4	gl_MultiTexCoords0	Read	Vertex texture coords
const	int	gl_MaxLights	Read	Maximum number of lights permitted
const	int	gl_MaxTextureCoords	Read	Maximum number of texture coordinates
uniform	mat4	gl_ModelViewMatrix	Read	The model position matrix
uniform	mat4	gl_ProjectionMatrix	Read	The projection matrix
uniform	struct	gl_LightSourceParameters	Read	Light source settings
	vec4	gl_Position	Write	Must be written by all vertex shaders
varying	vec4	gl_FrontColor	Write	Front facing color for fragment shader
varying	vec4	gl_BackColor	Write	Back color to pass to fragment shader
varying	vec4	gl_TexCoord[]	Write	Texture coordinates at vertex

Table 14.1. A subset of a vertex shader's input and output *built-in* variables.

Qualifier	Type	Variable	R/W	Information
	vec4	gl_FragCoord	Read	Window relative coordinates of fragment
	bool	gl_FrontFacing	Read	TRUE if fragment part of front-facing primitive
uniform	mat4	gl_ModelViewMatrix	Read	The model position matrix
uniform	mat4	gl_ProjectionMatrix	Read	The projection matrix
varying	vec4	gl_Color	Read	Fragment color from vertex shader
varying	vec4	gl_TexCoord[]	Read	Texture coordinates in fragment
	vec4	gl_FragColor	Write	Output of fragment color from shader
	vec4	gl_FragDepth	Write	Z depth, defaults to *fixed function* depth

Table 14.2. A subset of a fragment shader's input and output *built-in* variables.

Note that the = assignment operator is copying all the elements of a vector from source to destination. Tables 14.1 and 14.2 list the built-in variables that we have found to be the most useful. A full list and a much more detailed description on each can be found in [9] and [3].

```
OpenGL Version:  1.5.4454 Win2000 Release - Vendor :  ATI Technologies Inc.
OpenGL Renderer: RADEON 9600 XT x86/MMX/3DNow!/SSE
No of aux buffers 0  - GL Extensions Supported:
GL_ARB_fragment_program  GL_ARB_fragment_shader   ...
GL_ARB_vertex_program    GL_ARB_vertex_shader     ...

OpenGL Version:  2.0.0 - Vendor :  NVIDIA Corporation
OpenGL Renderer: Quadro FX 1100/AGP/SSE/3DNOW!
No of aux buffers 4 - GL Extensions Supported:
GL_ARB_fragment_program  GL_ARB_fragment_shader   ...
GL_ARB_vertex_program    GL_ARB_vertex_shader     ...
```

Figure 14.2. Output from the GLtest program, which interrogates the display hardware to reveal the level of feature supported. Extensions that begin with GL_ARB_ indicate that the feature is supported by the OpenGL Architecture Review Board and are likely to be included in all products. Those which begin with something like GL_NV_ are specific to a particular GPU manufacturer.

14.1.5   Upgrading to Version 2

Because the new functionality of OpenGL 2.0 is so tightly integrated with the hardware, an upgrade may entail replacing the graphics adapter. In terms of software, it is the responsibility of the device driver to support the extra functionality in the language (and therefore the adapter manufacturer's responsibility). From an application programmer's point of view, the OpenGL libraries are merely links to the device driver, and so long as the hardware vendor provides an OpenGL 2.0 driver for your operating system, it is not difficult to create a pseudo (or stub) library for the new Version 2 functions. OpenGL Version 1 remained stable for many years and forms what one might term *the gold standard*. Extensions have been gradually added ([10] documents the new features appearing in the minor revisions 1.1 through 1.5). To check what extension features a particular driver supports, OpenGL provides a function that can be called to list the extensions and version and give some indication of the hardware installed. Listing 14.2 shows a small program for testing the adapter's capability.

Its output, as shown in Figure 14.2, illustrates the adapter vendor, version, and what extensions are supported. The key supported extensions for OpenGL 2 are the fragment and vertex shader programs.

Once you know what version and features your hardware supports, it is much easier to debug an application program.

```c
#include <stdio.h>
#include <gl\glu.h>
#include <gl\glaux.h>

void main(int argc, char **argv){
 const char *p;
 GLint naux;
 auxInitDisplayMode(AUX_SINGLE | AUX_RGBA);
 auxInitPosition(0,0,400,400);
 auxInitWindow(argv[0]);
 printf("OpenGL Version: %s\n", glGetString(GL_VERSION));
 printf("OpenGL Vendor : %s\n", glGetString(GL_VENDOR));
 printf("OpenGL Renderer: %s\n", glGetString(GL_RENDERER));
 printf("\nGL Extensions Supported:\n\n");
 p=glGetString(GL_EXTENSIONS);
 while(*p != 0){putchar(*p); if(*p == ' ')printf("\n"); p++; }
 glGetIntegerv(GL_AUX_BUFFERS,&naux);
 printf("No of auxiliary buffers %ld\n",naux);
 glFlush();
}
```

Listing 14.2. List the OpenGL extensions supported by the graphics hardware.

14.1.6   Using OpenGL 2 in a Windows Program

There are two alternatives: use the GLUT (see Section 13.2) or obtain the
header file `glext.h` (available from many online sources), which contains
definitions for all the OpenGL constants and function prototypes for all
functions added to OpenGL since Version 1. Make sure you use the Version 2 `glext.h` file. Of course, providing prototypes for functions such as
`glCreateProgram()` does not mean that when an application program is
linked with the `OpenGL32.lib` Windows stub library for `opengl32.dll`,
the function code will be found. It is likely that a *missing function* error will
occur. This is where an OpenGL function that is unique to Windows come
in. Called `wglGetProcAddress(...)`, its argument is the name of a function whose code we need to locate in the driver. (Remember that most of the
OpenGL functions are actually implemented in the hardware driver and not
in the Windows DLL.)

Thus, for example, if we need to use the function `glCreateProgram()`,
we could obtain its address in the driver with the following C code:

```
typedef GLuint (* PFNGLCREATEPROGRAMPROC) (void);
PFNGLCREATEPROGRAMPROC glCreateProgram;
glCreateProgram = (PFNGLCREATEPROGRAMPROC) wglGetProcAddress(glCreateProgram);
```

After that, `glCreateProgram()` can be used in a program in the normal
way. This idea is so useful that in all our programs we created an additional
header file, called `gl2.h`, that can be included in every file of our programs. It
wraps around the extensions header file, called `glext.h`, and provides an initialization function that finds all the function addresses specific to OpenGL 2
in the driver. (Any program that wishes to use it makes a call to the function
`gl2Initialize()` once from the `WinMain()` entry function and defines
the symbol `_gl2_main_` in any one of the project files which have included
the header `gl2.h`.)

By following these ideas, our programs will have a good degree of future
proofing. Providing we have the latest version of the hardware driver and
the `glext.h` header file, it will not really matter what development tools or
Windows version we choose to use.[2]

With all this background preamble out of the way, we can look at an
example of how to use programmable shaders in practice.

---

[2]There is a function for UNIX systems, the GLX extension for X windows, that performs
an analogous task to `wglGetProcAddress`.

14.2   Using Programmable Shaders

There is a large body of literature on algorithms, designs and implementation strategies for programmable shaders. We can only briefly explore this vast topic, and would refer you to the specialist text by Olano et al. [5] and the SIGGRAPH course notes [6] for a much more rigorous and detailed explanation of the subject. The whole topic began with the work of Perlin [7], which had such a profound effect on algorithmic texturing that GLSL has built-in functions to generate Perlin noise.

However, GLSL has utility outside of the usual meaning of programmable shaders: bricks, marble, etc. Lighting, shading and image mapping are much improved by writing application-specific vertex and fragment shader programs. Phong shading, bump mapping as well as algorithmic textures are just a few things that dramatically improve the realism of the rendered images when compared with what *fixed functionality* can deliver. In the remainder of this chapter, we shall explore how to incorporate the use of programmable shaders into the program we discussed in Section 13.1 which visualizes triangular mesh models.

14.2.1   Modifying the Program

As we have already mentioned, the additional features of OpenGL 2 can be incorporated into any application by adding the special header file that allows us to give the illusion of using Version 2 functions in the traditional way. We must initialize this mechanism by a call to `gl2Initialize()`, and because we will be modifying the program from Chapter 13, it seems like a good idea to put most of the additional code in a single file and provide a few hooks that can be used at key points in the original code. These hooks take the form of three functions and an array of program identifiers which can be called to achieve the following:

- `void ShadersInit(void)`. This function initializes the OpenGL 2 function name pointers. It reads shader program sources from a particular folder (such as the same folder from where the application was loaded) and sets default values for any uniform or attribute shader variables.

- `extern GLuint progList[NPROGRAMS]`. This is a list of handles to the shader programs. It is made globally available so that the application program can, if necessary, set uniform or attribute values from within the rendering code.

- void UseShaderProgram(long id). By calling this function, the rendering code can switch to use one of the GPU shader programs and configure the shader with appropriate uniform or attribute values.

- void ShadersClose(void). This removes the shader programs from the GPU.

In addition, direct calls from within the rendering code will be made to three of the Version 2 functions:

- (GLuint)getUniLoc(GLuint, char*). This will get the address of a named shader program uniform variable.

- glUniform1f(GLuint, GLfloat). This function is one of several variants.[3] The first argument is the address of a named shader program uniform variable (returned by getUniLoc(), for example) and the second argument is the value of that variable which is to be passed to the shader program.

- glVertexAttrib3f(GLuint ,GLfloat,GLfloat,GLfloat). This sets vertex attributes in an analogous manner to the vertex and shader uniform variables.

It is possible (with minimal change in rendering code) to implement an extremely wide range of GPU programmed effects: lighting, shading, texturing, image mapping and animation. Before examining how each of these effects may be programmed into the GPU, a look at Listing 14.3 will illustrate the logic of our additional shader management file and Listing 14.4 shows how we must change the rendering function to match.

```
 //   include other main headers
..///
 //   define this symbol once - here in this file
#define __gl2_main_
 //   This is a wrapper header file for the "glext.h"
#include "gl2.h"
 //   and some utility functions.
```

Listing 14.3. GPU programmability can be added to an existing application by providing most of the functionality in a single file. All the shaders can be compiled and executed by using a few functions. This approach minizes use of the Version 2 extensions which need to be called directly from the application's rendering routine.

---

[3]For example, glUniform3f(..) sets a three-element vector, glUniform1i(..) an integer and so on.

```
   //   number of GPU shader programs to be loaded
#define NPROGRAMS 32
  //  list of shader program IDs (handles)
GLuint progList[NPROGRAMS];
static GLuint vertList[NPROGRAMS], fragList[NPROGRAMS];
static char  shader_location_path[] "\\programs\\shader\\";
static char *shader_names[NPROGRAMS] = { // list of shader program names
     "brick","marble", "A... "....};       // Read the shaders in pairs from files
  // XXX.vert and XXX.frag  (e.g. brick.vert brick.frag)
// Read the shader program's source codes from text tiles
// - the "gl2Read* " function is includedin the file "gl2.h".
static int readShaderSource(char *file, GLchar **vertexShader,
GLchar **fragmentShader){
  return gl2ReadShaderSource(file,shader_location_path,vertexShader,
fragmentShader); }
void ShadersInit(void){
 GLint i,status;
 char *vs1; char *fs1;
 GLchar *VertexShaderSource, *FragmentShaderSource;
   gl2Initialize();                     // this function is in file "gl2.h"
 for (i=0;i<NPROGRAMS;i++){     // read each shader program
 // read the shader Vertex/Fragment programs
   readShaderSource(shader_names[i], &vs1, &fs1);
 // create a vertex shader object
   vertList[i] = glCreateShader(GL_VERTEX_SHADER);
   fragList[i] = glCreateShader(GL_FRAGMENT_SHADER);
 // Set the shader object to get its source from our
   glShaderSource(vertList[i], 1, &vs1,NULL);
 // vertex and fragment shaders.
   glShaderSource(fragList[i], 1, &fs1,NULL);
 // compile the vertex shader code
   glCompileShader(vertList[i]);
 // check for error in compilation
   glGetShaderiv(vertList[i],GL_COMPILE_STATUS,&status);
   if(status == GL_FALSE)MessageBox(NULL,"Error in Vertex shader",NULL,MB_OK);
   glCompileShader(fragList[i]);
 // check for error in compilation
   glGetShaderiv(fragList[i],GL_COMPILE_STATUS,&status);
   if(status == GL_FALSE)MessageBox(NULL,"Error in Fragment shader",NULL,MB_OK);
 // make a shader program
   progList[i] = glCreateProgram();
 // add the compiled vertex shader to the program
   glAttachShader(progList[i],vertList[i]);
   glAttachShader(progList[i],fragList[i]);
 // ling the program with buit-in functions
   glLinkProgram(progList[i]);
 // NOW EVERYTHING IS READY TO GO
   if(i == ...    //
 // give attribute "names" an equivalent integer ID
     glBindAttribLocation(progList[i],1,"TextureA");
     ..//
   }
```

Listing 14.3. (continued).

```
  else if (i == ... // set some of the "uniform" variables for shader "i"
    (ones that don't need to change)
    glUniform3f(getUniLoc(progList[i], "LightColor"), 0.0, 10.0, 4.0);
    glUniform1f(getUniLoc(progList[i], "Diffuse"),0.45);
 // and make any other settings
    .. //
  }    // GO TO NEXT SHADER
}
void ShadersClose(void){
 int i; for(i=0;i<NPROGRAMS;i++){
    glDeleteShader(vertList[i]); glDeleteShader(fragList[i]);
    glDeleteProgram(progList[i]);
 }
}
void UseShaderProgram(long id){         // choose which shader program to run
 glUseProgram(id);                      // tell OpenGL
 if(id == 1){                           // set any parameters for the shader
    glUniform3f(getUniLoc(progList[id], "BrickColor"), 1.0, 0.3, 0.2);
 }
 else if ( ...) { //                    // do it for the others.
 }                                      // It may be necessary to set some of
}                                       // the attrubutes in the rendering
                                        // code itself.
```

Listing 14.3. (continued).

```
 ..//                      // other functions and headers as before
extern GLuint progList[];   // the list of GPU program handles
                            // (from file shaders.c)

void RenderScene(void){     // render the Mesh model - in response
                            // to WM_PAINT messag
 .. //
 UseShaderProgram(3);       // Phong shading program
 glEnable(GL_COLOR_MATERIAL);
 glBegin(GL_TRIANGLES);
 .. //
 fp=MainFp; for(i=0;i<Nface;i++,fp++){  //
    ..// render the polygon
    for(k=0;k<3;k++){ glMaterial*(..); glVertex(...); gNormal(...);  // etc/
 }
 glEnd();
 UseShaderProgram(0);       // switch back to fixed functionality
```

Listing 14.4. Changing the rendering code to accommodate GPU shaders. The code
fragments that appear here show the modifications that need to be made in Listings 13.11, 13.12 and 13.14 in Chapter 13.

```
if(Mapped && Nmap > 0)for(k=0;k<Nmap;k++){
 if(bBumpMapped){ //
   UseShaderProgram(5);  // bump
 }
 else if(bReflectionMapped){
   UseShaderProgram(4);  // environment
   if(MapsFp[k].bLoaded)
     glUniform1i(getUniLoc(progList[4], "RsfEnv"), GL_TEXTURE0+1+k);
 }
 else{
   UseShaderProgram(2);   // use image map
   if(MapsFp[k].bLoaded)
     glUniform1i(getUniLoc(progList[2], "Rsf"), GL_TEXTURE0+1+k);
 }
 glBegin(GL_TRIANGLES);
 fp=MainFp; for(i=0;i<Nface;i++,fp++){
   for(j=0;j<3;j++}{
     glVertex(...); gNormal(...);  // etc/
     .. // calculate mapping coordinates
     if(!bReflectionMapped)glTexCoord2f(alpha,beta);
     if(bBumpMapped){
       .. // calculate a tangent vector at vertex on surface
       glVertexAttrib3f(1,tangent[0],tangent[1],tangent[2]);
     }
   }
 }
 glEnd();
 UseShaderProgram(0);
 }
}
```

Listing 14.4. (continued).

There are a few comments we should make about the changes to the rendering code that become necessary to demonstrate the use of the example shaders we will explore shortly.

- We will not use display lists in the examples. Instead, the WM_PAINT message which calls Draw3dScene(..) (in Listing 13.11) will call RenderScene() as shown in Listing 14.4.

- The polygons are sorted so that all those that use the same shader program are processed together.

- Individual image maps are loaded into texture memory at the same time as the mesh is loaded. Each image map is bound to a different OpenGL texture identifier:

```
glActiveTexture(GL_TEXTURE0+textureID);
glBindTexture(GL_TEXTURE_2D,textureID+1);
```

Thus, when the shader program that carries out the texturing is executing, for each image map, a uniform variable (for example "RsfEnv") is set with a matching texture identifier, such as GL_TEXTURE0+1+k.

14.3   Some Basic Shader Programs

When vertex and fragment shader programs are installed, the fixed-functionality behavior for vertex and fragment processing is disabled. Neither ModelView nor ModelViewProjection transformations are applied to vertices. Therefore, at the very least, the shader programs must apply these transformations.

14.3.1   Emulating Fixed Functionality

It is possible to provide a complete emulation of the fixed-functionality behavior of the rendering pipeline in a pair of programmable shaders. The code required to do this is quite substantial, because transformations, mapping, lighting, fog etc. all have to be accounted for. Rost [9] provides a comprehensive example of fixed-functionality emulation. In this section, the examples will highlight how to implement a few of the most visually noticeable effects; that is, Phong shading and bump mapping.

14.3.2   Lighting and Shading

Lighting and shading are likely to be an essential part of any programmable shader. As we saw in Section 7.5.1 a lighting model is a particularly important stage in the rendering pipeline, and any shader we create should provide a simulation of diffuse and specular illumination. After lighting the next most important element is the Gouraud or Phong shading models as described in Section 7.5.2. Gouraud shading is provided by the fixed-functionality behavior, so it really is the ability to implement Phong shading that is the first real advantage of the programmability of the GPU. Figure 14.3 illustrates the difference that results when rendering a smooth sphere using the fixed-

```
// Vertex Shader -  Basic Gouraud shading
//
varying float Intensity; // pass light intensity to fragment shader

void main(void){
 // calculate the vertex position - in EYE coordinates
 vec3  Vpos  = vec3(gl_ModelViewMatrix * gl_Vertex);
 // a normalized vector in directon vertex to (0,0,0)
 vec3  vVec  = normalize(-Vpos);
 // calculate the vertex normal in - EYE
 vec3  Vnorm = normalize(gl_NormalMatrix * gl_Normal);
 // obtain a direction vector in direction of Light to vertex
 vec3   lVec  = normalize(vec3(gl_LightSource[0].position)-Vpos);
 // calcaulat the Diffise light component > 0
 float diff  = max(dot(lVec, vVec), 0.0);
 // Now obtain the specular light component using standard equations.
 vec3  rVec  = reflect(-lVec,Vnorm);
 vec3  vVec  = normalize(-Vpos);
 float spec  = 0.0;
 if(diffuse > 0.0){
    spec = max(dot(rVec, vVec), 0.0);
    spec = pow(spec, 18.0);
 }
 // combine diffuse and specular light component in 'simple 50/50' model
 Intensity = 0.5 * diffuse +  0.5 * spec;
 // Set the built-in variable vertex (All transformations
 // must be applied so use the SL function ftransform().
 gl_Position = ftransform();
 // Set the built in variable to be the current color.
 gl_FrontColor=gl_Color;
}
```

```
// Fragment Shader - Basic Gouraud shading
//
varying float Intensity;  // Input from vertex shader (interpolated)
                          // from the vertices bounding the polygon
                          // that gives rise to this fragment.
void main(void){
 // set output fragment color to input interpolated color
 // modulated by the interpolated light intensity
 gl_FragColor = gl_Color * Intensity;
}
```

Listing 14.5. A shader program pair that emulates very basic Gouraud shading for a single light source and plain colored material. The OpenGL "eye" coordinate system places the origin at $(0, 0, 0)$ with a direction of view of $(0, 0, 1)$ and an *up* direction of $(0, 1, 0)$. OpenGL's *y*-axis is vertical.

```
// Vertex Shader - Basic Phong smoothing model

varying vec4 PP;   // vertex location in EYE coordinates - passed to
                   // fragment shader
varying vec3 PN;   // vertex normal in EYE coordinates  - passed to
                   // fragment shader

void main(void){
 // copy the vertex position
 // compute the transformed normal
 PP            = gl_ModelViewMatrix * gl_Vertex;
 PN            = gl_NormalMatrix * gl_Normal ;
 gl_Position   = gl_ModelViewProjectionMatrix * gl_Vertex; // required
 gl_FrontColor = gl_Color;                                 // required
}
```

```
// FragmentShader - Basic Phong shading model

const float SpecularContribution=1.0;
const float DiffuseContribution=1.0;

// specular light color (WHITE) should be light color
const vec4 lc = (1.0,1.0,1.0,1.0);

varying vec3 PN;   // normal from vertex shader - interpolated for THIS fragment
varying vec4 PP;   // Position in EYE coordinates of this fragments location.
                   // interpolated from vertex coordinates in vertex shader.
void main(void){
    // compute a vector from the surface fragment to the light position
    vec3  lightVec  = normalize(vec3(gl_LightSource[0].position) - PP);
    // must re-normalize interpolated vector to get Phong shading
    vec3 vnorm=normalize(PN);
    // compute the reflection vector
    vec3  reflectVec = reflect(-lightVec, vnorm);
    // compute a unit vector in direction of viewing position
    vec3  viewVec   = normalize(-PP);
    // calculate amount of diffuse light based on normal and light angle
    float diffuse    = max(dot(lightVec, vnorm), 0.0);
    float spec       = 0.0;
    // if there is diffuse lighting, calculate specular
    if(diffuse > 0.0){
        spec = max(dot(reflectVec, viewVec), 0.0);
        spec = pow(spec, 32.0);
    }
    // add up the light sources - write into output build-in variable
    gl_FragColor = gl_Color * diffuse  + lc * spec;
}
```

Listing 14.6. The Phong shading program pair for a single light source.

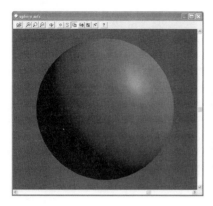 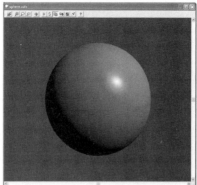

Figure 14.3. Simulating a smooth sphere using the built-in fixed-functionality Gouraud shader model (on the left) and Phong's model (on the right) programmed as a shader in the GPU.

functionality Gouraud shading model and our own programmable shaders for Phong shading.

Listing 14.5 shows a shader program pair which emulates a very basic Gouraud shading algorithm. Although simple, it shows the steps that are essential in setting up a vertex shader: Transformation, lighting calculation, surface color interpolation.

Phong's shading model delivers a much better illusion of smoothing, but at a price. The surface normal has to be interpolated across the polygon and the light intensity determined on a per-fragment basis in the fragment shader. Listing 14.6 shows how to implement Phong's model. An interesting comparison can be made with Listing 14.5, because for Gouraud shading, most of the code is in the vertex shader, whereas for Phong shading, it is in the fragment shader.

14.3.3 Image Mapping

OpenGL's fixed-functionality texture/image mapping is excellent, and the extensions introduced in minor Versions 1.1 through 1.5 which allowed several maps to be blended together is sufficient for most needs. Nevertheless, the shading language (GLSL) must facilitate all forms of image mapping so that it can be used (for example, with Phong's shading/lighting model) in all programs. In GLSL, information from texture maps is acquired with the use of a *sampler*. Rost [9] provide a nice example of image map blending with

```
// Vertex Shader  - for traditional image mapping

varying float Diffuse;   // diffuse light intensity
varying vec2  TexCoord;  // texture coordinate (u,v)

void main(void){
 // same as Gouraud shaded
 vec3 PP = vec3 (gl_ModelViewMatrix * gl_Vertex);
 // normal vector
 vec3 PN = normalize(gl_NormalMatrix * gl_Normal);
 // light direction vector
 vec3 LV = normalize(vec3(gl_LightSource[0].position)-PV);
 // diffuse illumination
 Diffuse = max(dot(LV, PN), 0.0);
 // Get the texture coordinates - choose the first -
 // we only assign one set anyway
 // pass to fragment shader
 TexCoord = gl_MultiTexCoord0.st;
 gl_Position = ftransform();          // required!
}
```

```
// Fragment Shader - simple 2D texture/image map, diffuse lighting

uniform sampler2D Rsf;  // this is the image sampler that matches
                        // the one in listing @ref?program:2gl2:shaders2`

varying float Diffuse;  // from vertex shader
varying vec2 TexCoord;  // from vertex shader

void main (void){
 // Get the fragment color by addressing the texture map pixel
 // date - the "texture2D" built in function takes the texure sampler "Rsf"
 // and fragment interpolated texture coorinates as arguments and the
 // ".RGB" (what SL calls a "swizzle") extracts the RGB values from the
 // texture2D() return value.
 vec3 color = (texture2D(Rsf, TexCoord).rgb * Diffuse);
 // promote the 3 element RGB vector to a 3 vector and assign to output
 gl_FragColor = vec4 (color, 1.0);
}
```

Listing 14.7. The image mapping shader program pair.

three maps. We will keep it simple in our example and show how a shader program pair can paint different maps on different fragments, which is a requirement for our small mesh model viewer application. The code for these

```
// Vertex Shader - Reflection mapping

varying vec3  Normal;  // surface normal - pass to fragment
varying vec3  EyeDir;  // vector from viewpoint (0,0,0) to vertex

void main(void) {
 vec4 pos     = gl_ModelViewMatrix * gl_Vertex;
 Normal       = normalize(gl_NormalMatrix * gl_Normal);
// use swizzle to get xyz (just one of many ways!)
 EyeDir       = pos.xyz;
 gl_Position = gl_ModelViewProjectionMatrix * gl_Vertex; // required
}
```

```
// Fragment Shader - reflection  mapping

const vec3 Xunitvec = vec3 (1.0, 0.0, 0.0);
const vec3 Yunitvec = vec3 (0.0, 1.0, 0.0);

uniform sampler2D RsfEnv;  // this is our sampler to identify the iamge map

varying vec3  Normal;     //  from vertex shader
varying vec3  EyeDir;     //  from vertex shader

void main (void){
    // Compute reflection vector - built-in function :-)
    vec3 reflectDir = reflect(EyeDir, Normal);
    // Compute the addess to use in the image map - we need an
    // address in the range [0 - 1 ] (angles of looking up /down)
    // see section @ref?section:1rendering:mapping`
    // in chapter @ref?chapter:1rendering` .
    vec2 index;
    index.y = dot(normalize(reflectDir), Yunitvec);
    reflectDir.y = 0.0;
    index.x = dot(normalize(reflectDir), Xunitvec) * 0.5;
    // get the correct range [0 - 1]
    if (reflectDir.z >= 0.0)   index = (index + 1.0) * 0.5;
    else{
        index.t = (index.t + 1.0) * 0.5;
        index.s = (-index.s) * 0.5 + 1.0;
    }
    // copy the addressed texture pixel to the output
    gl_FragColor = texture2D(RsfEnv, index);
}
```

Listing 14.8. The reflection map shading program pair.

shader programs is given in Listing 14.7. These are the GLSL programs used in conjunction with the mapping part of the code in Listing 14.4.

14.3.4   Reflection Mapping

The mesh model viewer program must also be able to simulate reflection by using a mapping technique. Fixed functionality provided an easy way to do this by automatically generating texture coordinates based on an *environment map* (as was seen in Section 7.6.2). To write a GLSL program to simulate reflection/environment mapping, our code will also have to do that. Listing 14.8 shows the outline of a suitable shader pair. For reflection maps, we do not want to use any surface lighting models, and so this is omitted.

14.3.5   Bump Mapping

Our portfolio of shader programs would not be complete without the ability to use an image texture as a source of displacement for the surface normal, i.e., bump mapping. Bump mapping requires us to use a `sampler2D` and provide a tangent surface vector. The same tangent surface vector that we will use when discussing procedural textures (see Section 14.3.6) will do the job. You can see how this is generated from the image map parameters if you look at the application program code for which Listing 14.4 is an annotated outline. Vertex and fragment shaders for bump mapping are given in Listing 14.9. A few of comments are pertinent to the bump-mapping procedure:

- To minimize the work of the fragment shader, the vertex shader calculates vectors for light direction and eye (viewpoint) direction in a coordinate frame of reference that is in the plane of the polygon which will be rendered by the fragment shader (this is called *surface-local space*).

  Figure 14.4 illustrates the principle. In the world (global) coordinate frame of reference, the eye is at the origin. The polygon's normal **n**, tangent **t** and bi-normal **b** are calculated in the global frame of reference. Transformation of all vectors to a surface-local reference frame allows the fragment shader to perform position-independent lighting calculations because the surface normal is always given by the vector $(0, 0, 1)$.

  By doing this, we ensure that the surface normal vector is always given by coordinates $(0, 0, 1)$. As such, it is easily displaced (or bumped) in the fragment shader by adding to it the small changes Δu and Δv obtained from the bump map.

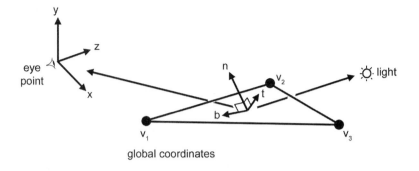

global coordinates

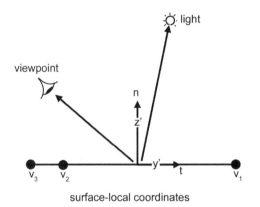

surface-local coordinates

Figure 14.4. Surface-local coordinates.

To obtain the surface-local coordinates, we first set up a rectilinear co-ordinate system with three basis vectors. We have two of them already: the surface normal $\mathbf{n} = (n_x, n_y, n_z)$ and the tangent vector $\mathbf{t} = (t_x, t_y, t_z)$ (passed to the vertex shader as a vector attribute from the application program). The third vector $\mathbf{b} = (b_x, b_y, b_z)$ (known as the *bi-normal*) is defined as $\mathbf{b} = \mathbf{n} \times \mathbf{t}$. Vectors $\mathbf{d}_w$ (such as viewpoint and light directions) are transformed from world coordinates to the surface-local frame by $\mathbf{d}_s = [T_{ws}]\mathbf{d}_w$, where $[T_{ws}]$ is the 3×3 transformation matrix:

$$[T_{ws}] = \begin{bmatrix} t_x & t_y & t_z \\ b_x & b_y & b_z \\ n_x & n_y & n_z \end{bmatrix}. \qquad (14.1)$$

```
// Vertex Shader -  Bump mapping
//

// These two vectors pass the direction from vertex to light
varying vec3 LightDir;
// and viewpoint (IN A COORD REFERENCE FRAME of the polygon!!!)
varying vec3 EyeDir;
// this passed surface texture coords to fragment shader
varying vec2  TexCoord;

attribute vec3 Tangent;  // tangent vector from application program

void main(void) {
    // do the usual stuff
    EyeDir          = vec3 (gl_ModelViewMatrix * gl_Vertex);
    gl_Position     = ftransform();
    gl_FrontColor=gl_Color;
    // pass on texture coords to fragment shader
    TexCoord        = gl_MultiTexCoord0.st;
    // Get normal, tangent and third vector to form a curviliner coord frame
    // of reference that lies in the plane of the polygon for which we are
    // processing one of its vertices.
    vec3 n = normalize(gl_NormalMatrix * gl_Normal);
    vec3 t = normalize(gl_NormalMatrix * Tangent);
    vec3 b = cross(n, t);
    // get the light direction and viewpoint (eye) direction in this new
    // coordinate frame of reference.
    vec3 v;
    vec3 LightPosition=vec3(gl_LightSource[0].position);
    v.x = dot(LightPosition, t);
    v.y = dot(LightPosition, b);
    v.z = dot(LightPosition, n);
    LightDir = normalize(v);
    v.x = dot(EyeDir, t);
    v.y = dot(EyeDir, b);
    v.z = dot(EyeDir, n);
    EyeDir = normalize(v);
}
```

```
// Fragment Shader  - Bump mapping

varying vec3 LightDir;  // input from vertex shader
varying vec3 EyeDir;
varying vec2 TexCoord;

uniform sampler2D Rsf;
```

Listing 14.9. Bump-mapping shader programs.

```
// fraction of light due to specular hightlights
const float SpecularFactor = 0.5;
const float Gradient = 4.0; // range [4.0 - 0.1]  bump gradient [large - small]

void main (void){
 vec3 SurfaceColor = vec3(gl_Color);  // get the color
 // get the surface normal perturbation vector -
 // it is encoded in the texture image
 vec2 grad = (texture2D(Rsf, TexCoord).rg);
 // calculate the new surface normal - perterb in (x,y) plane
 vec3 normDelta=normalize(vec3((grad.x-0.5)*Gradient,
 (grad.y-0.5)*Gradient,1.0));
 // now apply the lighting model
 vec3 litColor = SurfaceColor* max(dot(normDelta, LightDir), 0.0);
 vec3 reflectDir = reflect(LightDir, normDelta);
 float spec = max(dot(EyeDir, reflectDir), 0.0);
 spec *= SpecularFactor;
 litColor = min(litColor + spec, vec3(1.0));
 gl_FragColor = vec4(litColor, 1.0);
}
```

Listing 14.9. (continued).

- A fragment shader can have no knowledge of any adjacent fragments or vertices, so the normal displacement has to be encoded in the bump map. We do this by pre-processing the image pixel map after it is read/decoded and before it is copied into texture memory. We first convert it to a gray scale and then make a gradient map by storing the $\frac{\partial \mathbf{n}_u}{\partial u}$ as the *red* pixel component and the v gradient ($\frac{\partial \mathbf{n}_v}{\partial v}$) as the *green* component. The gradient map is copied to texture memory and thus the surface normal displacement vector is easily extracted from the texture by $\Delta \mathbf{n} = (\Delta n_1, \Delta n_2, 0) = (M_r(u, v), M_g(u, v), 0)$ (where $M_r(u, v)$ is the red component of map M at texture coordinate (u, v)). Because we are working with a normal now relative to the surface-local coordinates, $\mathbf{n} = (1, 0, 0)$ and the coordinates of the displaced normal ($\bar{\mathbf{n}} = \mathbf{n} + \Delta \mathbf{n}$) vector are $(\Delta n_1, \Delta n_2, 1)$.

- An alternative method of generating the displacement vector allows it to be calculated within the fragment shader program. This eliminates the pre-processing step at the expense of a little extra GPU processing in the fragment shader. By sampling the texture map at four points that have a small texture coordinate offset, the image's brightness gradient can be obtained. The following shader code fragment illustrates the idea:

```
const float d=0.01;
// Assumes gray R=G=B and get value
float y1 = (texture2D(Rsf, TexCoord+vec2( 0.0, d)).r);
// above and below the sample point.
float y2 = (texture2D(Rsf, TexCoord+vec2( 0.0,-d)).r);
// and to left
float x1 = (texture2D(Rsf, TexCoord+vec2( d, 0.0)).r);
// and right
float x2 = (texture2D(Rsf, TexCoord+vec2(-d, 0.0)).r);
vec3 normDelta=normalize(vec3((y2-y1)*Gradient,(x1-x2)*Gradient,1.0));
```

14.3.6   Procedural Textures

It is in the area of procedural textures that the real power of the programmable GPU becomes evident. More or less *any* texture that can be described algorithmically can be programmed into the GPU and rendered much faster than with any software renderer. This is an incredible step forward!

```
// Vertex Shader - procedural bumps

varying vec3 LightDir;       // vectors to be passed
varying vec3 EyeDir;         // to fragment shader

attribute vec3 Tangent;      // input vector from the application program

void main(void) {
 // usual stuff feed input in the OpenGL machione
 gl_Position    = ftransform();
 gl_FrontColor=gl_Color;
 gl_TexCoord[0] = gl_MultiTexCoord0;
 // get a coordinate system based in plane of polygon
 vec3 n = normalize(gl_NormalMatrix * gl_Normal);
 vec3 t = normalize(gl_NormalMatrix * Tangent);
 vec3 b = cross(n, t);
 // Calculate the direction of the light and direction of
 // the viewpoint from this VERTEX using a coord system
 // based at this vertex and in the plane of its polygon.
  EyeDir        = vec3 (gl_ModelViewMatrix * gl_Vertex);
  vec3 LightPosition=vec3(gl_LightSource[0].position);
  vec3 v;
```

Listing 14.10. Procedural bump mapping, illustrating how to generate little bumps or dimples completely algorithmically.

```
    v.x = dot(LightPosition, t);
    v.y = dot(LightPosition, b);
    v.z = dot(LightPosition, n);
    LightDir = normalize(v);
    v.x = dot(EyeDir, t);
    v.y = dot(EyeDir, b);
    v.z = dot(EyeDir, n);
    EyeDir = normalize(v);
}
```

```
// Fragment Shader  - procedural bumps

varying vec3 LightDir;   // from Vertex shader
varying vec3 EyeDir;     // from Vertex shader

// 8 bumps will span a texture coord range [0,1]
const float BumpDensity  = 8.0;
const float BumpSize     = 0.15;   //
const float SpecularFactor = 0.5; // fraction of the light that is specular
void main (void){
    vec2 c = BumpDensity * gl_TexCoord[0].st; // where are we in the texture
    vec2 p = fract(c) - vec2(0.5);            // scale bump relative to center
    float d, f;
    // calculate the normal perturbation based on distance from center of "bump"
    d = p.x * p.x + p.y * p.y;
    f = 1.0 / sqrt(d + 1.0);
    if (d >= BumpSize){ p = vec2(0.0); f = 1.0; }
    // surface normal is always aligned along "z"
    // so just male bump in (x,y) plane
    vec3 normDelta = vec3(p.x, p.y, 1.0) * f;
    // get the surface color
    vec3 SurfaceColor = vec3(gl_Color);
    // work out its lighting based on local (bumped)normal
    // and viewpoint direction
    vec3 litColor = SurfaceColor* max(dot(normDelta, LightDir), 0.0);
    vec3 reflectDir = reflect(LightDir, normDelta);
    float spec = max(dot(EyeDir, reflectDir), 0.0);
    spec *= SpecularFactor;
    litColor = min(litColor + spec, vec3(1.0));
    gl_FragColor = vec4(litColor, 1.0);  // pass it on dowm the pipeline
}
```

Listing 14.10. (continued).

As a taste of the power of the procedural shader, we offer an example of a pair of shaders to create some little bumps, or dimples depending on your point of view, similar to those described in [2]. Again, two short shader programs deliver the goods (see Listing 14.10). We will not say much more about it here because the essential ideas are the same as those used in bump mapping, and the comments in the listing speak for themselves.

14.4   Programmable Shaders in Direct3D

We couldn't finish this chapter without mentioning that Direct3D has an equally rich and powerful GPU programmable shading system, HLSL (High Level Shader Language). This has been around a lot longer than OpenGL Version 2. Like GLSL, its syntax is based on C, but despite being around for a longer time, it seems to change rapidly. The earliest version of a shading language for Direct3D required you to program in assembly language for the GPU, not an alluring prospect. The procedure has evolved through Shader Model 1, Shader Model 2 and, at the time of writing, Shader Model 3. It may be Shader Model 4 when you read this. To us, OpenGL just always seems to be more stable.

Now, OpenGL 2 and Direct3D are not the only shader systems around. RenderMan [8] is extremely respected with its long association to the computer animation business. There are some wonderful RenderMan shaders, and in a lot of cases, the algorithms they contain should be modifiable for real-time use. Because GPU-enabled hardware preceded both OpenGL 2 and HLSL, chip manufacturers devised their own shading languages. NVIDIA's Cg [4] matches its GPU hardware very well, and ATI's RenderMonkey [1] is a friendly environment for developing shader programs (see Figure 13.7).

14.5   Summary

This chapter has introduced the concept of the programmable GPU. It showed how OpenGL Version 2's *high-level-like* shading language can be put into practice to enhance the realistic appearance of objects and scenes whilst still maintaining the real-time refresh rates so vital for VR applications.

14.5.1 Afterthought

There really is an almost inexhaustible supply of procedural shaders that could be created to enhance the visual appearance of a virtual world: wood, brick, stone and rough metal are just a few that immediately spring to mind. Neither are we limited to using the GPU programmability for surface shaders. Volume shaders—cloud, hair, fur and atmospherics—are well within the realm of real-time rendering now. OpenGL 2 even comes with the Perlin noise generator [7] (that lies at the heart of so many beautiful textures) built in. The extent to which you explore, develop and use programmable shaders is only limited by your imagination and skill in designing suitable algorithms.

Bibliography

[1] Advanced Micro Devices, Inc. "RenderMonkey Toolsuite". http://ati.amd.com/developer/rendermonkey/, 2006.

[2] R. S. Ferguson. *Practical Algorithms for 3D Computer Graphics.* Natick, MA: A K Peters, 2001.

[3] J. Kessenich et al. *The OpenGL Shading Language, Language Version 1.10, Document Revision 59.* http://oss.sgi.com/projects/ogl-sample/registry/ARB/GLSLangSpec.Full.1.10.59.pdf, 2004.

[4] NVIDIA Corporation. "Cg Toolkit 1.5". http://developer.nvidia.com/object/cg_toolkit.html, 2006.

[5] M. Olano et al. *Real-Time Shading.* Natick, MA: A K Peters, 2002.

[6] M. Olano et al. *Real-Time Shading.* SIGGRAPH 2004 Course Notes, Course #1, New York: ACM Press, 2004.

[7] K. Perlin. "An Image Synthesizer". *Computer Graphics* 19:3 (1985) 287–296.

[8] S. Raghavachary. *Rendering for Beginners: Image Synthesis using RenderMan.* Boston, MA: Focal Press, 2004.

[9] R. Rost. *OpenGL Shading Language.* Reading, MA: Addison-Wesley Professional, 2006.

[10] D. Shreiner, M. Woo, J. Neider and T. Davis. *OpenGL Programming Guide, Fifth Edition.* Reading, MA: Addison-Wesley Professional, 2005.

15

Using Multimedia in VR

No VR environment would be complete without the ability to play movies, DVDs, feed live video, work with video cameras or carry out multichannel videoconferencing and indulge in a bit of distributed or augmented reality.

As soon as one hears the terms *live video*, DVD movies etc., one shudders at the thought of the complexity and size of the software requirements. Just to write the decoding functions for a DVD player application is a daunting task. New encoding and compressing algorithms[1] are cropping up all the time, and it is quite impossible to write every application so that it can cope with every codec that it is likely to encounter.

It doesn't help either when you realize that there is a different aspect to the meaning of *real-time* when it comes to constructing players for video and audio material. We have been discussing real-time 3D graphics, and if we can render our view of the 3D scene in less than 25 ms, for all intents and purposes, it is real time. But what if we can't meet that target? It plays a little slower; 15 frames per second is still acceptable. After that, we have to compromise and reduce the quality of the rendering or the complexity of the scene. For audio and video work, slow-motion replay is unacceptable. Imagine watching a whole movie in slow motion. Or, have you ever listened to a slowed down tape recording? In these technologies, the best thing one can do is *drop frames* or leave out bits of the audio. We need a different approach to real-time programming, where essentially what one is doing is keeping in sync with a real-time clock. If your DVD player program takes a few milliseconds longer to render the frame, it will have to adapt by, say,

---

[1] In the context of TV, video, DVDs, AVI files, streaming movies etc. the universally adopted term for the computer code that implements the encoding and decoding process is *codec*.

omitting the occasional frame. If your player is running on a faster processor, it must not play the movie as if in fast forward, either; it must pause and wait to synchronize with the real-time clock.

The good news is that every PC made today has the processing power to easily play back movies at the required real-time speed, so the only complication we have is to ensure that they play smoothly and don't slow down or speed up due to some big calculation being done in the back ground.

Gathering these thoughts together brings us to the subject of this chapter, DirectShow. We mentioned this briefly in Chapter 12 when discussing application program tools and templates. DirectShow is a powerful and versatile API for multimedia applications. It facilitates acquiring data from sources, processing the data and delivering it to a destination. It also has a nomenclature, a *jargon*, in which one can discuss and describe what is going on and what one would like to do. DirectShow jargon speaks of signals coming from sources passing through filters en route from sources to sinks. The signal path may split to pass through different filters or to allow multiple outputs. For example, when playing a movie, the video goes to the screen, and the audio goes to the loudspeakers. DirectShow uses a neat graphical metaphor, called a *FilterGraph*, that depicts the filter components and their connections (such as those shown in Figure 15.1). It takes this metaphor further, to the level of the API, where it is the job of the application program to use the COM interfaces of the DirectShow API to build a FilterGraph to do what the application wants to do (play a movie, control a DV camcorder etc.) and send the signals through it. DirectShow had a slow start. It began life under the title ActiveMovie, but in the early days, it was not popular, because multimedia application developers could do everything they wanted using either the media control interface (MCI) or video for Windows (VfW). It wasn't until the multimedia explosion with its huge video files and the easy attachment of video devices to a PC, none of which really worked in the MCI or VfW approach, that DirectShow came into its own. Now, and especially in the case of VR where there are few off-the-shelf packages available, when we need to incorporate multimedia into an application, DirectShow offers the best way to do so.

15.1   A Brief Overview of DirectShow

Meeting DirectShow for the first time, it can seem like an alien programming concept. The terms *filter*, *signal* and *FilterGraph* are not traditional programming ideas, and there are other complicating issues.

In DirectShow, an application program does not execute its task explicitly. Instead, it builds a conceptual structure. For example, the *FilterGraph* contains components which are called the *filters*. The application then connects the filters to form a chain from a *source* to a *sink*. The route from source to sink is essentially a signal path, and in the context of a multimedia application, the signal consists of either audio or video samples. Once the FilterGraph is built, the application program instructs it to move samples through the filters from source to sink. The application can tell the graph to do many things to the data stream, including pausing the progress of the samples through the filter chain. But just what is appropriate depends on the nature of the source, sink and filters in between.

To make the idea of a FilterGraph and the signal processing analogy a little more tangible, we will quote a few examples of what might be achieved in an application program using DirectShow FilterGraphs:

- The pictures from a video camera are shown on the computer's monitor.

- A DVD movie is played on the computer's output monitor.

- The output from a video camera is saved to a multimedia file on disk.

- A movie or DVD is decoded and stored on the computer's hard drive using a high compression scheme.

- A series of bitmap pictures is combined to form a movie file.

- Frames from a movie are saved to bitmap image files on disk.

- An audio track is added to or removed from a movie file.

Figure 15.1 shows diagrammatically how some of these FilterGraphs might look. Indeed, the DirectShow SDK includes a helpful little application program called *GraphEdit* which allows you to interactively build and test FilterGraphs for all kinds of multimedia applications. The user interface for GraphEdit is illustrated in Figure 15.2. The four filters in Figure 15.1 perform four different tasks. In Figure 15.1(a), the output of a digital video recorder is compressed and written to an AVI file. The signal is split so that it can be previewed on the computer's monitor. In Figure 15.1(b), a video signal from a camera is mixed with a video from a file and written back to disk. A preview signal is split off the data stream. The mixer unit can insert a number of effects in real time. In Figure 15.1(c), a video stream from an AVI

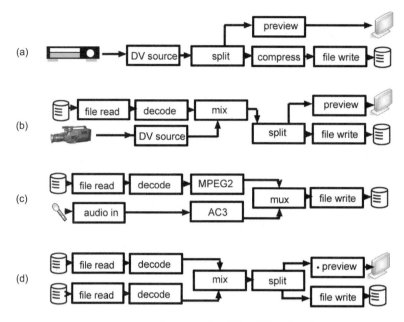

Figure 15.1. An example of four FilterGraphs.

file is combined with a live audio commentary. The signals are encoded to the DVD standard, combined and written to DVD. In Figure 15.1(d), two video sources from disk files are mixed and written back to file with simultaneous preview.

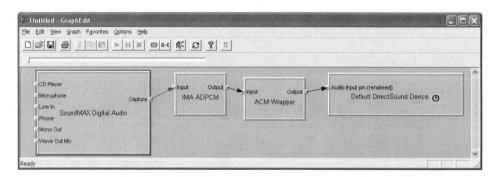

Figure 15.2. The GraphEdit interactive tool for testing DirectShow FilterGraphs. It allows the program's author to use any of the registered filters available on the computer, connect them together and experiment with their action.

We could continue to list many more applications of DirectShow. The fact that all of them arise out of mixing and matching a small (but extensible) set of FilterGraph components illustrates the power of the concept. We can also see a pattern emerging. Every FilterGraph starts with a source that generates a stream of samples: video frames from a camera, images in a set of bitmap files etc. It ends with a sink: a window on the monitor, some additional bytes appended to a recompressed movie file etc. In between the source and sink, the samples pass through a filter where they can be altered in some way. For example, the DVD MPEG2 encoded data stream might be decoded or the color balance in a video signal compensated for poor lighting conditions etc.

DirectShow identifies three types of filter:

1. *Source filters.* These could be a video camera, a movie file, a microphone or an audio file.

2. *Sink filters.* A video renderer. A renderer filter is a filter that displays video pictures or plays audio, or in fact, any filter that presents multimedia content. A file writer filter stores the media samples to disk.

3. *In-place filters.* These change the data stream in some way. They may even split the data stream so that for example the output of a video camera can be previewed on the monitor at the same time as it written to an AVI file.

When filters are linked together, media samples are said to travel from an upstream filter (such as a source) to a downstream filter (such as a sink). Although we have classified both video cameras and movie file readers as source filters, there are some significant differences. At the other end of the filter chain, the types of output filters have their differences, too. A file writer is different from a window presenting live video grabbed from a webcam, for example. Rather than thinking about a subclass of filters, DirectShow introduces the idea of a graph as conforming to a *push* model or a *pull* model. In a push mode of operation, the source or upstream filter is the driver; it sets the timing. In a pull FilterGraph, the downstream filter sets the timing and controls the flow of samples. A typical pull graph application would be a media file player (movie or audio), because the presentation to the viewer (done by the filter at the end of the chain) must appear as smooth as possible and not play at a speed dictated by the speed at which the file can be read. A push model would apply, for example, when capturing live video from a camera to an AVI file.

In multimedia applications, timing is important, and all FilterGraphs are designed so that they can be synchronized with a real-time clock. Whether playing back video files or recording video from a USB webcam, all programs need to keep up with this clock. If the hardware or application software cannot deliver the samples through the graph fast enough then some samples will have to be discarded.

> For example, a video recorder application is saving AVI movie files from a DV camera with a compression filter in the chain. For NTSC video, it must read, compress and write 30 samples (frames) per second. But suppose the compression filter can only compress 25 samples per second. The camera cannot be slowed down, nor are we allowed to only write 25 frames per second into the AVI file while still claiming the AVI is a 30-fps movie file.

> DirectShow cleverly solved the problem for us, without any work on the part of the program designer. It does this (in our example) by dropping 5 out of every 30 samples, say samples 6, 12, 18, 24, 30, before the compression filter stage. After the compression filter stage it adjusts the sample timings in the AVI file so that either 30 samples (with some duplicates) are written per second or 25 samples are written per second with the timing changed so that a player will think each sample should be displayed for 40 ms instead of 33 ms.

> Similar things happen in movie-player applications. If the video source is supposed to play at 30 fps but the renderer filter can only keep up at 25 fps then the DirectShow FilterGraph drops five out of every 30 frames read from the file. It may even skip reading samples from the source file itself.

Another term you will read about, with respect to DirectShow filters, is the *pin*. A pin is a point of connection on a filter. Source filters have one or more pins. Output filters generally have one pin. In-place filters have at least two pins. Pins are designated either as being an *input* pin or an *output* pin. The FilterGraph is build by connecting one filter's output pin to another filter's input pin. Thus, a FilterGraph can be something as simple as a source filter and a sink filter, with the source filter's output pin connected to the sink filter's input pin.

15.1.1 Threading

A great deal of emphasis has been put on the importance of timing in Direct-Show FilterGraphs. In addition, the FilterGraph seems almost to be an entity working independently and only communicating with its driving application from time to time. DirectShow is a multi-threaded technology. The Filter-Graph executes in a separate thread[2] of execution. Programming in a multi-threaded environment bring its own set of challenges. But to achieve the real-time throughput demanded by multimedia applications, our programs must embrace multi-threading concepts.

Nevertheless, this book is not focused on multi-threading or even about writing multi-threaded code, so we will gloss over the important topics of deadlocks, interlocks, synchronization, critical sections etc. In our example programs, we will use only the idea of the *critical section*, in which blocks of code that access global variables must not be executed at the same time by different threads; say one thread reading from a variable and another thread writing to it. We suggest that if you want to get a comprehensive explanation of the whole minefield of multi-threaded programming in Windows, consult one of the specialized texts [1] or [3].

There is one aspect of multi-threaded programming that we cannot avoid. Applications *must* be linked with the multi-threaded C++ libraries.

15.2 Methodology

In this section, we will summarize what an application program must do in order to use DirectShow. The flexibility of the FilterGraph concept means that all multimedia applications follow these same simple rules:

- Create a FilterGraph with a COM FilterGraph object and its interfaces.

- Create instances of the filter objects and add them to the FilterGraph. Usually there will be at least one source and one sink filter.

- Link the pins of the filters together to form the desired filter chain. This can be done by creating a `CaptureGraphBuilder2` COM object, and its COM interfaces. The `CaptureGraphBuilder2` object is an extremely useful tool because it implements what is known as *intelligent*

---

[2]Threads are in essence processes or programs executing independently and concurrently from one another, but they use the same address space and have access to the same data.

connect. Without intelligent connect, building a FilterGraph would be difficult indeed. To see why, consider a video player application. Two filters are loaded into the FilterGraph, one to read the file and the other to show the movie on the screen. They must be connected before the graph can run. However, an AVI movie file will more likely than not use data compression. If a matching filter is not added to the graph and inserted between reader and renderer, the graph cannot be built. This is were intelligent connect comes in. It can detect what filters are required to match the output of the file reader to the input of the screen renderer. It can also load those filters and put them into the chain. So using the `CaptureGraphBuilder2` interface methods considerably simplifies and shortens the code required to build the FilterGraph.

- Acquire appropriate control interfaces to the FilterGraph; that is, attach whatever parts of the application program require it to the Filter-Graph. For example, tell the FilterGraph to render its video output in the application program's window.

- Start, stop, rewind, pause etc. the FilterGraph as required.

- Destroy the FilterGraph when finished and exit the application.

Figure 15.3 shows diagrammatically the programming strategy for all applications in this book that use DirectShow.

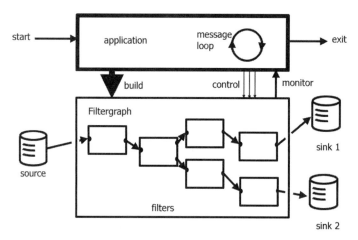

Figure 15.3. A DirectShow application builds and controls a FilterGraph to carry of the bulk of the work of the program.

Note the mention of the COM. As we discussed in Section 12.4, DirectShow's API is a prime example of an API specification conforming to the component object model. Any program hoping to build and use a Filter-Graph must do so using COM objects and interfaces. In the next few sections, we will write a few basic DirectShow programs that cover a broad spectrum of multimedia applications. Later, we will embellish and adapt them for use as some interesting and specialized VR tools.

15.3 A Movie Player

As a first application, we will look at a very basic use of DirectShow: playing an AVI movie file. Its simplicity is the very reason why DirectShow can be so useful. The FilterGraph we require for this application is shown in Figure 15.4.

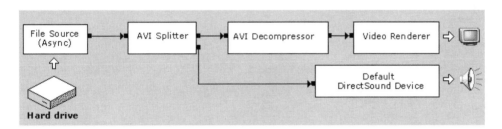

Figure 15.4. A movie player FilterGraph.

The application has an entry point, message loop and window message handling function, as illustrated in Listing 15.1.

All the other DirectShow applications that we will examine in this and subsequent chapters use an overall skeleton structure as outlined in Listing 15.1. Each example will require some minor modifications and additions to each of the functions listed. We will not reprint this code in each program listing and only refer to additions if they are very significant. Using our annotated printed listings, it should be possible to gain an understanding of the program logic and follow the execution path.

```
#include <windows.h>         // This is the minimal set of header files required
#include <atlbase.h>         // for a DirectShow appluication.
#include <dshow.h>

HWND  ghApp=0;               // our application window's handle
HINSTANCE ghInst=0;          // the application's instance handle
char gFileName[MAX_PATH];    // string for filename

IGraphBuilder *pGB = NULL;   // pointer to the graph builder interface
IMediaControl *pMC = NULL;   // pointer to the media control interface

LRESULT CALLBACK WndMainProc(HWND hWnd,UINT message,
WPARAM wParam,LPARAM lParam){
 switch(message){
  case WM_COMMAND:           // menu command messages
    switch(wParam) {
     case ID_FILE_OPENCLIP:
      OpenClip(NULL); break;  // open a movie file - this builds the graph
     case ID_FILE_EXIT:
      CloseClip();           // this stops the movie playing
                             // and destroys the graph
      PostQuitMessage(0) break;
     case: ..//              // other menu commands
     default: break;
    }
    break;
  case WM_DESTROY:
   PostQuitMessage(0);
   break;
  ..//                       Other messages
  default: break
 return DefWindowProc(hWnd, message, wParam, lParam);
}



int PASCAL WinMain(HINSTANCE hInstC, HINSTANCE hInstP,
LPSTR lpCmdLine, int nCmdShow){
 .. // declare local variables and fill window class "wc"
 CoInitialize(NULL);                                   // tell Windows we use COM
 ghInst = hInstC;
 // register the window class
 RegisterClass(&wc);
 // create the application's window
 ghApp = CreateWindow(CLASSNAME, APPLICATIONNAME, ...
 while(GetMessage(&msg,NULL,0,0)){
   TranslateMessage(&msg);                              // message loop
   DispatchMessage(&msg);
 }
 CoUninitialize();                                      // release COM
 return (int)0;
}
```

Listing 15.1. The outline of the movie player application.

In any DirectShow application, there are a couple of essential function calls required so that it can use the COM:

```
CoInitialize(NULL); // tell Windows we use COM
...
...
CoUninitialize(); // tell Windows we have finished with COM
```

The DirectShow FilterGraph is created when a specific movie is opened for presentation. We need pointers to the graph builder (`IGraphBuilder *pGB;`) interface and the media control interface (`IMediaControl *pMC;`), which allows us to play the movie, stop it or pause it. All the actions in the program are tied to menu commands calling the functions in Listing 15.2.

The four functions in this listing do the following:

- `OpenClip()`: Obtain the video filename using a *FileOpen* dialog.

- `CloseClip()`: Stop the running graph and delete it.

- `CloseInterfaces()`: Destroy the graph by releasing all the interfaces it uses.

- `PlayMovieInWindow()`: This is the *key* step in playing the movie. It simply builds the FilterGraph object gets pointers to the graph's graph builder interface and its media control interface. The graph builder interface's `RenderFile()` method does all the work of building the necessary graph using intelligent connect. Finally, the `run()` method from the media control interface plays the movie.

Given that this short program plays video files (with and without a sound track) and can decode the video stream using any codec installed on the computer, the 100 or so lines of code it takes to write the whole program serve to illustrate the power of DirectShow. However, the behavior of the program is still a little untidy. For one thing, the movie plays in a separate pop-up window. This is because the intelligent connect mechanism has inserted one of DirectShow's internal video renderer filters, and it will create a window of its own in which to present the video. This may be acceptable, but we can do better, because DirectShow offers a way for an application program to tell the renderer filter[3] to draw its output in one of its own windows.

---

[3]One can of course write one's own renderer filter. We shall do this later and also in Chapter 16 when we write some programs for stereoscopic work.

```
void OpenClip(void){
  HRESULT hr;
  if (! GetClipFileName(gFileName))return;      // get the movie filename
  hr = PlayMovieInWindow(g_szFileName);         // build the graph and play it
  if (FAILED(hr)) CloseClip();                  // FAILED is DS macro
}

void CloseClip(void){
    HRESULT hr;
    if(pMC) hr = pMC->Stop();                   // Stop media playback
    CloseInterfaces();                          // Free DirectShow interfaces
}


HRESULT PlayMovieInWindow(LPTSTR szFile){
  USES_CONVERSION; // 8 bit char to wide char macros are used
  WCHAR wFile[MAX_PATH];
  HRESULT hr;
  if (!szFile) return E_POINTER;
  // Convert filename to wide character string
  wcsncpy(wFile, A2W(szFile), NUMELMS(wFile)-1);
  // Get the interface for DirectShow's GraphBuilder
  hr=CoCreateInstance(CLSID_FilterGraph, NULL, CLSCTX_INPROC_SERVER,
          IID_IGraphBuilder, (void **)&pGB);
  // Have the graph builder construct the appropriate graph automatically
  hr=pGB->RenderFile(wFile, NULL);
  // QueryInterface for DirectShow interfaces
  hr=pGB->QueryInterface(IID_IMediaControl, (void **)&pMC);
  // Complete window initialization
  // Run the graph to play the media file
  hr=pMC->Run();
  return hr;
}

void CloseInterfaces(void){                     // release the interfaces
    SAFE_RELEASE(pMC);                          // DS macro which checks
    SAFE_RELEASE(pGB);                          // pointer before freeing.
}
```

Listing 15.2. These functions build and execute the movie player graph.

This is done by using two interfaces. The IVideoWindow interface tells DirectShow's renderer filter where and how to draw its output. We will also need to use the IMediaEventEx interface, since it tells DirectShow to route notification of events (such as the graph stopping when the movie is over) through the application's message handler functions. When events pass through a message handler routine, we can act on them to do such things as

```
   .. // add these lines of code to PlayMovieInWindow(..)
   RECT rect;
   GetClientRect(ghApp, &rect);
   // get an interface for passing commands to the video output window
   pGB->QueryInterface(IID_IVideoWindow, (void **)&pVW);
   // tell DirectShow to render into main window's client area
   pVW->put_Owner((OAHWND)ghApp);
   pVW->put_WindowStyle(WS_CHILD | WS_CLIPSIBLINGS | WS_CLIPCHILDREN);
   pVW->SetWindowPosition(rect.left, rect.top, rect.right, rect.bottom);
   // Have the graph signal events
   pME->SetNotifyWindow((OAHWND)ghApp, WM_GRAPHNOTIFY, 0));
   .. //

   ..//add these lines to the code for CloseInterfaces(..)
   hr = pVW->put_Visible(OAFALSE);   // Relinquish ownership (IMPORTANT!)
   hr = pVW->put_Owner(NULL);        // no wnser
   // Disable event callbacks
   hr = pME->SetNotifyWindow((OAHWND)NULL, 0, 0);
   SAFE_RELEASE(pME);                // Release DirectShow interfaces
   SAFE_RELEASE(pVW);                // SAFE_RELEASE is a DS macro
   ..//

   ..// add these functions to the code
   void MoveVideoWindow(void){    // new function to handle size changes
    // Track the movement of the window being drawn into and resize if needed
     if(pVW){
       RECT client;
       GetClientRect(ghApp, &client);
       hr = pVW->SetWindowPosition(client.left,client.top,
       client.right,client.bottom);
     }
   }

   HRESULT HandleGraphEvent(void){ // new functions to handle graph events
     LONG evCode, evParam1, evParam2;
     HRESULT hr=S_OK;
     if (!pME) return S_OK;        // Otherwise process all queued events
     while(SUCCEEDED(pME->GetEvent(&evCode,&evParam1,&evParam2,0))){
        //
        // handle any event codes here !!!!!!!!!!!!!!!
        //
        // Free memory associated with callback - required behaviour
           hr = pME->FreeEventParams(evCode, evParam1, evParam2);
        }
        return hr;
   }
```

Listing 15.3. Changes required to make the renderer filter play the movie within the application's client area.

rewind the movie back to the start or load another one. To put this functionality into the application, the code in Listings 15.1 and 15.4 is modified by adding some instructions, global variables and extra functions as follows.

Two extra interface pointers are required:

```
IVideoWindow  *pVW    = NULL;
IMediaEventEx *pME    = NULL;
```

In the `WndMainProc()` message handler, we intercept graph event messages and act upon a change of client area size so that the video window always fits inside the client area of the container window. (We are really asking another thread, the one running the renderer filter, to make sure it draws inside the application's window.) This is done with the following code fragment:

```
case WM_GRAPHNOTIFY:
  HandleGraphEvent();    break;
case WM_SIZE:
 if ((hWnd == ghApp))MoveVideoWindow();  break;
```

We tell DirectShow to route all messages through our application's window with the `pVW->NotifyOwnerMessage(...)` method of the video window interface. Listing 15.3 shows other modifications we need to make to render the output into the client area of the main window.

Despite the appearance that the video output is now in the main window's client area, it is still being drawn by a system renderer. In most circumstances, this is irrelevant, but if you try to capture a frame from the movie by reading pixels from the window's client area, all you will get is a blank rectangle. Similarly, if you try and copy the full screen to the clipboard then the part of the window where the movie is rendered will again appear blank. The reason for this is that the DirectShow renderer bypasses the GDI drawing functions and communicates directly with the graphics hardware so as to maximize performance.

15.3.1   A Final Touch: Drag and Drop

As a final touch for this application, we want to show you how to make your movie player accept files that have been *dragged and dropped* into the window, from Windows Explorer, for example. Drag and drop is really standard stuff but can be hard to find out how to do it from the Windows API docu-

```
#include <windows.h>                        // necessary header files
#include <tchar.h>
#include <shlobj.h>
#include <oleidl.h>

// our class calls back to this function in the  program
extern void OpenDraggedClip(char *);

class CDropTarget : public IDropTarget{  //
public:
 STDMETHODIMP QueryInterface(REFIID riid, void **ppv);
 STDMETHODIMP_(ULONG) AddRef();
 STDMETHODIMP_(ULONG) Release();
 STDMETHODIMP DragEnter(IDataObject *, DWORD, POINTL, DWORD *);
 STDMETHODIMP DragLeave(void);
 STDMETHODIMP DragOver(DWORD, POINTL, DWORD *);
 STDMETHODIMP Drop(IDataObject *,DWORD, POINTL, DWORD *);
 CDropTarget(HWND );
 ~CDropTarget();
private:
 LONG m_cRef;
 int m_int1;
};

CDropTarget::CDropTarget(HWND hWnd){      // constructor
 if(RegisterDragDrop(hWnd,this) != S_OK)MessageBox(NULL,
 "D&D setup fail",NULL,MB_OK);
}
CDropTarget::~CDropTarget(){;}            // destructor
static CDropTarget *pT=NULL;             // This is our drag and drop object

void  SetUpDragAndDrop(HWND hWnd){       // Call this method to initialize D&D
 pT = new CDropTarget(hWnd);
 return;
}

void CloseDragAndDrop(void){             // call at end of application
  if(pT)delete pT;
}
```

Listing 15.4. This class implements drag and drop.

mentation. Because it is nicely self-contained, the facility can be added to any
Windows application with little difficulty. We won't go into great detail, since
it takes us away from our main theme, but Listings 15.4 and 15.5 present a
C++ class and its methods to implement drag and drop. If you put this code
into a separate file, you can include it with all your programs and make use of
it by adding the code from Listing 15.6 in the appropriate places.

```
STDMETHODIMP CDropTarget::QueryInterface(REFIID riid, void **ppv){
 if(ppv == NULL)return E_POINTER;
 if(riid == IID_IDropTarget || riid == IID_IUnknown) {
   AddRef();
   *ppv=this;
   return S_OK;
 }
 *ppv = NULL;
 return E_NOINTERFACE;
}

STDMETHODIMP_(ULONG) CDropTarget::AddRef() {
   return InterlockedIncrement(&m_cRef);
}

STDMETHODIMP_(ULONG) CDropTarget::Release(){
 LONG lRef = InterlockedDecrement(&m_cRef);
 if (lRef == 0)delete this;
 return lRef;
}

STDMETHODIMP CDropTarget::DragEnter(
  IDataObject * pDataObject,
                    //Pointer to the interface of the source data
                    // object
  DWORD grfKeyState, //Current state of keyboard modifier keys
  POINTL pt,         //Current cursor coordinates
  DWORD * pdwEffect  //Pointer to the effect of the drag-and-drop
                    // operation
){ return S_OK; };

STDMETHODIMP CDropTarget::DragLeave(void){ return S_OK;}

STDMETHODIMP CDropTarget::DragOver(
  DWORD grfKeyState, //Current state of keyboard modifier keys
  POINTL pt,         //Current cursor coordinates
  DWORD * pdwEffect  //Pointer to the effect of the drag-and-drop
                    // operation
){ return S_OK;}
```

Listing 15.5. Methods of the drag-and-drop class.

```
STDMETHODIMP CDropTarget::Drop(
  IDataObject * pDataObject,
                        //Pointer to the interface for the source data
  DWORD grfKeyState, //Current state of keyboard modifier keys
  POINTL pt,         //Current cursor coordinates
  DWORD * pdwEffect  //Pointer to the effect of the drag-and-drop
                        // operation
){
  FORMATETC ff;
  ff.cfFormat=CF_HDROP;
  ff.ptd=NULL;
  ff.dwAspect=DVASPECT_CONTENT;
  ff.lindex= -1;
  ff.tymed=TYMED_HGLOBAL;
  STGMEDIUM pSM;
  pDataObject->GetData(&ff,&pSM);
  HDROP hDrop = (HDROP)(pSM.hGlobal);
  TCHAR name[512];
  int i=DragQueryFile(hDrop,0xFFFFFFFF,name,512);
  DragQueryFile(hDrop,0,name,512);
  // now that we've got the name as a text string use it  ///////////
  OpenDraggedClip(name); // this is the application defined function
  ////////////////////////////////////////////////////////////////////
  ReleaseStgMedium(&pSM);
 return S_OK;
}
```

Listing 15.5. (continued).

15.4   Video Sources

The other major use we will make of DirectShow concerns data acquisition. Essentially, this means grabbing video or audio. In this section, we will set up some example FilterGraphs to acquire data from live video sources. These could be digital camcorders (DV) connected via the FireWire interface or USB cameras ranging in complexity from a simple webcam using the USB 1 interface to a pair of miniature cameras on a head-mounted display device multiplexed through a USB 2 hub. It could even be a digitized signal from an analog video recorder or S-video output from a DVD player.

To DirectShow, all these devices appear basically the same. There are subtle differences, but provided the camera/digitizer has a WDM (windows driver model) device driver, DirectShow has a source filter which any application program can load into its FilterGraph and use to generate a stream of video samples. We shall see in this section that using a live video source in a

```
OleInitialize(NULL);          // Drag and drop requires OLE not jsut COM so
                              // replace CoInitialize(NULL) with this.
SetUpDragAndDrop(hWnd)         // make the application D&D compatible

Close DragAndDrop();           // destroy the D&D object
OleUninitialize();             //

void OpenDraggedClip(TCHAR *filename){ // this is called with file is dropped
  if (g_psCurrent != Init)CloseClip();
  OpenClip(T2A(filename));  // convert text string to basic "char" format
}
```

Listing 15.6. Modifications required in an application so that it can handle drag and drop. When a file is dropped into the application window, the function `OpenDraggedClip()` is called. Note also the change to `OleInitialize()` instead of `CoInitialize()`.

DirectShow FilterGraph is as easy as using a file-based source. As a result, our code listings will be relatively short and only complicated by the need to provide code to enable the application programs to select the desired input source when more than one possible device is connected to the host computer.

There are, of course, some differences between the types of video signal sources that we must be aware of as we plan our application design. We note three of significance:

1. *Data rate.* DV camcorders will typically provide 30 samples (or video frames) per second (fps). A USB 1 webcam may only be able to deliver 15 fps.

2. *Sample size or frame resolution.* DV frame sizes in pixels are typically 720×576 for PAL regions and 640×480 in NTSC regions. A webcam resolution might be 320×240.

3. *Pixel data format.* In most applications, we would like the source to deliver a 3-byte-per-pixel RGB24 data stream, but some video source devices provide a YUV2 format. DirectShow's intelligent connect mechanism inserts appropriate conversion filters so that if we want to use RGB24 (or RGB32 for that matter), a YUV2-to-RGB conversion filter will be loaded and connected into the graph.

A special DirectShow filter, called the *grabber filter*, is often put into the filter chain by an application program. This not only lets an application program peek into the data stream and grab samples, but it is also used to tell the upstream source filters that they must deliver samples in a particular format.

As we build the code for a couple of examples using live video source filters, you will see that the differences from the previous movie player program are quite small. The longest pieces of code associated with video sources relate to finding and selecting the source we want. In the context of FilterGraphs using video sources, the `ICaptureGraphBuilder2` interface comes into its own by helping to put in place any necessary intermediate filters using intelligent connect. Without the help of the `ICaptureGraphBuilder2` interface, our program would have to search the source and sink filters for suitable input and output pins and link them. Some of the project codes on the CD and many examples from the DXSDK use this strategy, and this is one of the main reasons why two DirectShow programs that do the same thing can look so different.

15.4.1 Viewing Video from a DV or USB Camera

For the most part (the `WinMain` entry, message handler function etc.), the code for this short program is very similar to the movie player code, so we will not list it here. Instead, we go straight to look at the differences in the structure of the FilterGraph given in Listing 15.7. A couple of functions merit a comment:

1. `g_pCapture->RenderStream()`. This `ICaptureGraphBuilder2` method carries out a multitude of duties. Using its first argument, we select either a preview output or a rendered output. This gives an application program the possibility of previewing the video while capturing to a file at the same time. The second argument tells the method that we need video samples. The last three arguments—`pSrcFilter`, NULL, NULL—put the graph together. They specify:

   (a) The source of the video data stream.
   (b) An optional intermediate filter between source and destination; a compression filter for example.
   (c) An output filter. If this is NULL, the built-in renderer will be used.

   If we are recording the video data into a file using a file writing filter (say called `pDestFilter`), the last three arguments would be (`pSrcFilter`,

```
ICaptureGraphBuilder2 * g_pCapture=NULL;// declare global pointer

 // this is main function to build/run the capture graph
HRESULT CaptureVideo(){
 HRESULT hr;
 // this will point to the Video source filter
 IBaseFilter *pSrcFilter=NULL;
 hr = GetInterfaces();                     // Get DirectShow interfaces
 // Tell the capture graph builder to do its work on the graph.
 hr = g_pCapture->SetFiltergraph(g_pGraph);
 // Use the system device enumerator and class enumerator to find
 // a video capture/preview device, such as a desktop USB video camera.
 hr = FindCaptureDevice(&pSrcFilter);// see next listing
 // Add Capture filter to our graph.
 hr = g_pGraph->AddFilter(pSrcFilter, L"Video Capture");
 // Render the preview pin on the video capture filter
 // Use this instead of g_pGraph->RenderFile
 hr = g_pCapture->RenderStream(&PIN_CATEGORY_PREVIEW,&MEDIATYPE_Video,
                               pSrcFilter, NULL, NULL);
 // Now that the filter has been added to the graph and we have
 // rendered its stream, we can release this reference to the filter.
 pSrcFilter->Release();
 hr = SetupVideoWindow();               // Setup video window
 hr = g_pMC->Run();                     // Start previewing video data
 return S_OK;
}

 // very similar to movie player - (Do check for errors in hr!!!
HRESULT GetInterfaces(void){
 HRESULT hr;
 // Create the filter graph
 hr = CoCreateInstance (CLSID_FilterGraph, NULL, CLSCTX_INPROC,
                        IID_IGraphBuilder, (void **) &g_pGraph);
 // Create the capture graph builder
 hr = CoCreateInstance (CLSID_CaptureGraphBuilder2 , NULL, CLSCTX_INPROC,
                        IID_ICaptureGraphBuilder2, (void **) &g_pCapture);
 // Obtain interfaces for media control and Video Window
 hr = g_pGraph->QueryInterface(IID_IMediaControl,(LPVOID *) &g_pMC);
 hr = g_pGraph->QueryInterface(IID_IVideoWindow, (LPVOID *) &g_pVW);
 hr = g_pGraph->QueryInterface(IID_IMediaEvent, (LPVOID *) &g_pME);
 // Set the window handle used to process graph events
 hr = g_pME->SetNotifyWindow((OAHWND)ghApp, WM_GRAPHNOTIFY, 0);
 return hr;
}
```

Listing 15.7. Build FilterGraph thatpreviews live video from a DV/USB camera. Note the call to FindCaptureDevice(), which finds the video source filter. Its code is given in Listing 15.8.

NULL, pDestFilter). If the graph needs to use a compression filter (i.e., pCompress), the whole chain can be established in one call using the last three arguments (pSrcFilter, pCompress, pDestFilter).

2. FindCaptureDevice(). This is the most important function, and its code is given in Listing 15.8.

```
HRESULT FindCaptureDevice(IBaseFilter ** ppSrcFilter){
  HRESULT hr;
  IBaseFilter * pSrc = NULL;
  CComPtr <IMoniker> pMoniker =NULL;
  ULONG cFetched;
  // Create the system device enumerator
  CComPtr <ICreateDevEnum> pDevEnum =NULL;
  hr = CoCreateInstance (CLSID_SystemDeviceEnum,NULL,CLSCTX_INPROC,
        IID_ICreateDevEnum,(void **)&pDevEnum);
  // Create an enumerator for the video capture devices
  CComPtr <IEnumMoniker> pClassEnum = NULL;
  // we are looking for a video input source
  hr = pDevEnum->CreateClassEnumerator(CLSID_VideoInputDeviceCategory,
  &pClassEnum,0);
  // If there are no enumerators for the requested type, then
  // CreateClassEnumerator will succeed, but pClassEnum will be NULL.
  if (pClassEnum == NULL)  {
    MessageBox(ghApp,"No video capture device was detected",NULL, MB_OK);
    return E_FAIL;
  }
  // Use the first video capture device on the device list.
  if (S_OK == (pClassEnum->Next (1, &pMoniker, &cFetched))) {
    // Bind Moniker to a filter object
    hr = pMoniker->BindToObject(0,0,IID_IBaseFilter, (void**)&pSrc);
    return hr;
    }
  }
  else return E_FAIL;
  // Copy the found filter pointer to the output parameter.
  // Do NOT Release() the reference, since it will still be used
  // by the calling function.
  *ppSrcFilter = pSrc;
  return hr;
}
```

Listing 15.8. Enumerating and selecting the video capture device. In this case, we take the first source found. Note the use of the ATL CComPtr pointer class, which handles our obligation of releasing interfaces when we are finished with them.

In all DirectX technology—Direct3D, DirectInput and DirectShow— finding a suitable device (or filter) is a significant task. However, the same code can usually be used repeatedly. In `FindCaptureDevice()` (see Listing 15.8), the strategy adopted is to enumerate all devices and pick the first suitable one. This requires some additional interfaces and a system *moniker*. The DirectX SDK contains a comprehensive discussion on device enumeration.

Listing 15.8 may be extended to find a second video source for stereo work (see Chapter 16) or make a list of all attached video devices so that the user can select the one she wants.

15.4.2   Capturing Video

Recording the video signal into an AVI file requires only two minor additions to the video preview example (see Listing 15.9), both of which require the use of methods from the `ICaptureGraphBuilder2` interface:

1. `g_pCapture->SetOutputFileName(..)`. This method returns a pointer to a filter that will send a video data stream to a number of different file formats. In this example, we choose an AVI type.

2. `g_pCapture->RenderStream(...)`. This method is used twice. The first call is to render to an onscreen window linked to a preview pin on the video source filter. A second call requests a `PIN_CATEGORY_CAPTURE` pin on the source filter and links it to the file writer filter directly. The last three arguments (`pSrcFilter,NULL,g_pOutput`) specify the connection.

We have considered file sources and sinks, video sources and a variety of display renderers. This only leaves one element of DirectShow to take a look at in order to have a comprehensive set of components which we will be able to put together in many different combinations. We now turn to the last element in DirectShow of interest to us, the in-place filter.

15.5   Custom Filters and Utility Programs

In this section, we will look a bit closer at the key element in DirectShow that makes all this possible—the *filter*—and how it fits into a FilterGraph to play its part in some additional utility programs. We will also make more general comments and give overview code listings. and defer details for specific applications that will be covered in subsequent chapters.

```
IBaseFilter * g_pOutput=NULL;

HRESULT CaptureVideo(){
    HRESULT hr;
    IBaseFilter *pSrcFilter=NULL;
    // Get DirectShow interfaces
    hr = GetInterfaces();
    // Attach the filter graph to the capture graph
    hr = g_pCapture->SetFiltergraph(g_pGraph);
    CA2W pszWW( MediaFile );
    hr = g_pCapture->SetOutputFileName(
        &MEDIASUBTYPE_Avi,   // Specifies AVI for the target file.
         pszWW,
         &g_pOutput,          // Receives a pointer to the output filter
         NULL);
    // Use the system device enumerator and class enumerator to find
    // a video capture/preview device, such as a desktop USB video camera.
    hr = FindCaptureDevice(&pSrcFilter);
    // Add Capture filter to our graph.
    hr = g_pGraph->AddFilter(pSrcFilter, L"Video Capture");
    // Render the preview pin on the video capture filter to a preview window
    hr = g_pCapture->RenderStream(
                    &PIN_CATEGORY_PREVIEW,
                    &MEDIATYPE_Video,
                    pSrcFilter,
                    NULL, NULL);
    // Render the capture pin into the output filter
    hr = g_pCapture->RenderStream(
                    &PIN_CATEGORY_CAPTURE,  // Capture pin
                    &MEDIATYPE_Video,       // Media type.
                    pSrcFilter,             // Capture filter.
                    NULL,                   // Intermediate filter
                    g_pOutput);             // file sink filter.
    // Now that the filter has been added to the graph and we have
    // rendered its stream, we can release this reference to the filter.
    pSrcFilter->Release();
    hr = g_pMC->Run();
    return S_OK;
}
```

Listing 15.9. Capturing from a live video source to a file requires only minor changes to the FilterGraph used previously for previewing a video source.

There are two strategies one can adopt in writing a DirectShow filter:

1. Build the filter as a distinct and independent component. This is what one might do if the filter has wide applicability or is to be made available commercially. A good example of this type of filter is a video codec.

2. Include the filter as part of the application program code. In research and development work, this has the advantage that it is easier to tweak and change the filter's interfaces and to quickly implement things such as user configuration of filter properties. Any filters that we create will fall into this category.

In either case, the functional part of the filter will not be too dissimilar. To build a DirectShow filter, we follow the following five steps:

1. Select the filter type: source, sink or in-place.

2. Find an example of a similar filter in one of the templates provided with the DirectX SDK. Use them as a guide proceed to the next step.

3. Specify a C++ class derived from the DirectShow base class for the type of filter.

4. Supply code for the virtual functions of the base class (these implement the majority of the filter's behavior) and override other class methods as required.

5. For filters that are going to be used as true COM objects and made available as individual distributable components, add the necessary code and install the filter in Windows.

All the filters we will use in our programs are relatively specialized. There is no advantage to be gained from writing them as fully independent components, so we will simply implement them as objects representing instances of C++ classes. Before looking at the outline of some custom filters, it is useful to see how the applications we have considered so far could use a custom built component filter, one that has been installed and registered with Windows.

15.5.1   Using Video Filters

Video filters play a vital part in playing and compiling multimedia content. A common use of an in-place filter is a compressor between camera and file writer. If we wanted to use a filter of our own creation, one that has been registered, an instance of it would be created and installed with the code in Listing 15.10.

Since we are not going to register our filter as a Windows component, its object is created in the C++ style with the keyword new. Listing 15.11 shows how an in-place filter would be created and added to a FilterGraph in such cases. Several examples of filters designed to be used in this way (both source and renderer filters) will be used in Chapter 16 on stereopsis programming.

```
.. // add this code into the filter graph to use an
   // optional image processing filter
   IBaseFilter *pProcessFilter=NULL;
   // get a pointer to filter
   GetProcessorFilter(&pProcessFilter);
   // add it to graph.
   hr=g_pGraph->AddFilter(pProcessFilter,L"Image Processor");
   hr = g_pCapture->RenderStream (&PIN_CATEGORY_PREVIEW, &MEDIATYPE_Video,
                                  pSrcFilter, pProcessFilter, NULL);
..//
// the installed filter is identified by its GUID
struct
   __declspec(uuid("{B5285CAC-6653-4b9a-B607-7D051B9AB96A}"))CLSID_CUSTOM_IPF;
// Create an instance of the filter and return a pointer to it.
HRESULT GetProcessorFilter(IBaseFilter ** ppSrcFilter){
    HRESULT hr=S_OK;
    IBaseFilter * pSrc = NULL;
    hr = CoCreateInstance (__uuidof(CLSID_CUSTOM_IPF), NULL, CLSCTX_INPROC,
                           IID_IBaseFilter,
                           (void **) &pSrc);
    if (FAILED(hr)){
        MessageBox(NULL,"Cound not create filter",NULL,MB_OK);
        return hr;
    }
    *ppSrcFilter = pSrc;
    return hr;
}
```

Listing 15.10. Adding a custom filter to the FilterGraph is done by creating an instance of it. Because the filter has been registered with Windows, the CoCreateInstance() function is used. The correct filter is identified through its GUID/UUID.

```
.. // add this code into the filter graph to use an
   // (in-process)1 image processing filter
   IBaseFilter *pProcessFilter=NULL;
   // create an instance of the filters object.
   pProcessFilter = (IBaseFilter *) new CImageProcessFilter(NULL, &hr);
   hr=g_pGraph->AddFilter(pProcessFilter,L"Image Processor"); // add it to graph.
   hr = g_pCapture->RenderStream (&PIN_CATEGORY_PREVIEW, &MEDIATYPE_Video,
                                  pSrcFilter, pProcessFilter, NULL);
..//
```

Listing 15.11. Adding an in-place filter when the filter is implemented within the application alone. The class CImageProcessFilter will have been derived from a DirectShow base filter class.

15.5.2   Image Processing with an In-Place Filter

One application of an in-place filter that comes to mind is video overlay. For example, you might want to mix some real-time 3D elements into a video data stream. Other possibilities might be something as trivial as producing a negative image or, indeed, any image-processing operation on a video stream. The key to processing an image in an in-place filter is to get access to the pixel data in each frame. This requires reading it from the upstream filter, changing it and sending it on to the downstream filter. We won't give full details here but instead outline a suitable class declaration and the key class methods that do the work. Listing 15.12 shows the class declaration, and Listing 15.13 covers the overridden `Transform(...)` method that actually performs the pixel processing task.

Like the other filters in DirectShow, when we are customizing one of them, we can get by with some minor modifications to one of the examples of the SDK. Most of the original class methods can usually be left alone to get on with doing what they were put there to do. Looking at a few features in the code, one can say:

- To implement an image-processing effect, all the action is focused in the `Transform()` method.

```
class CIPProcess : public CTransformFilter, // derive from base classes
    public IIPEffect,                       // and interfaces
    public ISpecifyPropertyPages,
    public CPersistStream
{
public:
  ..// other public methods
  //// These methods are overridden from CTransformFilter base class
  //// All our image processing work can be done in this function
  HRESULT Transform(IMediaSample *pIn, IMediaSample *pOut);
  ..// other overrides will be here
private:
  ..// other private methods go here
  HRESULT CopySample(IMediaSample *pSource, IMediaSample *pDest);
  HRESULT LocalTransform(IMediaSample *pMediaSample);
  ..// private member variables go here
};
```

Listing 15.12. Class declaration for an in-place video stream processing function.

```
// Copy the input sample into the output sample - then transform the output
// sample 'in place'.
HRESULT CIPProcess::Transform(IMediaSample *pIn, IMediaSample *pOut){
  // Copy the properties across
 CopySample(pIn, pOut);     // copy the sample from input to output
 return LocalTransform(pOut);  // do the transform
}

HRESULT CIPProcess::CopySample(IMediaSample *pSource, IMediaSample *pDest){
  // Copy the sample data
  BYTE *pSourceBuffer, *pDestBuffer;
  long lSourceSize = pSource->GetActualDataLength();
  pSource->GetPointer(&pSourceBuffer);
  pDest->GetPointer(&pDestBuffer);
  CopyMemory( (PVOID) pDestBuffer,(PVOID) pSourceBuffer,lSourceSize);
  ..// Other operations which we must do but for which we can
  ..// just use the default code from a template:
  ..// Copy the sample times
  ..// Copy the Sync point property
  ..// Copy the media type
  ..// Copy the preroll property
  ..// Copy the discontinuity property
  ..// Copy the actual data length
  long lDataLength = pSource->GetActualDataLength();
  pDest->SetActualDataLength(lDataLength);
  return NOERROR;
}

HRESULT CIPProcess::LocalTransform(IMediaSample *pMediaSample){
  BYTE *pData;        // Pointer to the actual image buffer
  long lDataLen;       // Holds length of any given sample
  int iPixel;        // Used to loop through the image pixels
  RGBTRIPLE *prgb;     // Holds a pointer to the current pixel
  AM_MEDIA_TYPE* pType = &m_pInput->CurrentMediaType();
  VIDEOINFOHEADER *pvi = (VIDEOINFOHEADER *) pType->pbFormat;
  pMediaSample->GetPointer(&pData); // pointer to RGB24 image data
  lDataLen = pMediaSample->GetSize();
  // Get the image properties from the BITMAPINFOHEADER
  int cxImage  = pvi->bmiHeader.biWidth;
  int cyImage  = pvi->bmiHeader.biHeight;
  int numPixels  = cxImage * cyImage;
  // a trivial example - make a negative
  prgb = (RGBTRIPLE*) pData;
  for (iPixel=0; iPixel < numPixels; iPixel++, prgb++) {
      prgb->rgbtRed = (BYTE) (255 - prgb->rgbtRed);
      prgb->rgbtGreen = (BYTE) (255 - prgb->rgbtGreen );
      prgb->rgbtBlue  = (BYTE) (255 - prgb->rgbtBlue);
  }
  return S_OK;
}
```

Listing 15.13. The overridden `Transform()` method copies the input sample from input pin to output pin and then modifies the pixel data in the output buffer.

- The first thing the filter does is copy the video sample from the input buffer to the output buffer.

- The filter also copies all the other properties about the video sample.

- The filter performs its changes on the video sample in the output buffer.

- The filter assumes that the video data is in RGB24 bit format. Any graph using this filter must ensure (by using intelligent connect, for example) that the samples are in this format.

- The filter finds the dimensions of the video frame and gets a pointer to the RGB24 data. It can then modify the pixel values in a manner of its choosing. Listing 15.13 demonstrates a filter that makes the image negative.

- Consult the SDK example projects to see other examples of in-place transform filters.

15.5.3   Making Movies from Images

Building a movie file from a collection of images requires a custom source filter. In Section 16.7, we will look in detail at the structure of a filter of this class, which takes a sequence of pairs of images and make a stereoscopic movie. We will defer until then any further discussion about this. It would only take a very small modification to the stereo code to do the same thing with single (mono) image sequences.

15.5.4   Saving Frames from a Video Stream

There are two ways to save frames from a video stream. The most conventional way is to insert the DirectShow *grabber* filter into the filter chain, use it to request the samples be delivered in RGB24 format and write the sampled bitmaps into individual files. This approach is discussed in [2]. Alternatively, a renderer filter could save each frame to an image file at the same time as they are previewed onscreen. We will look at an implementation of this second strategy in Chapter 16.

The wonderful flexibility of the FilterGraph approach to processing multimedia data streams means that anything we have suggested doing to a video source could equally well be applied to a file source.

15.6 Playing Movies or Live Video into Textures

In this last section of the chapter, we shall investigate a topic to which we will return in Chapter 18. In VR, placing live video into a synthetic environment is fundamental to such things as spatial videoconferencing, in which a 3D mesh model of a monitor could have its screen textured with the video feed to give the illusion of a TV monitor. We will also see that by rendering a movie into a texture applied to a deformable mesh, it is possible to distort a projected image to correct for display on a non-flat surface, as for example in the immersive system outlined in Chapter 4. The same technique can be adapted to project images onto small shapes to give the illusion that the item is sitting on the table in front of you.

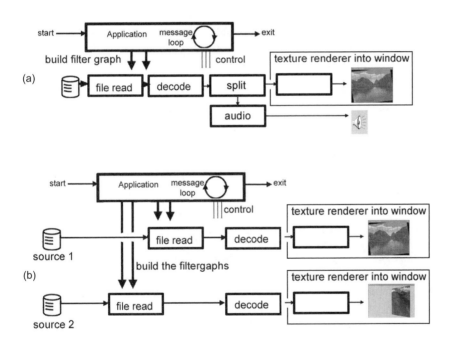

Figure 15.5. Playing movies into textures. (a) Using a custom renderer filter at the end of the FilterGraph chain that writes into a texture on a mesh which is rendered as a 3D object using Direct3D or OpenGL. (b) Using an application that builds and executes two FilterGraphs to play two movies at the same time using both OpenGL and Direct3D.

We have looked (see Chapter 12) into the texturing and rendering of 3D meshes in real time using both the OpenGL and Direct3D APIs. Either of these powerful libraries has the potential to acquire one or more of its texture image sources from the output of a DirectShow FilterGraph. Turning the problem on its head, so to speak, what we can do is write custom rendering DirectShow filters that render movies (or live video sources) into memory buffers and use these memory buffers as the source image of texture maps applied to 3D mesh models. To provide the illusion of video playing in the 3D scene, one simply arranges to render the scene continuously, say with a Windows timer, so that it updates the textures from their sources every time a frame is rendered.

Figure 15.5 shows how these filters might be arranged in a typical application and how we could use them in a couple of contrived examples to illustrate the principles involved:

1. A program that plays two movies at the same time using an OpenGL textured mesh in one window and a Direct3D textured mesh in the other. To achieve the result requires running two FilterGraphs at the same time. We will show how this is achieved by using two threads of execution.

2. A program that plays the same movie into two windows. The same texturing effect is achieved as in the first program, but this time only one FilterGraph is required, with the output from our custom renderer filter being simultaneously presented using the OpenGL and Direct3D windows.

The appearance of both these programs is illustrated in Figure 15.6. The logical structure of the program consists of:

A WinMain() function creates and displays a small application window. Commands from the menu attached to the application window call two functions. One will render a movie using OpenGL and the other will use Direct3D. Each function will:

1. Create and display a child window in which to render a 3D image-mapped polygonal mesh using OpenGL/Direct3D.

2. Construct and run a DirectShow FilterGraph with a custom filter at the end of the chain to render the video frames into the OpenGL/Direct3D textures.

main window

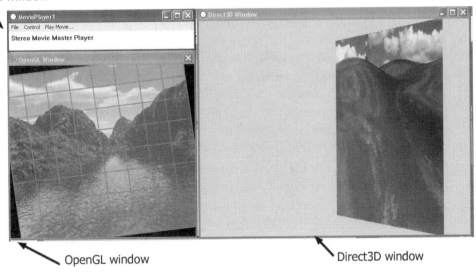

OpenGL window   Direct3D window

Figure 15.6. Rendering video into a texture (twice). A small main window carries a menu to control the display windows. Two child windows display the video (from movie files) by playing them into a texture and rendering that texture onto mesh models. The window on the left illustrates this being done using OpenGL and the one on the right using Direct3D.

Commands from the menu also stop/start the movie, close the child windows and delete the FilterGraphs.

We have already discussed how to implement the application framework, initialize OpenGL and Direct3D and render textured meshes in both system. So we go straight on to discuss the rendering filters and how they copy the video signal into OpenGL/Direct3D mesh textures. Whilst the rendering filters for OpenGL and Direct3D are nearly identical in concept, there is one subtle difference and one feature which applies to all 3D textures that we must highlight:

- DirectShow uses multi-threading; therefore, the filters in the chain execute in a separate thread under the control of the FilterGraph manager. On the other hand, our visualization (the rendering of the textured mesh) runs in the same thread as the application program; the windows are updated in response to WM_PAINT messages. We must find a way to avoid a conflict between writing/creating the texture and reading/rendering it. In fact, we must use slightly different approaches when rendering with Direct3D and OpenGL:

1. Direct3D texture objects have an interface method which can do exactly what we need. Called `LockRect( ..)`, the method collaborates with the Direct3D rendering device interface method `BeginScene()` to prevent reading and writing from a texture at the same time. Thus, in the case of Direct3D, our rendering filter will write directly into the Direct3D texture object.

2. To use OpenGL (it not being a Microsoft product) requires a little more work. However, the same result can be achieved by setting up an intermediate memory buffer that is protected by the standard thread programming technique of *critical sections* [1, 3]. So, in the case of OpenGL, our DirectShow rendering filter will copy the video sample data into a system RAM buffer (pointed to by the C++ pointer called `Screen3`). The window's `WM_PAINT` message handler (running in the main application thread) will read from the RAM buffer, make the texture and render the mesh.

- Any image used as a mesh texture must have dimensions[4] that are a power of two. Of course, a movie frame is very unlikely to have a width and height with pixel dimensions of 64, 128, 256 or other powers of two. To get around this problem, we need to either scale the map or adjust the texture-map coordinates. Scaling an image (i.e., changing its dimension) can be quite slow unless hardware-accelerated, and so we will adjust the surface image map coordinates to compensate.

For example, a video frame of resolution 720×576 is to be used as a texture for a rectangular mesh. In both OpenGL and Direct3D, the texture coordinates (u, v) at the four corners of the mesh should be $(0, 0), (0, 1), (1, 1)$ and $(1, 0)$, as shown in Figure 15.7. If our hardware requires that a texture has dimensions which are a power of two, we must copy the pixel data into a memory block whose size is the next

---

[4]OpenGL has always had this restriction, but it is now being relaxed in version 2.0. Direct3D can support texture dimensions that are not a power of two, but an application program should not rely on this being available and therefore should still be able to adjust the source dimensions accordingly.

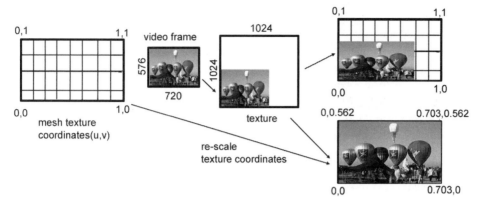

Figure 15.7. Adjusting texture coordinates for image-map textures with image dimensions that are not powers of two requires the (u, v) texture coordinates to be scaled so that the part of the texture containing the image covers the whole mesh.

highest power of two exceeding 720×576. This is 1024×1024, but then the image will only occupy the bottom left hand corner of the texture rectangle. So for the image to still cover the whole rectangular mesh, we must change the (u, v) coordinates at the four corners to be $(0, 0)$, $(0, \frac{720}{1024})$, $(\frac{720}{1024}, \frac{576}{1024})$ and $(\frac{576}{1024}, 0)$.

15.6.1 The Renderer Filters

To write a custom renderer filter, we will take advantage of the examples that ship with the DirectX SDK. It doesn't have an example that exactly matches our requirement, but the *rendering into a texture* example is pretty close and provides useful hints to get us started. To implement the filter, we derive a class from the DirectShow API base class CBaseVideoRenderer. This powerful class offers all the features, such as the interfaces that handle *connecting pins* and *media type* requests. By overriding three of the class's methods, we will get the renderer filter we need for our FilterGraph. Listing 15.14 presents the derived class details.

To be precise, we are not going to write a complete DirectShow filter, because we don't need to! The filter will not be used in other applications, and we want to keep the code as short and self-contained as possible. So we will pass over such things as component registration and class factory implementation. By deriving the CRendererD3D class from CBaseVideoRenderer, all the standard behavior and interfaces which the base class exhibits will be present in the derived one. Listing 15.15 gives the constructor and destructor.

```
// Although a GIUD for the filter will not be used we must still declare
// one because the base class requires it.

struct
__declspec(uuid("{7B06F833-4F8B-48d8-BECF-1F42E58CEAD5}"))CLSID_RendererD3D;

class CRendererD3D : public CBaseVideoRenderer
{
public:
    CRendererD3D(LPUNKNOWN pUnk,HRESULT *phr);
    ~CRendererD3D();

public:  // these methods must be overidden
         // - see individual listings for details
         // Is the Video Format acceptable?
    HRESULT CheckMediaType(const CMediaType *pmt );
         // Video format notification
    HRESULT SetMediaType(const CMediaType *pmt );
         // New video sample (THIS FILLS THE BUFFER
    HRESULT DoRenderSample(IMediaSample *pMediaSample);

    LONG m_lVidWidth;    // Video/movie width
    LONG m_lVidHeight;   // Video/movie height
    LONG m_lVidPitch;    // Pitch is the number of bytes in the video buffer
                         // required to step to next row
};
```

Listing 15.14. The derived class for the Direct3D texture renderer filter. The member variables record the dimension of the movie/video frames, and the *pitch* specifies how the addresses in RAM changes from one row to the next. This allows the rows in an image to be organized into memory blocks that are not necessarily the same size as the width of the frame.

To complete the filter, it only remains to override the following three methods from the base class:

1. CheckMediaType(). Listing 15.16 presents the code for this method. Its job is to tell DirectShow whether or not the renderer can use an input media sample[5] in a particular format. It is called repeatedly until the FilterGraph manager gives up or a suitable format is offered. Our version of the function rejects all nonvideo samples and frame formats that are not 24-bit RGB.

---

[5]A media sample is a convenient term to describe a unit that will be rendered. In this case, a media sample is the image which makes up a video frame. In another instance, it could be an audio sample.

```
CRendererD3D::CRendererD3D(LPUNKNOWN pUnk,HRESULT *phr )       // Constructoe
          : CBaseVideoRenderer( __uuidof(CLSID_RendererD3D),
                               NAME("D3D Texture Renderer"), pUnk, phr)
{
          *phr = S_OK;
}

CRendererD3D::~CRendererD3D() {       // Do nothing
}
```

Listing 15.15. The Direct3D video texture renderer class constructor and destructor.
The class constructor has very little work to do; it simply initializes the base class and
one of the member variables.

2. SetMediaType(). This function is called to inform the filter about
    the media samples; their frame width and height, for example. Our
    filter uses the method to create the Direct3D texture. In the OpenGL
    filter, it is used to allocate a RAM buffer in which to store the video
    frames. The method is called once each time the filter starts playing a
    new movie. Listing 15.17 presents the code.

3. DoRenderSample(). This does most of the work. It is called each time
    a new frame is to be rendered. It picks up a pointer to the video sample

```
HRESULT CRendererOpenGL::CheckMediaType(const CMediaType *pmt){
 // only allows RGB video sample formats
     HRESULT   hr = E_FAIL;
     VIDEOINFO *pvi=0;

 // Reject this option if not a video type
     if( *pmt->FormatType() != FORMAT_VideoInfo ) return E_INVALIDARG;
     pvi = (VIDEOINFO *)pmt->Format();
     if( IsEqualGUID( *pmt->Type(), MEDIATYPE_Video)){
 // Only accept  RGB 24 bit formatted frames
         hr = S_OK;
         if( IsEqualGUID( *pmt->Subtype(), MEDIASUBTYPE_RGB24) ){; }
         else hr = DDERR_INVALIDPIXELFORMAT;
     }
     return hr;
}
```

Listing 15.16. The check media type method returns S_OK when a suitable video
format has been enumerated. The DirectShow mechanism will put in place whatever
(decoder etc.) is needed to deliver media samples in that format.

```
HRESULT CRendererD3D::SetMediaType(const CMediaType *pmt){
    HRESULT hr;  UINT uintWidth = 2,uintHeight = 2;
    // Retrieve the size of this media type
    D3DCAPS9 caps;
    VIDEOINFO *pviBmp;                        // Bitmap info header
    pviBmp = (VIDEOINFO *)pmt->Format();
    m_lVidWidth  = pviBmp->bmiHeader.biWidth;
    m_lVidHeight = abs(pviBmp->bmiHeader.biHeight);
    m_lVidPitch  = (m_lVidWidth * 3 + 3) & ~(3); // We are forcing RGB24
    ZeroMemory( &caps, sizeof(D3DCAPS9));
    hr = g_pd3dDevice->GetDeviceCaps( &caps );
    if( caps.TextureCaps & D3DPTEXTURECAPS_POW2 ){
        while( (LONG)uintWidth < m_lVidWidth )uintWidth = uintWidth << 1;
        while( (LONG)uintHeight < m_lVidHeight )uintHeight = uintHeight << 1;
        // if power of two we must update the mesh texture map coordinates
        UpdateMappingD3D(m_lVidWidth, uintWidth, m_lVidHeight, uintHeight);
    }
    else{ // use actual video frame dimensions
        uintWidth = m_lVidWidth;
        uintHeight = m_lVidHeight;
    }
    // Create the texture that maps to the 24 bit image
    hr = g_pd3dDevice->CreateTexture(uintWidth, uintHeight, 1, 0,
                D3DFMT_X8R8G8B8,D3DPOOL_MANAGED, &g_pTexture, NULL);
    D3DSURFACE_DESC ddsd;                    // get format used and store
    ZeroMemory(&ddsd, sizeof(ddsd));
    CComPtr<IDirect3DSurface9> pSurf;   // automatically released
    if (SUCCEEDED(hr = g_pTexture->GetSurfaceLevel(0, &pSurf)))
    pSurf->GetDesc(&ddsd);
    g_TextureFormat = ddsd.Format;      // Save format info
    return S_OK;
}
```

Listing 15.17. The `SetMediaType()` method extracts information about the movie's video stream and creates a texture to hold the decoded frame images.

being delivered by the DirectShow filter chain and copies the image data into the texture. See Listing 15.18.

Overriding these three methods is all we need to do to complete the renderer filter. If we wanted, we could easily turn it into a fully fledged DirectShow filter, install it as a Windows component and made it amenable to construction via the `CoCreateInstance(...)` mechanism.

The code for the OpenGL renderer filter is virtually identical, so we will not examine its code here. Chapter 16 will look in some detail at a custom OpenGL DirectShow renderer filter for presenting stereoscopic movies.

```
HRESULT CRendererD3D::DoRenderSample( IMediaSample * pSample ) {
    BYTE   *pBmpBuffer, *pTxtBuffer; // Bitmap buffer, texture buffer
    LONG   lTxtPitch;                // Pitch of bitmap, texture
    BYTE   * pbS = NULL;
    BYTE   * pdD = NULL;
    UINT row, col, rgb;
    pSample->GetPointer( &pBmpBuffer );     // video frame pixels
    // Lock the Texture
    D3DLOCKED_RECT d3dlr;
    // lock the texture
    if (FAILED(g_pTexture->LockRect(0, &d3dlr, 0, 0))) return E_FAIL;
    pTxtBuffer = static_cast<byte *>(d3dlr.pBits);        // get texture pointer
    lTxtPitch = d3dlr.Pitch;                              // the texture pitch
    // we only do this format
    if (g_TextureFormat == D3DFMT_X8R8G8B8)      {
        for(row=0;row<m_lVidHeight;row++){ // copy the pixel data
    // we have to do it this way because the
            pdS = pBmpBuffer;
    // video and texture may have different pitches
            pdD = pTxtBuffer;
            for(col=0;col<m_lVidWidth;col++){
                for(rgb=0;rgb<3;rgb++)*pdD++ = *pdS++;
            }
            pBmpBuffer  += m_lVidPitch;
            pTxtBuffer += lTxtPitch;
        }
    }
    // Unlock the Texture
    if (FAILED(g_pTexture->UnlockRect(0)))return E_FAIL;
    return S_OK;
}
```

Listing 15.18. The `DoRenderSample()` method is overridden from the base class to gain access to the video samples. In the case of OpenGL, the video pixel data is copied to a RAM buffer.

The Application

The program uses the standard Windows template code to establish a small window and menu bar. In response to user commands, the application creates windows to render the OpenGL output and builds the geometry. It does the same thing using Direct3D. Listing 15.19 shows the code to build and configure the Direct3D output so that it appears in a floating (child) window. The code in Listing 15.20 achieves the same thing using OpenGL.

```
// global pointers to Direct3D objects
LPDIRECT3D9            g_pD3D        = NULL; // Used to create the D3DDevice
LPDIRECT3DDEVICE9      g_pd3dDevice  = NULL; // Our rendering device
LPDIRECT3DVERTEXBUFFER9 g_pVB        = NULL; // Buffer to hold vertices
LPDIRECT3DTEXTURE9     g_pTexture    = NULL; // Our texture
HWND                   g_hWnd        = 0;    // global handle to D3D window

HRESULT InitD3D( HWND hWnd ){
  .. // code is based in standard template
}

HRESULT InitGeometryD3D(){
  .. // build the mesh - code is based on standard template
}

void  CleanupD3D(){.. } // release Direct3D objects and tear down filter graph

LRESULT WINAPI D3DMsgProc( HWND hWnd, UINT msg, WPARAM wParam, LPARAM lParam ){
    switch( msg )      { // render
        .. // other messages
        case WM_TIMER:
        case WM_PAINT:
            RenderDirect3D();         // Update the main window when needed
            break;
        break;
    }
    return DefWindowProc( hWnd, msg, wParam, lParam );
}
INT WinD3D(HINSTANCE hInst){  //   Create the window to receive the D3D output
    UINT uTimerID=0;
    // Register the window class
    WNDCLASSEX wc = { sizeof(WNDCLASSEX), CS_CLASSDC, D3DMsgProc, 0L, 0L,
                      GetModuleHandle(NULL),
                      LoadIcon(hInst, MAKEINTRESOURCE(IDI_TEXTURES)),
                      NULL, NULL, NULL, CLASSNAME, NULL };
    RegisterClassEx( &wc );
    // Create the application's window
    g_hWnd = CreateWindow( CLASSNAME, TEXT("Direct3D Window "),
                        WS_OVERLAPPEDWINDOW, 100, 100, 300, 300,
                        GetDesktopWindow(), NULL, wc.hInstance, NULL );
    if( SUCCEEDED(InitD3D(g_hWnd))){    // Initialize Direct3D
        if( SUCCEEDED(InitGeometry())){// Create the scene geometry
            // start a timer to render the scene every 15 - 20 ms
            uTimerID = (UINT) SetTimer(g_hWnd, TIMER_ID, TIMER_RATE, NULL);
    }   }

    return 0;
}
```

Listing 15.19. Initializing the window to receive the Direct 3D output requires setting up the geometry and creating the Direct3D device and other objects. The initialization code is based on the standard templates in Listings 13.16 and 13.17.

```
void WinD3Dclose(HINSTANCE hInst){
    KillTimer(g_hWnd, TIMER_ID);
    DestroyWindow(g_hWnd);
    CleanupD3D();
    UnregisterClass(CLASSNAME,hInst);
}
```

Listing 15.19. (continued).

```
unsigned char   *Screen3; // pointer to temporary buffer
long            X,Y;      // video frame size
long            BUF_SIZEX=64;        // Map size nearest power of 2
long            BUF_SIZEY=64;        // greater than image size
HWND  hWndGL;   // handle to OpenGL window
static GLfloat maxx=1.0,maxy=1.0;  // texture size
CCritSec g_cs;   // critical section for thread synchronization
HWND GLWindowPr(HINSTANCE hInstance, HWND hWndParent){
  WNDCLASSEX wc1 = { sizeof(WNDCLASSEX), CS_CLASSDC, GLProcessWndProc, 0L, 0L,
                    GetModuleHandle(NULL),
                    LoadIcon(hInstance, MAKEINTRESOURCE(IDI_ICON1)),
                    NULL, NULL, NULL, CLASSNAME1, NULL };
 RegisterClassEx(&wc1));
 hwndGL = CreateWindow( // window
     CLASSNAME1,"OpeGL Window",
     WS_POPUP | WS_CAPTION | WS_SYSMENU | WS_THICKFRAME,
     50, 150, 300, 350, // default size and position
     hWndParent,NULL,hInstance,NULL);
 }
 uTimerID = (UINT) SetTimer(hwnd, TIMER_ID, TIMER_ID_RATE, NULL);
 return hWndGL;
}
LRESULT WINAPI GLProcessWndProc( HWND hWnd, UINT msg,
WPARAM wParam, LPARAM lParam ){
    static int deltacount=0;
    switch( msg )    {
        .. // other  messages set up the OpenGL environment
        case WM_TIMER:
        case WM_PAINT:
        DrawSceneOpenGL(hWnd);
        break;
        default:
        break;
    }
    return DefWindowProc( hWnd, msg, wParam, lParam );
}
```

Listing 15.20. Initializing the OpenGL output window. Initialization makes use of
code from the basic template in Listings 13.1 and 13.2.

Like initialization, rendering (in response to a WM_PAINT message) follows closely the examples we have considered before. A minor difference between the Direct3D and OpenGL program logic occurs because the Direct3D geometry is created during initialization, whereas the OpenGL mesh geometry is defined in the DrawSceneOpenGL() function.

15.6.2   Rendering using Direct3D

Rendering the textured mesh with Direct3D is accomplished by the code in Listing 15.21. Function RenderDirect3D() is called in response to a Windows *timer*, and on each call a *world coordinate* rotation matrix is applied to the mesh. The angle of rotation is calculated as a function of the elapsed time since the program started.

The texture image is updated automatically by the DoRenderSample() method of the renderer filter. However, when the texture is first created, it

```
VOID SetupMatrices(){
   ..// other matrices are the same as in the template code
   D3DXMATRIX matWorld;
   D3DXMatrixIdentity( &matWorld );
   D3DXMatrixRotationY( &matWorld,  // rotate slowly about vertical axis
       (FLOAT)(timeGetTime()/1000.0 - g_StartTime + D3DX_PI/2.0));
   hr = g_pd3dDevice->SetTransform( D3DTS_WORLD, &matWorld );
}

VOID RenderDirect3D(){
    HRESULT hr = S_OK;
    if( g_bDeviceLost ){ .... // rebuild the direct 3D if necessary
    // Clear the backbuffer and the zbuffer
    hr = g_pd3dDevice->Clear( 0, NULL, D3DCLEAR_TARGET|D3DCLEAR_ZBUFFER,
                          D3DCOLOR_XRGB(bkRed,bkGrn,bkBlu), 1.0f, 0 );
    // Begin the scene
    hr = g_pd3dDevice->BeginScene();
    // Setup the world, view, and projection matrices
    SetupMatrices();
    hr = g_pd3dDevice->SetTexture( 0, g_pTexture );
    hr = g_pd3dDevice->SetStreamSource( 0, g_pVB, 0, sizeof(CUSTOMVERTEX) );
    hr = g_pd3dDevice->SetVertexShader( NULL );
    hr = g_pd3dDevice->SetFVF( D3DFVF_CUSTOMVERTEX );
    hr = g_pd3dDevice->DrawPrimitive( D3DPT_TRIANGLESTRIP, 0, 2*nGrid-2 );
    hr = g_pd3dDevice->EndScene();
    hr = g_pd3dDevice->Present( NULL, NULL, NULL, NULL );
}
```

Listing 15.21. Render the mesh (with texture, if it exists) using direct3D.

```
HRESULT UpdateMappingD3D( LONG lActualW, LONG lTextureW,
                         LONG lActualH, LONG lTextureH ){
    HRESULT hr = S_OK;
    if(0 == lTextureW || 0 == lTextureH) return E_INVALIDARG;
    FLOAT tuW = (FLOAT)lActualW / (FLOAT)lTextureW;
    FLOAT tvH = (FLOAT)lActualH / (FLOAT)lTextureH;
    // Alter the texture coordinates to comply with image map dimensions that
    // are powers of two
    CUSTOMVERTEX* pVertices;  // get a pointer to the vertex data
    if (FAILED(hr = g_pVB->Lock(0, 0, (void**)&pVertices, 0))) return E_FAIL;
    for(DWORD i=0; i<nGrid; i++){ // go throug every vertex in grid
      pVertices[2*i+0].tu        = tuW * ((FLOAT)i)/((FLOAT)nGrid-1.f);
      pVertices[2*i+0].tv        = 0.0f;
      pVertices[2*i+1].tu        = tuW * ((FLOAT)i)/((FLOAT)nGrid-1.f);
      pVertices[2*i+1].tv        = tvH;
    }
    g_pVB->Unlock();  // we are down making our changes - release the lock
    return S_OK;
}
```

Listing 15.22. Updated texture coordinates to account for texture sizes that must be of dimension 2^x.

may be necessary (as we have already discussed) to alter the vertices' texture-mapping coordinates. This is done by the function UpdateMappingD3D() of Listing 15.22.

15.6.3 Rendering using OpenGL

To render the textured mesh using OpenGL, the code in Listing 15.23 is used. Function DrawGeometry() draws a rectangular mesh of quadrilateral polygons with texture coordinates applied. The texture itself is created from the pixel data stored in the memory buffer pointed to by Screen3 and placed there by the renderer filter in the movie player's FilterGraph. In the OpenGL rendering thread, visualization is accomplished by calling functions glBindTexture() and glTexImage2D(), which copy the image pixels from host RAM into texture memory in the GPU.

To construct an OpenGL version of the renderer filter, the following two methods from the CBaseRenderer base class are overridden:

- CRendererOpenGL::SetMediaType(). This method, called once when the graph is started, uses the size of the video frame to allocate

```
static void DrawGeometry(void){
   glBindTexture(GL_TEXTURE_2D,2);
   glEnable(GL_TEXTURE_2D);
   glBegin(GL_QUADS);
   int i,j,n,m; float x,y,dx,dy;
   n=8,m=8;
   dx=1.0/(float)n;   dy=1.0/(float)m;
 // draw mesh of polygons to fill unit square
   for(i=0,x=0.0;i<n;i++){
 // and with texture coordinates to match texture scale
       for(j=0,y=0.0;j<m;j++){
           glTexCoord2f(x*maxx,y*maxy);         glVertex3f(x,y,0.0);
           glTexCoord2f((x+dx)*maxx,y*maxy);    glVertex3f(x+dx,y,0.0);
           glTexCoord2f((x+dx)*maxx,(y+dy)*maxy);glVertex3f(x+dx,y+dy,0.0);
           glTexCoord2f(x*maxx,(y+dy)*maxy);    glVertex3f(x,y+dy,0.0);
           y += dy;        }
       x += dx;     }
   glEnd();
   glDisable(GL_TEXTURE_2D);
}

void DrawSceneOpenGL(HWND hWnd){
 // lock the texture image Screen3 so others can't write
 CAutoLock lock(&g_cs);
 // to it until we are finished
 glClear(GL_COLOR_BUFFER_BIT | GL_DEPTH_BUFFER_BIT);
 if(Screen3 != NULL){                    // make the texture
   glBindTexture(GL_TEXTURE_2D,2);       // use texture ID=2
   glTexImage2D(GL_TEXTURE_2D,0,3,BUF_SIZEX,BUF_SIZEY,0,
            GL_RGB,GL_UNSIGNED_BYTE,(GLvoid *)Screen3);
   static GLfloat aa=0.0;  // incrementable rotation angle
   glPushMatrix();
   glTranslatef(0.5,0.5,0.0);
   glRotatef(aa,0.0,0.0,1.0);
   glTranslatef(-0.5,-0.5,0.0);
   DrawGeometry();           // draw the mesh
   glPopMatrix();
   aa += 1.0;
   glFlush();
   glFinish();               // finish rendering
 }
 return;
}
```

Listing 15.23. Drawing in OpenGL. The WM_PAINT message handler calls
DrawSceneOpenGL() and swaps the front and back buffer, as described in Section 13.1.2.

a RAM buffer for the pixel data (see Listing 15.24). The size of the buffer is set to the smallest size suitable to accommodate a texture of dimension 2^x (see Figure 15.7). Global variables *maxx* and *maxy* contain the necessary mapping coordinate scaling factors, and BUF_SIZEX and BUF_SIZEY define the texture size.

- CRendererOpenGL::DoRenderSample(). This method copies the pixel data from the video sample to the memory buffer (Screen3). Care must be taken to ensure the address offset between subsequent rows (the pitch) is applied correctly. Since this method is accessing Screen3, which could potentially be in use by another thread (i.e., the main application thread executing the DrawSceneOpenGL() function), a *critical section* synchronization object (AutoLock lock(&g_cs);) is deployed to prevent conflict (see Listing 15.25).

```
HRESULT CRendererOpenGL::SetMediaType(const CMediaType *pmt){
    HRESULT hr = S_OK;
    VIDEOINFO *pviBmp = NULL;    // Bitmap info header
        unsigned char *S;
    // Retreive the size of this media type
    pviBmp = (VIDEOINFO *)pmt->Format();
    X = m_lVidWidth  = pviBmp->bmiHeader.biWidth;
    Y = m_lVidHeight = abs(pviBmp->bmiHeader.biHeight);
    m_lVidPitch = (m_lVidWidth * 3 + 3) & ~(3); // We are forcing RGB24
    AllocateScreenBuffersD(X, Y);
    return hr;
}

void AllocateScreenBuffersD(long x, long y){
 int i;
 if(x < 1 || y < 1){DeAllocateScreenBuffersD(); return; }
 for(i=1;i<4096;i*=2){if(x <= i){BUF_SIZEX = i; break;}}
 for(i=1;i<4096;i*=2){if(y <= i){BUF_SIZEY = i; break;}}
 Screen3=(unsigned char *)malloc(BUF_SIZEX*BUF_SIZEY*3*sizeof(unsigned char));
 maxx=(GLfloat)x/(GLfloat)BUF_SIZEX;
 maxy=(GLfloat)y/(GLfloat)BUF_SIZEY;
 return;
}
```

Listing 15.24. The OpenGL custom DirectShow rendering filter's SetMediaType() method.

```
HRESULT CRendererOpenGL::DoRenderSample( IMediaSample * pSample ){
    CAutoLock lock(&g_cs);    // criutical section sync object
    HRESULT hr = S_OK;
    BYTE * pSampleBuffer = NULL;
    unsigned char *S = NULL;
        UINT row;    UINT col;
    LONG lTexturePitch;    // Pitch of texture
    LONG yy;
        if( !pSample ) return E_POINTER;
        // Get the video bitmap buffer
        hr = pSample->GetPointer( &pSampleBuffer );
        if( FAILED(hr)){return hr;}
        S=Screen3;
                if(X > 0 && Y > 0 && S != NULL){
          yy=Y;
          for(row = 0; row < yy; row++ ) {
            BYTE *pBmpBufferOld = pSampleBuffer;
            for (col = 0; col < X; col++)   {
              *S++ = pSampleBuffer[2];
              *S++ = pSampleBuffer[1];
              *S++ = pSampleBuffer[0];
                      pSampleBuffer += 3;
            }
            S += (BUF_SIZEX-X)*3;
            pSampleBuffer   = pBmpBufferOld + m_lVidPitch;
                    }
        }
    return hr;
}
```

Listing 15.25. The OpenGL custom DirectShow rendering filter's method
`DoRenderSample()`.

15.6.4 Using Two FilterGraphs

In this example, where we will use two FilterGraphs in the one program, each
of them is built up in the same way as the other single FilterGraph examples
we have already written. The only new feature relating to this example is the
way in which our custom render filters are inserted in the chain. Listing 15.26
shows the main steps in FilterGraph construction in the case of the filter that
renders into a Direct3D texture.

The OpenGL filter differs only in two lines: Connect the source to a different AVI file and create an instance of an object of the `CRendererOpenGL()` class, i.e.:

```
hr=g_pGB->AddSourceFilter(MediaFileForOpenGL,L"Source File for OGL",
&pSrcFilter);
pRenderer = new CRendererOpenGL(NULL, &hr);
```

```
CComPtr<IGraphBuilder>  g_pGB;            // GraphBuilder
CComPtr<IMediaControl>  g_pMC;            // Media Control
CComPtr<IMediaPosition> g_pMP;            // Media Position
CComPtr<IMediaEvent>    g_pME;            // Media Event
CComPtr<IBaseFilter>    g_pRenderer;      // our custom renderer
extern TCHAR filenameF3DMovie;           // string with movie name

 // build a filter graph to play movie file
HRESULT InitDShowRendererD3D(){
 HRESULT hr = S_OK;
 CComPtr<IBaseFilter>   pFSrc;            // Source Filter
 CComPtr<IPin>          pFSrcPinOut;      // Source Filter Output Pin
 if (FAILED(g_pGB.CoCreateInstance(CLSID_FilterGraph, NULL,
 CLSCTX_INPROC)))return E_FAIL;
 // Create our custom texture renderer filter
 g_pRenderer = new CRendererD3D(NULL, &hr);
 // add it to the graph
 if (FAILED(hr = g_pGB->AddFilter(g_pRenderer, L"RendererD3D")))return hr;
 // add the source filter
 hr = g_pGB->AddSourceFilter (ffilename, L"SOURCE", &pFSrc);
 // find an output on on the source filter
 if (FAILED(hr = pFSrc->FindPin(L"Output", &pFSrcPinOut)))
 // The intelligent connect mechanism will link the source filter's output pin
 // and our custom renderer filter since they are the only source
 // and sink filters in the graph. It will also add any necessary decoder
 // filters in the chain.
 if (FAILED(hr = g_pGB->Render(pFSrcPinOut)))return hr;
 // Get the graph's media control, event & position interfaces
 g_pGB.QueryInterface(&g_pMC);
 g_pGB.QueryInterface(&g_pMP);
 g_pGB.QueryInterface(&g_pME);
 // Start the graph running;
 if (FAILED(hr = g_pMC->Run()))return hr;
   return hr;
 }
```

Listing 15.26. Build and run a FilterGraph to play a movie into a Direct3D texture by using our custom renderer filter, i.e., an instance of class `CRendererD3D()`.

15.6.5   Rendering Two Outputs from One Filter

As we approach the end of this chapter, we hope you can see the versatility of the DirectShow FilterGraph concept and how its SDK and the useful C++ base classes it provides can be used to build a comprehensive array of multimedia application programs. But before we leave this section, we couldn't resist including a little program that removes one of the FilterGraphs from our dual-texture movie player and adapts the other's renderer filter to play the same movie into the two textures at the same time. Have a look at the accompanying code for full detail. Listing 15.27 shows the small changes that need to be made to the Direct3D renderer filter so that it takes the image source from the Screen3 buffer used by the OpenGL rendering filter.

```
HRESULT CDualRenderer::SetMediaType(const CMediaType *pmt){
 .. // Include all the code from this method in the Direct3D filter
 .. // class CRendererD3D.
    // modify the lines below
    // get dimensions of OGL texture
 X = m_lVidWidth  = pviBmp->bmiHeader.biWidth;
 Y = m_lVidHeight = abs(pviBmp->bmiHeader.biHeight);
    // add this line
 AllocateScreenBuffersD(X, Y);  // allocate RAM for the Screen3 buffer
 .. //
}

HRESULT CDualRenderer::DoRenderSample( IMediaSample * pSample ){
 .. // Include all the code from this method in the Direct3D filter
 .. // class CRendererD3D.

    // Add the code that copies from the sample buffer to the memory
    // buffer pointed to by  Screen3 in this method from the CRendererOpenGL
    // class.
 CAutoLock lock(&g_cs);
 hr = pSample->GetPointer( &pSampleBuffer );
 S=Screen3;
 for(row = 0; row < yy; row++ ) {
    for (col = 0; col < X; col++)  {
 .. // etc. etc.
 .. //
}
```

Listing 15.27. Filter modifications to render, simultaneously, the same video samples into both Direct3D and OpenGL textured meshes.

15.7 Summary

In this chapter, we have explored the basics of the DirectShow development library and given a number of fundamental examples that use it to acquire, store and process signals from a wide variety of multimedia sources. The key concepts of the signal, filter and FilterGraph were explained and demonstrated.

In the next chapter, we will build on these FilterGraph concepts, and in particular the use of custom OpenGL-based rendering filters, to allow us to play, build and modify stereoscopic movies.

Bibliography

[1] J. Hart. *Windows System Programming,* Third Edition. Reading, MA: Addison-Wesley Professional, 2004.

[2] M. Pesce. *Programming Microsoft DirectShow for Digital Video and Television.* Redmond, WA: Microsoft Press, 2003.

[3] B. E. Rector and J. M. Newcomer. *Win32 Programming.* Reading, MA: Addison-Wesley Professional, 1997.

16 Programming Stereopsis

Chapter 10 described and explained the concept of stereopsis, its significance and its theoretical basis. In this chapter, we will look at how to implement a few *must have* stereoscopic application programs.

Approaching the problem from the software side, there are two initial considerations:

1. How are the stereoscopic sources formatted?

2. What are the requirements for driving the stereoscopic display hardware?

Before looking at the format of stereoscopic sources, we will consider how to drive the display hardware. Section 16.1 explains how projection display devices are driven by a suitable graphics adapter. To program the stereo-ready graphics adapter, we have two choices: use either OpenGL or Direct3D. We choose to use OpenGL because it offers a number of advantages:

- OpenGL has a much longer history of supporting stereoscopic display technology and is very robust—and does not just work on Windows PCs.

- Individual window areas of the screen can be configured for stereoscopic use while other windows and the background remain unaffected. Thus, a program can divide up the screen, with some areas of it displaying non-stereoscopic text, for example. The only examples we have seen of stereoscopic rendering using Direct3D required that the whole display be switched to a full-screen stereo mode.

433

- There is no need for special configuration settings, as happens in some types of computer games, for example.

- It is easy to use OpenGL's multiple image buffers to display stereo images and stereo movies with a degree of device independence.

The only disadvantage we can think of for OpenGL stereo is that the graphics adapters which fully support stereo by giving a left/right frame synchronization signal are still not in widespread use, and as a consequence they tend to be quite costly.[1] But in our opinion, the advantages of using OpenGL significantly outweigh the minor disadvantage.

Shortly, we will write some software tools to display stereoscopic images, play stereoscopic movies and view 3D mesh models in stereo. But before doing that, we need to explain a little more about how the display adapter hardware produces the stereoscopic output and how the existing image/movie file formats are enhanced to include stereoscopic information.

16.1 Display Adapter Hardware for Stereopsis

This section will concentrate on briefly reviewing the hardware options for driving the different display options; for example, with a CRT screen or DLP projector. However, we found that the CRT devices were the most useful in terms of delivering stereo images. But of course, we also need to consider the display adapters, of which there are two main types: stereo-ready and non-stereo-ready.

16.1.1 Stereo-Ready Adapters

Unfortunately, stereo-ready adapters tend to be expensive because they are mainly intended for professional CAD work. NVIDIA,[2] ATI[3] and 3Dlabs[4] manufacture families of graphics processors that are stereo-ready. To be stereo-ready, an adapter should be able to generate a signal for shutter eyewear systems that opens the left shutter when the left image is active and vice versa for

---

[1] At the time of writing, NVIDIA's FX class GPU-based adapters support all the OpenGL 2 functionality with stereo output, and the lower-end models are relatively inexpensive.

[2] http://www.nvidia.com/.

[3] http://ati.amd.com/.

[4] http://www.3dlabs.com/.

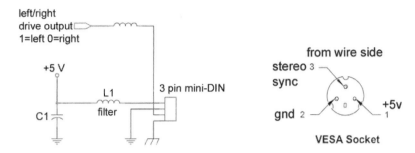

Figure 16.1. Circuit and pin configuration layout for stereo-ready graphics adapters.

the right. There is an agreed standard for this signal and its electrical interface through the VESA standard connector. The output, shown in Figure 16.1, is a circular three-pin mini-DIN jack.

- *Pin 1.* +5 Volt power supply capable of driving a current of at least 300mA.

- *Pin 2.* Left/right stereo sync signal (5 volt TTL/CMOS logic level) high (5V) when the left eye image is displayed; low (0 V) when the right eye image is displayed. When not in stereo mode, the signal must be low.

- *Pin 3.* Signal and power ground.

Surprisingly, even through there are many graphics adapter vendors, the number of suppliers of the key processor chip is much smaller. For example, the same NVIDIA FX1100 GPU chip is included in adapters from HP[5] and PNY.[6] So, there are not only stereo-ready graphics adapters but also stereo-ready graphics processors. For those adapters that don't use a stereo-ready GPU, a stereo sync signal must be generated in the display electronics. One way to do this is to clock a D-type flip-flop on the leading edge of the vertical sync pulses being sent to the monitor, as illustrated in Figure 16.2. The software driver for such adapters must ensure that the D-input is set to indicate a left or right field. It can do this by hooking up the D-input to a system port on the CPU. The driver's vertical sync interrupt service routine (ISR) writes

---

[5] http://www.hp.com/.

[6] http://www.pny.com/.

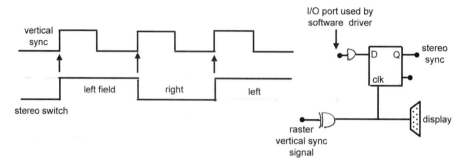

Figure 16.2. Generating stereoscopic sync signals in stereo ready-adapters that do not have stereo-enabled GPUs can be achieved by adding a flip-flop with one input from the vertical sync signal and another from the operating system driver using a system I/O port to indicate left or right field.

to this port to indicate a left or right field. Details of the circuitry can be obtained in the StereoGraphics Hardware Developer's Kit documentation [3].

The OpenGL 3D rendering software library (Section 13.1) fully supports stereo rendering and can report to an application program whether or not the adapter has this capability. It is the job of the manufacturer's device driver to interface with the hardware sync signal generator. To comply with the *stereo certification process,* a display adapter must implement quad buffering (see Section 16.4), allow for dual Z-buffers, fully support OpenGL and provide a three-pin mini-DIN connector. So if you have a certified *stereo-ready* adapter, it will work with all our example stereoscopic programs.

16.1.2   Non-Stereo-Ready Adapters

Graphics adapters that are not advertised as *stereo-ready* can still be used for stereoscopic work. Computer game players use many stereoscopically enabled products, but since they are not prepared to pay *stereo-ready* prices, a different approach is adopted. With non-stereo-enabled adapters, the application programs themselves must support specific stereo delivery hardware, be that shutter-eye or anaglyph. In many cases, the software product vendor will have to negotiate with the display adapter manufacturer to build support into the driver to allow the non-stereo hardware to work with specific applications. This is less satisfactory for application developers, since any program will have to be rewritten to support specific stereo technology. Nevertheless, there are one or two general ideas that can be accommodated. StereoGraphics

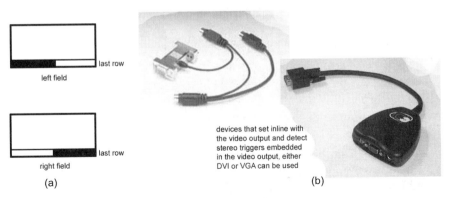

left field

last row

right field

last row

(a)

devices that set inline with
the video output and detect
stereo triggers embedded
in the video output, either
DVI or VGA can be used

(b)

Figure 16.3. (a) For non-stereo-ready graphics adapters, the last row in the display raster can be used to indicate with a brightness level whether it is from the left or right image. (b) Special inline hardware detects the brightness level and synthesizes the sync signals.

Corporation[7] (now part of Real D) and eDimensional[8] provide shutter eye systems that include a device that is *inlined* between the traditional analog VGA output of display adapter and the computer's monitor. Devices such as these check on the video signals and examine the video brightness on one or two lines at the bottom of the video raster or in the left and right images themselves.

To give a definitive example, let us say that the last row of pixels in a raster of resolution 1024×768 provides the stereo cue. On that row, we draw the first 512 pixels at peak white brightness and the second 512 pixels at the black level; call this the right frame. On the next frame, we draw the first 512 pixels at black level and then go to peak white. The external device detects the brightness level in the video waveform and tells from its magnitude whether the next raster is a left or right one. The device blocks out this row from the display and forms a left/right sync signal for the eyewear to use. Figure 16.3 illustrates this principle and typical inline devices currently in use. One major drawback of the non-stereo-ready display approach is that the applications must operate in a full screen-mode because the hardware cannot detect a left field/right field indication at the bottom of a normal overlapped desktop window. For games and some VR applications, this is less likely to be a handicap as it might be for CAD software.

---

[7]http://www.reald-corporate.com/scientific.

[8]http://www.edimensional.com/.

16.2   File Formats for Storing Stereoscopic Images

Since stereoscopic pictures are basically a combination of two separate pictures, it is possible to adapt any of the single image formats for stereo use. There are a number that we have heard of—JPS, GIS, BMS, PNS—but there may be others, too. Each of these is simply an extension of the well-known JPEG, GIF, BMP, and PNG standards respectively. Simply by changing the filename extension and using an implied layout for the images, a display program can separate out the left and right parts and present them to the viewer in the correct way. However, many image file formats allow applications to embed extra information in what are termed *chunks* as part of the standard file headers, and we can use one of these to store information about the stereo format. The chunks typically give details of such things as image width and height. The GIF format has what is know as the *GIF application extension* to do the same thing. The GIS stereo version of the GIF-encoded image can be used in just this way. We shall not explore the GIS specification any further, because as far as photographic pictures are concerned, the JPEG image format (see Section 5.4.1) reigns supreme. The JPEG file specification is a member of a class of file formats called Exif, and this allows third-party information chunks to be inserted with the image data that are known as application markers (APP0, APP1 etc.).

For stereo work, two pictures positioned side by side or one above the other appear to the JPEG encoding and decoding algorithms as one single image, twice as wide or twice as high as the non-stereo image. You've probably spotted a small problem here: how do you know which of the two alternatives are being used? In fact, there are other possibilities too (interlaced, for example). These alternatives are presented graphically in Figure 16.4. Since

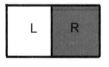
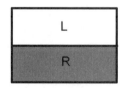
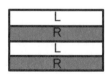

Figure 16.4. A stereoscopic image file format can combine the left and right images by laying them out: side by side, over and under or interlaced.

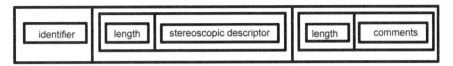

Figure 16.5. The structure of an APP3 segment containing a JPS stereoscopic descriptor.

the JPEG file format allows any number of proprietary information chunks to be embedded (as we have seen in Section 5.4.1 where the APP3 marker was defined), it is possible to insert details of the way an image is to be handled

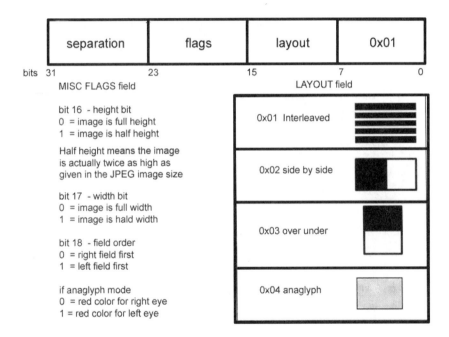

Figure 16.6. The 32-bit JPS stereoscopic descriptor is divided up into four 8-bit fields. The *separation* field in bits 24–31 gives the separation in pixels between the left and right images. There is little reason to make this non-zero. The *misc flags* and *layout* fields are the most informative. The *type* field contains 0x01 to indicate a stereoscopic image.

when interpreted as a stereo pair. This is exactly what happens in the JPS [4] stereo picture format, where an APP3 application tag is embedded in a JPEG file. In this case, the body of the APP3 is structured as shown in Figure 16.5. (It follows the two-byte APP3 chunk identifier 0xFF3E and the 16-bit integer that gives the length of the remaining chunk, minus the two-byte ID and length fields of four bytes.) The eight-byte identifier should always contain the character pattern "_JPSJPS_". The 16-bit length field is the length of *that* block. The stereoscopic descriptor is the most important part of the APP3; it is detailed in Figure 16.6. The remaining part of the APP3 can contain anything, such as the stereo camera type or eye separation. Specific details on the stereo format are given in Figure 16.6.

One final comment on the JPS format: if the stereoscopic APP3 marker is not present in the file then display programs are expected to assume that the image is in what is called *cross-eyed* format, where the left and right images are in side-by-side layout with the right image on the left! The name may give you a clue as to why it is done in this way. Almost unbelievably, there is a way to view such images without any special hardware or glasses, but it takes practice.

To view cross-eyed stereo images:

Display the image in a full-screen mode (right eye image on the left half of the monitor). Sit about 60 cm away from the screen. Close your right eye, put your first finger 15 cm in front of your nose and line it up so that it appears below some feature in the image on the right of the screen.

Now, close your left eye and open your right eye; your finger should line up with the same feature in the left side of the screen. If it doesn't then iterate. Swap back and forth, opening left and right eyes and moving your finger about until the alignment is almost correct.

The last step is to open both eyes. Start by focusing on your finger. Now take it out of the field of view and defocus your eyes so that the image in the background drifts into focus. This step is quite hard, because you must keep looking at the point where your finger was while you defocus.

It works, and it isn't as hard to do as viewing those Magic Eye stereograms!

16.3   A File Format for Stereo Movies

We are not aware of any formats commonly used for recording stereo movies. On approach we've tried successfully uses two separate files for the left and right frame sequences. It works well for the lossless compression FLC [2] format, and there is no reason why the same strategy should not be used for AVI, MPEG etc. But why go to all the trouble of synchronizing two separate input streams or having to specify two source files when we can use the same strategy as in the JPS format for photographs? Since there is no recognized stereo movie MPEG format, we will just define our own simple one.

> Use standard AVI (all codecs are applicable) and assume that each individual frame is recored in a *left-over-right* layout.

We choose, and urge you, to adopt *left-over-right* as opposed to *left-beside-right* because it will make the coding of any program for playing back the movies much simpler and more efficient. There is a good reason for this, which becomes much more significant when writing a program to play stereo movies as opposed to one displaying images. It has to do with the row organization of the memory buffer used to store the decompressed movie frame before display. In ordering the images in a left-above-right order, all the pixels in the left image are stored in consecutive memory addresses. Consequently, the left and right images can be copied to video RAM in one block transfer. If the images were stored left-beside-right, it would be necessary to copy one row at a time, so that the first half of the row is copied to the left frame buffer etc. With left-over-right storage, it is easy to manufacture C pointers to the memory occupied by left and right images. The two pointers behave as if they pointed to independent buffers, even though they are pointing to different locations in the same buffer.

With this in mind, the program described in Section 16.5 will assume the height of a movie frame as defined in the AVI file is twice the actual height and the left image is in the top half of the frame.

16.4   Displaying Stereoscopic Images

Displaying images and pictures is at the core of graphical user interfaces. Windows' native drawing API (the GDI, or graphics device interface) offers a number of functions for drawing pictures, but none of them are designed to support stereoscopic hardware. For the greatest flexibility, OpenGL is the ideal programming environment in which to do stereoscopic work. We shall use it to render 3D scenes in stereo and even as part of a stereoscopic movie player. It might seem a little strange to use an API known primarily for 3D work to display images and movies, but OpenGL works just as well for pixel manipulations. It can therefore easily deliver the performance required to present 30 or more pictures per second, either within a desktop window or across the full screen. Using OpenGL to render pictures and movies has a few surprisingly useful side effects, some of which helped us build the driving software for a large-scale virtual-reality theater similar to that described in Chapter 4.

To construct the code, we will start with the standard Windows application framework. An image display application should have a fairly user friendly interface, for example with the following features:

- File selection dialog

- Show image in a resizable window

- Stretch the image to fill the screen

- Run in slideshow mode through all the files in a folder.

You can find the full code for such an application on the CD. In the remainder of this section, we will concentrate on looking at the key steps in adapting the basic OpenGL template in Listing 13.2.

The way OpenGL handles stereoscopy is to extend the double-buffer concept to *quad buffering*; that is, double buffers for left and right stereoscopic pairs. So we have back left, back right, front left and front right frame buffers. The application program renders the image for the left eye into the left back buffer, then it renders an image for the right eye into the back right buffer. When rendering is complete, both back and both front buffers are swapped. Using quad buffering will double the rendering time, which may cause a little concern for some big 3D scenes, but in simple pixel image rendering, frame rates of 50 Hz or more are still attainable.

Starting with the basic OpenGL template, preparing a window for stereoscopic use only requires setting an extra flag in the PIXELFORMATDESRCIPTOR:

```
PIXELFORMATDESCRIPTOR pfd = {
    ..
    };
if (Stereo){ // we want to render in stereo
  pfd.dwFlags |= PFD_STEREO;
}
```

Naturally, a robustly written program should check to see that the application has accepted the request for stereoscopic display. If not, it should fall back into the default mono mode. There are a wide variety of freely available libraries and code for reading images stored with various file formats. So here we will assume that somewhere in our application there are a collection of routines that extract width, height, stereoscopic format data and image pixel data into global program variables and RAM memory buffers. (Suitable code is included along with our own code.) The image pixel data in the RAM buffer will be assumed to be stored in 24-bit (RGB) format and ordered in rows beginning at the top left corner of the image and running left-to-right and top-to-bottom. A global pointer (unsigned char *Screen;) will identify the location of the pixel data buffer. Two long integers (long X,Y;) store the width and height of the image. In the case of stereoscopic pictures, our program will make sure that the pixel ordering in the buffer is adjusted so that the correct images are presented to left and right eyes.

Section 16.2 provided details of the stereoscopic format for image data organized in JPS files. Since the image part of the JPS format is encoded using the JPEG algorithm, we can use any JPEG decoder to obtain the pixel data and fill the RAM buffer. To acquire the stereoscopic format data, the program must seek out an APP3 application marker and use its information to determine whether the left and right images are stored side by side or one above the other. So that our application is not restricted to displaying JPS files, we make the assumption that for all non-stereo file formats, the image is stored in left-above-right order.

When reading a file containing implied stereo images in a left-over-right format, both images may be loaded to RAM in a single step, say to a buffer pointed at by: char *Screen;. Given that the width is int X; and the height is int Y;, we can immediately make it look as if we have independent image buffers by setting Y = Y/2;, ScreenL = Screen; and ScreenR = (Screen+3*X*Y);. With no further work required, the program can proceed to render the left and right images using these pointers and dimensions. For cases involving other stereo image layouts, the program will have to shuffle the data into the left-over-right organization. We will not show this code explicitly but instead carry on to use OpenGL to display the stereo image pairs.

There are two alternative ways in which one can render a pixel array (an image) using OpenGL:

1. Use the pixel-drawing functions. This is the fastest way to display an image. The 3D pipeline is bypassed and the pixels are rendered directly into the frame buffer; only being processed for scaling and position. It is possible to use logical operations on a per-pixel basis to achieve blending or masking effects, too. The OpenGL pixel drawing functions expect the pixels to be ordered by rows starting at the *bottom* of the image. This will mean that when we use this method to render pixels from a RAM buffer, it will be necessary to turn the image upside down before passing it to OpenGL.

```
static GLvoid initializeGL(HWND hWnd){       // Set up display
  glClearColor(0.0,0.0,0.0,1.0);             // dark background
  glClearDepth(1.0);                         // set Z buffer to furthest
  glClear(GL_COLOR_BUFFER_BIT |              // Clear color and Z
          GL_DEPTH_BUFFER_BIT);              // buffers.
  glMatrixMode(GL_PROJECTION);               // Viewpoint is NON
  glLoadIdentity();                          // persoective and uses a unit
  glOrtho(0.0,1.0,0.0,1.0,-1.0,1.0);         // square [0,0] to [1,1]
  glMatrixMode( GL_MODELVIEW );              // position the bitmap
  glRasterPos2i(0,0);                        // put it at bottom left
  RECT rc;
  GetClientRect(hWnd,&rc);                   // get the windows size
  if(X > 0 && Y > 0)                         // Scale the pixels so that
   glPixelZoom((GLfloat)rc.right/(GLfloat)X,// the iamge fills the window
              (GLfloat)rc.bottom/(GLfloat)Y);
  else glPixelZoom((GLfloat)1.0,(GLfloat)1.0);
  glViewport(0,0,X,Y);                       // viewport fill the window
  return;
}
```

Listing 16.1. Initialize the viewing window by setting the screen origin for the image pixel bitmap (bottom left corner) and any pixel scaling that we wish to employ. Note that OpenGL is still in essence a 3D rendering system, so we need to set up a view, in this case an orthographic view. Otherwise, the little rectangle of pixels would be distorted and not fill the window.

2. Use the image pixel data to make a texture (image) map, apply the map to a few primitive polygons and render an orthographic view that fills the viewport. This second approach is not as crazy as it sounds, because it can lead to some really interesting effects. For example, you can render a video source into a texture and apply that texture to dynamically deforming objects to achieve some very interesting video mixing and blending effects. We will use this in some of the projects in Chapter 18.

The first approach will be used in this example. It requires a small amount of additional initialization code to define the viewport and set the pixel scaling. Listing 16.1 provides the detail.

```
void DrawScene(HWND hWnd){      // draw the image bitmaps for left and right eye
 HDC hDC;
 PAINTSTRUCT  ps;
 hDC = BeginPaint(hWnd, &ps);
 glClear(GL_COLOR_BUFFER_BIT
     | GL_DEPTH_BUFFER_BIT); // all buffers cleared
  // if stereo draw into back left buffer
 if(bStereoDraw)glDrawBuffer(GL_BACK_LEFT);
 else           glDrawBuffer(GL_BACK);
 if(ScreenL != NULL)glDrawPixels((GLsizei)X,(GLsizei)Y,
                 GL_RGB,GL_UNSIGNED_BYTE, // format of pixels and byte order
                 (GLvoid *)ScreenL);      // buffer with pixels
 if(bStereoDraw){
   if(ScreenR != NULL){         // ther is a right image so draw it
     glFlush ();
     glClear (GL_DEPTH_BUFFER_BIT);
     glDrawBuffer (GL_BACK_RIGHT);
     glDrawPixels((GLsizei)X,(GLsizei)Y,
                 GL_RGB,GL_UNSIGNED_BYTE,
                 (GLvoid *)ScreenR);
 }   }
  // flush OpenGL - finish rendering and swap buffers
 glFlush();
 glFinish();
 hDC = wglGetCurrentDC();
 SwapBuffers(hDC);
 EndPaint(hWnd, &ps);
 return;
}
```

Listing 16.2. Rendering stereoscopic images is accomplished by writing the left and right image pixels into the left and right back buffers.

```
case WM_SIZE:                          // respond to window size changes
// window only changes size in NON full screen mode
  if(!FullScreen){
    glViewport(0,0,(GLfloat)LOWORD(lParam), (GLfloat)HIWORD(lParam));
// change pixel size and aspect ration to fill window
    if(X > 0 && Y > 0)
       glPixelZoom((GLfloat)LOWORD(lParam),/(GLfloat)X,
                   (GLfloat)HIWORD(lParam),/(GLfloat)Y);
    else                               // in case X or Y are zero
       glPixelZoom((GLfloat)1.0,(GLfloat)1.0);
    InvalidateRect(hWnd,NULL,FALSE);  // redraw the OpenGL view
  }
  break;
```

Listing 16.3. Handling the WM_SIZE message is only required when rendering into a resizable window.

Rendering the pixel buffer is just a matter of calling the function glDrawPixels(...); (provided we have arranged that the images in it are upside down). For stereo images, the correct back buffer (left or right) must be activated for writing, but once both back buffers have been filled, a swap with the front ones makes the image visible. Listing 16.2 introduces the code detail for the rendering process (a process initiated in response to a WM_PAINT message). For a windowed display, the application will have to handle the WM_SIZE message so that the pixel scaling and viewport can be changed to match the size of the window's client area (see Listing 16.3).

There is little more to be said about displaying stereoscopic images. The full code listing of this example program can be consulted, and its logic should be clear to follow using the key points discussed earlier as a guide.

16.5 A Stereoscopic Movie Player

The move player that we are going to put together in this section will build on a number of other examples we have already discussed. Before getting into details, it is useful to consider what alternatives we have at our disposal:

- *Rendering*. There is no practical alternative for flexibly rendering stereo movies other than to use OpenGL in the same way that we have just done for stereoscopic images. We can use the same software design and build the rest of the application so that it interfaces to the stereoscopic rendering routines via a pointer to a RAM buffer storing the pixel data from the video frame and global variables for width and height.

- *Decompression.* To write a file parser and decompression routine to support all the *codecs* one might encounter is impractical. As we discussed in Section 5.5.2, the philosophy of providing extensible components allows any player program to call on any installed codec, or even to search the Internet and download a missing component. Under Windows, the simplest way to achieve this is to use the DirectShow API.

- *Audio.* A movie player that only produces pictures just isn't good enough. Nor is one that doesn't manage to synchronize the video and audio tracks. Again, we have seen in Chapter 15 that DirectShow delivers everything we need in terms of audio processing.

- *Timing.* Synchronization between sound and vision is not the only timing issue. The movie must play at the correct speed, and if it looks like the rendering part of a program can't keep up then frames must be skipped, not slowed down. Again, DirectShow is designed to deliver this behavior, and if we were to write a custom rendering filter using OpenGL with stereoscopic output, we would gain the benefit of DirectShow's timing.

- *Stereo file format.* We will assume that each frame in the AVI file is in the format discussed in Section 16.3. This makes sense because the left-over-right organization allows for the most efficient use of memory copy operations, thus maximizing the speed of display.

Having examined the alternatives the conclusion is obvious:

> Use DirectShow to decode the video data stream and play the audio. Use OpenGL to render the stereoscopic video frames.

In Section 15.3, we coded a simple DirectShow movie player application that used the built-in video rendering filter at the end of the chain. For the stereo movie player, we will use as much of that program as possible but replace the built-in DirectShow renderer filter with a custom one of our own that formats the video information as 24-bit RGB pixel data in bottom-to-top row order and writes it into a frame buffer in system RAM. A

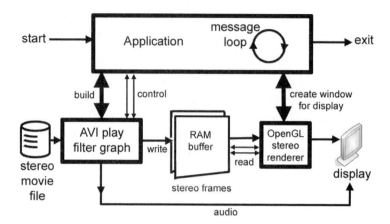

Figure 16.7. The main features and data flow of the stereo-movie player application.

small modification to the stereo image display program will be all that is required to provide the visualization of the video using OpenGL pixel drawing functions. Figure 16.7 illustrates the block structure and data flow for the program.

This program uses the stereo-image display program's code as its starting point. The application contains a message processing loop and it has a small main window (as before). It also creates a child window to display the contents of the RAM frame buffer using OpenGL. A timer is used to post WM_PAINT messages 50 times per second to the display window and hence render the video frames in the RAM buffer. At this point in the code, if the FilterGraph has not been built or is not running then no movie is playing the RAM buffer will be empty and the window (or screen) will just appear blank.

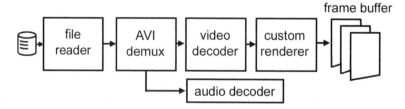

Figure 16.8. The FilterGraph for the AVI stereo movie player. Only the custom rendering filter at the end of the chain is unique to this player.

In response to a command to *play a movie*, the application program will use DirectShow to build and run an appropriate FilterGraph. By combining the FilterGraph (which also renders the audio) and the OpenGL display, we will have an application program that takes a single AVI file with left-over-right stereoscopic frames, encoded using any codec (so long as that codec is installed on the computer) and plays it back in either a window or by filling the screen. Figure 16.8 shows the filter components of the DirectShow graph that we need.

Next we turn to the design of the custom renderer filter.

```
// Although a GIUD for the filter will not be used we must still declare
// one because the base class requires it.
struct
  __declspec(uuid("{C7DF1EAC-A2AB-4c53-8FA3-BED09F89C012}")) CLSID_OpenGLRenderer;

class CRendererOpenGL : public CBaseVideoRenderer
{
public:   // Class constuctor and  destructor
// pointer to application object
    CRendererOpenGL(CMoviePlayer *pPres,
                    LPUNKNOWN pUnk,HRESULT *phr);        // standard COM pointers
    ~CRendererOpenGL();

public:   // these methods must be overidden -
          // see individual listings for details
// Is the Video Format acceptable?
    HRESULT CheckMediaType(const CMediaType *pmt );
// Video format notification
    HRESULT SetMediaType(const CMediaType *pmt );
// New video sample (THIS FILLS THE BUFFER
    HRESULT DoRenderSample(IMediaSample *pMediaSample);
                                                     // USED BY OPENGL
private:               // member variables
// pointer to a C++ object we will use to encapsulate the whole Filter graph
    CMoviePlayer* m_pCP;
    LONG m_lVidWidth;   // Video/movie width
    LONG m_lVidHeight;  // Video/movie height
// Pitch is the number of bytes in the video buffer required to step to next row
    LONG m_lVidPitch;
};
```

Listing 16.4. The derived class for the renderer filter. The member variables record the dimension of the movie/video frames and the *pitch* specifies how the addresses in RAM change from one row of pixels to the next. This allows the rows in an image to be organized into memory blocks that are not necessarily stored contiguously.

16.5.1   The Renderer Filter

If you aren't familiar with DirectShow and its filter concept, now would be a good time to read Chapter 15 and in particular Section 15.6, where we describe a monoscopic version of this filter.

```
CRendererOpenGL::CRendererOpenGL(                          // constructor
// pointer to application object
   CMoviePlayer *pPres,
   LPUNKNOWN pUnk,
   HRESULT *phr )              //
   : CBaseVideoRenderer(__uuidof(CLSID_OpenGLRenderer),  // configure base class
                  NAME("OpenGL Renderer"), pUnk, phr)
// initialize member variable
   , m_pCP( pPres){
// simply notify - everything OK
 *phr = S_OK;
}

CRendererOpenGL::~CRendererOpenGL(){                        // do nothing
}
```

Listing 16.5. The video renderer class constructor and destructor. The class constructor has very little work to do. It simply initializes the base class and one of the member variables.

```
HRESULT CRendererOpenGL::CheckMediaType(const CMediaType *pmt){
 // only allows RGB video sample formats
   HRESULT   hr = E_FAIL;
   VIDEOINFO *pvi=0;
 // Reject this option if not a video type

   if( *pmt->FormatType() != FORMAT_VideoInfo ) return E_INVALIDARG;
   pvi = (VIDEOINFO *)pmt->Format();
   if( IsEqualGUID( *pmt->Type(), MEDIATYPE_Video)){
 // Only accept  RGB 24 bit formatted frames
      hr = S_OK;
      if( IsEqualGUID( *pmt->Subtype(), MEDIASUBTYPE_RGB24) ){; }
      else hr = DDERR_INVALIDPIXELFORMAT;
   }
   return hr;
}
```

Listing 16.6. Check media type: return S_OK when a suitable video format has been enumerated.

To implement the filter, we derive a class, CRendererOpenGL, from the base class CBaseVideoRenderer and override three of its methods. Listing 16.4 presents the derived class details.

By deriving the CRendererOpenGL class from CBaseVideoRenderer, all the standard behavior and interfaces which the base class exhibit, will be present in the derived one. Listing 16.5 shows the class's constructor and destructor methods.

To implement the OpenGL rendering filter, it only remains to override the following three base class methods:

1. CheckMediaType(). Listing 16.6 presents the code for this method. Its only function is to determine whether the renderer can use an input media sample in a particular format.

```
 // gives us data about video
HRESULT CRendererOpenGL::SetMediaType(const CMediaType *pmt){
    HRESULT hr = S_OK;
 // BitmapInfo header format
    VIDEOINFO *pviBmp = NULL;
  // get the data
        pviBmp = (VIDEOINFO *)pmt->Format();
 // extract width and height
    X = m_lVidWidth  = pviBmp->bmiHeader.biWidth;
    Y = m_lVidHeight = abs(pviBmp->bmiHeader.biHeight);
 // calculate Pitch - must be even byte
    m_lVidPitch  = (m_lVidWidth * 3 + 3) & ~(3);
 // if exists remove it since this is new
    if(ScreenL != NULL)free(ScreenL); ScreenL=NULL;
    if((ScreenL = (unsigned char *)malloc(3*X*Y)) == NULL){
        hr = E_UNEXPECTED;                                   // can't do it
        return hr;
        }
 // stereo video - left over right
    if(ScreenL != NULL){
 // calculate address of Right image
        ScreenR =(ScreenL + ((Y/2) * X));
 // Double height in file
        if(bStereoDraw)Y /= 2;
    }
    return hr;
}
```

Listing 16.7. Set media type: extract information about the movie's video stream and allocation RAM buffers ready to hold the decoded frame images.

2. `SetMediaType()`. Our filter uses this method to allocate a RAM buffer in which to store the video frames. Listing 16.7 presents the code.

3. `DoRenderSample()`. This does most of the work. It picks up a pointer to the video sample being delivered by the DirectShow filter chain and copies the video pixel data into the RAM buffer identified by the pointer `ScreenL` (see Listing 16.8).

That completes the implementations of the stereo renderer filter. Apart from the custom rendering filter, all the other components in the FilterGraph are the same as we encountered in the player discussed in Section 15.3.

```
HRESULT CRendererOpenGL::DoRenderSample( IMediaSample * pSample ){
 HRESULT hr = S_OK;
 BYTE * pSampleBuffer = NULL, *S = NULL;
 UINT row,col,yy;
 if( !pSample ) return E_POINTER;
 hr = pSample->GetPointer( &pSampleBuffer );// get pointer to image pixels
 if( FAILED(hr)){return hr;}                 // cannot get image pixels
// pointer to RAM buffer to be rendered by OpenGL
 S=ScreenL;
 if(X > 0 && Y > 0 && S != NULL){
// stereo drawing so the iamge is 2 times as high
   if(bStereoDraw)yy=Y*2; else yy=Y;
// Copy the image  - we need to do it pixel by
   for(row = 0; row < yy; row++ ) {
// by pixel because the order of RGB in the video
     BYTE *pBmpBufferOld = pSampleBuffer;
// stream is reversed. Versions of OpenGL > 1.0
     for (col = 0; col < X; col++)   {
// allow different byte orderings which would mean we
       *S++ = pSampleBuffer[2];
// could copy the samples one row at a time but
       *S++ = pSampleBuffer[1];
// because of the PITCH we cannot copy the whole
       *S++ = pSampleBuffer[0];
// image as a single block.
       pSampleBuffer += 3;
     }
     pSampleBuffer  = pBmpBufferOld +  m_lVidPitch;
 }   }

 return hr;
}
```

Listing 16.8. The `DoRenderSample()` method is overridden from the base class to gain access to the video samples.

```
#include " .... // Header files

HINSTANCE     hInstance = NULL;             // global variables
unsigned char *ScreenL=NULL,*ScreenR=NULL;  // Left and Right buffers
long          X=0,Y=0;                      // Movie size
 HWND         hWndMain=NULL,hWndDisplay=NULL; // Window handles
 // Windows classes for main and child windows
WNDCLASSEX wc1 = {...}, wc2 = { ... };

INT WINAPI WinMain(...){         // application entry point
    ..
    CoInitialize (NULL);          // initialize the COM mechanism
    RegisterClass(&wc1);
    hWndMain = CreateWindow(...  // main window
    while(1){
     .. // Message loop
    }
    ..
    CoUninitialize();             // finished with COM
    return 0L;
}

LRESULT WINAPI MainWindowMsgProc(...){ // Handle Menu commands to load movie
 switch( msg )    {                     // and display output etc.
   case WM_COMMAND:
     if (wParam == ID_FILE_QUIT)PostQuitMessage(0);
     else if(wParam == ID_DISPLAY_STEREO_OUTPUT)
 // display the output  !!!!!!!!!KEY
       hWndDisplay=DisplayMovie(hInstance,hWnd);
 // run the movie        !!!!!!!!!KEY
     else if(wParam == ID_LOAD_RUN_MOVIE)RunMovie(hWnd);
     else if(wParam == ID_CONTROL_ ....)ControlMovie( ...);
     else if ....
   default: break;
 }
 return DefWindowProc( hWnd, msg, wParam, lParam );
}

 // Create OpenGL window
HWND DisplayMovie(HINSTANCE hInstance, HWND hWndParent){
   RegisterClass(&wc2);
   if(!FullScreen)hwndGL=CreateWindowEx( ... //windowed
   else          hwndGL=CreateWindowEx( ... //full screen
   uTimerID=(UINT)
 // set a timer to re-draw the video 25 fps.
     SetTimer(hwndGL,TIMER_ID,TIMER_ID_RATE,NULL);
   return hWndGL;
}
```

Listing 16.9. An overview of the key execution steps in the stereo movie player application. This is a *skeletal view* of the key points only, so that you can follow the execution flow in the full code.

```
LRESULT WINAPI OpenGLWndMSgProc(...){
  switch( msg )     {
  // fall through  to the paint message to render the video
    case WM_TIMER:
    case WM_PAINT: DrawScene(hWnd);
       break;
    case WM_CREATE: ..                  //  WM_SIZE , WM_DESTROY  and other messages.
    default: break;
  }
  return DefWindowProc( hWnd, msg, wParam, lParam );
}
```

Listing 16.9. (continued).

16.5.2   Putting the Application Together

We turn now to the remainder of the application. There are a number of ways in which we could implement it. We choose to follow the framework outlined in Listing 16.9 because it is minimal and should allow easy customization and adaption. It follows the general design philosophy for all the programs in this book:

The application opens in WinMain(). A window callback function, MainWindowMsgProc(), is identified through a Windows class. The program then goes into a message loop.

Two of the main window's menu commands are of note.

The DisplayMovie() function brings about the creation of a child window to contain the rendered stereo movie. This allows either windowed or full-screen display. The child window's message handling function, OpenGLWndMSgProc(), processes the WM_PAINT message and renders the output of the pixel memory buffer ScreenL/R.

A *timer* posts a WM_TIMER message to OpenGLWndMSgProc() every few milliseconds to re-render the frame buffer using OpenGL. If no movie is playing into this buffer (i.e., ScreenL=NULL;), the window simply remains blank.

The function RunMovie() is responsible for building the DirectShow FilterGraph and running it. The graph will use our custom renderer filter to play a movie in a loop into the screen buffer, identified by pointer ScreenL.

Now let's focus on the movie player code in function RunMovie();.

16.5.3   The Movie Player Function and Application FilterGraph

The movie player is implemented as an instance of an application-specific C++ class. We do this because in subsequent examples, we will run two instances of the FilterGraph in separate threads of execution.

When a command is received to play a movie, an instance of the class `CMoviePlayer` is created, and its method which builds and runs the Filter-Graph is executed. If the user wants to pause or stop the movie, a command is forwarded to the object. Listing 16.10 illustrates this linking code.

```
// build - and run a filter graph to play a movie
void RunMovie(HWND hWnd){
 if(g_pMoviePlay)StopMovie();    // if a movie is playing stop it
 g_pMoviePlay=new CMoviePlayer();// Create an instance of the movie
                                 // player object
 if(g_pMoviePlay){
    g_pMoviePlay->m_hwnd=hWnd;    // tell it the parent window and .....
    g_pMoviePlay->BuildMoviePlayerFilterGraph(); // run the graph
} }

void StopMovie(void){           // executed by menu command from the main window
 if(g_pMoviePlay)  {
    delete g_pMoviePlay;        // Unload the object, this will stop the movie
    g_pMoviePlay = NULL;        // and release everything
} }

void ControlMovie(int ID){    // pass commands to the object - pause - start etc.
   if( g_pMoviePlay != NULL)g_pMoviePlay->Control(ID);
}
```

Listing 16.10. Building a movie's presentation FilterGraph is handled by an object of the `CMoviePlayer` class.

The `CMoviePlayer` class (Listing 16.11) allows us to conveniently represent all the elements which go to make a DirectShow FilterGraph as a single object.

Special note should be taken of the use of the Active Template Library's (ATL) COM interface pointer `CComPtr<>`, for example in the statements:

```
    CComPtr<ICaptureGraphBuilder2>  m_pCG;
    CComPtr<IGraphBuilder>          m_pGB;
```

These exceptionally helpful pointers automatically release the interfaces once the application program ceases to use them. As a result, it is not

```
class CMoviePlayer {   // class object which encapsulates all the elements
public:                // of an object that will load and play
    CMoviePlayer( );
    ~CMoviePlayer();
    // object methods
    HRESULT BuildMoviePlayerFilterGraph();
    void CheckMovieStatus(void);
    void Control(int);
    void Cleanup(void);
    HWND    m_hwnd;     // video window handle.
private:
    // DirectShow Interface pointers used by the object. NOTE: These use
    // very helpful ATL COM template class which is exceptionally helpful
    // because it automatically releases any interfaces when they
    // are no longer required.
    // Helps to render capture graphs
    CComPtr<ICaptureGraphBuilder2>  m_pCG;
    CComPtr<IGraphBuilder>          m_pGB;          // GraphBuilder object
    CComPtr<IMediaControl>          m_pMC;          // Media Control
    CComPtr<IMediaEvent>            m_pME;          // Media Event
    CComPtr<IMediaSeeking>          m_pMS;          // Media Seeking
    CComPtr<IMediaPosition>         m_pMP;          // Media Position
};
```

Listing 16.11. The movie player class.

necessary to release an interface explicitly. Even if you forget to do it, any memory occupied by the COM objects will be freed when the application exits.

The efficiency of the DirectShow API is evident here, because even with the movie class, the custom renderer filter and a simple application framework, it takes only a few lines of code to build the stereo movie player. (We refer you back to Section 15.1 to review the procedure for putting together FilterGraphs and loading filters.) Listing 16.12 gives the code to build the filters, connect them and run the graph. One line in the code is responsible for instancing our custom stereo rendering filter:

```
pRenderer = (IBaseFilter *)new CRendererOpenGL(this, NULL
            &hr);
```

It takes this form because we chose to stop short of making a filter that can be registered with Windows as a system component, but in every other respect it acts like any other DirectShow filter.

```
HRESULT CMoviePlayer::BuildMoviePlayerFilterGraph(){
 // ALL hr return codes should be checked with if(FAILED(hr)){Take error action}
 HRESULT hr = S_OK;
 CComPtr<IBaseFilter> pRenderer;   // custom renderer filter
 CComPtr<IBaseFilter> pSrcFilter;  // movie source reader
 CComPtr<IBaseFilter> pInfTee;     // filter to split data into two
 CComPtr<IBaseFilter> pAudioRender;
 CRendererOpenGL  *pCTR=0;    // Custom renderer
  hr = m_pGB.CoCreateInstance(CLSID_FilterGraph, NULL, CLSCTX_INPROC);
  // Get the graph's media control and media event interfaces
  m_pGB.QueryInterface(&m_pMC); // media Control Interface
  m_pGB.QueryInterface(&m_pME); // media Event Interface
  m_pGB.QueryInterface(&m_pMP); // media Position Interface

  // Create the OpenGL Renderer Filter !
  pRenderer = (IBaseFilter *)new CRendererOpenGL(this, NULL, &hr);
  // Get a pointer to the IBaseFilter interface on the custom renderer
  // and add it to the existing graph
  hr = m_pGB->AddFilter(pRenderer, L"Texture Renderer");
  hr = CoCreateInstance (CLSID_CaptureGraphBuilder2 , NULL, CLSCTX_INPROC,
       IID_ICaptureGraphBuilder2, (void **) &(m_pCG.p));
  // Attach the existing filter graph to the capture graph
  hr = m_pCG->SetFiltergraph(m_pGB);
  USES_CONVERSION;                 // Character conversion to UNICODE
  hr=m_pGB->AddSourceFilter(A2W(MediaFile),L"Source File",&pSrcFilter);
  // split the data stream into video and audio streams
  hr = CoCreateInstance (CLSID_InfTee, NULL, CLSCTX_INPROC,
       IID_IBaseFilter, (void **) &pInfTee);
  hr=m_pGB->AddFilter(pInfTee,L"Inf TEE");
  hr = m_pCG->RenderStream(0,0,pSrcFilter,0,pInfTee);
  IPin *pPin2;                      // get output pin on the splitter
  hr = m_pCG->FindPin(pInfTee,PINDIR_OUTPUT,0,0,TRUE,0,&pPin2);
  /////////////////// audio section
  hr = CoCreateInstance (CLSID_AudioRender, NULL, CLSCTX_INPROC,
       IID_IBaseFilter, (void **) &pAudioRender);
  hr=m_pGB->AddFilter(pAudioRender,L"AudioRender");
  hr = m_pCG->RenderStream (0, &MEDIATYPE_Audio,
       pSrcFilter, NULL, pAudioRender);
  /////////////////// end of audio
  hr = m_pCG->RenderStream (0, &MEDIATYPE_Video,  // Connect all the filters
        pPin2, NULL, pRenderer);                  // together.
  hr = m_pMC->Run();               // Start the graph running - i.e. play the movie
 return hr;
}
```

Listing 16.12. Build the FilterGraph using the DirectShow GraphBuilder object and
its associated CaptureGraphBuilder2 COM object. Note that every hr return code
should be checked to make sure that it is has returned an S_OK code before proceed-
ing. If an error code is returned, appropriate error action should be taken.

16.5.4   Control

The final listing of this section (Listing 16.13) presents the remaining methods of the CMoviePlayer class which are concerned with controlling the presentation of the movie: playing it, restarting it or positioning the playback point. Most of these tasks are handled through the IMediaControl, IMediaPosition and IMediaSeeking interfaces, which offer a large selection of helpful methods, only a few of which are explictly listed in our code. Full information is available in the DirectX SDK (DXSDK).

```
CMoviePlayer::CMoviePlayer() : m_hwnd( NULL ){}  // constructor

CMoviePlayer::~CMoviePlayer(){ Cleanup();}       // close down smoothly

void CMoviePlayer::Cleanup(void){
  if( m_pMC )m_pMC->Stop();                      // shut down the graph
}

void CMoviePlayer::Control(int id){              // integers indicate commands
  HRESULT hr = S_OK;
  if(id == 0){                                   // Run the movie
    hr = m_pMC->Run();
  }
  else if(id == 1){                              // Check movie status
    CheckMovieStatus();                          // restart if necessary.
  }
  else if(id == 2){                              // stop the movie
    if( m_pMC )m_pMC->Stop();
  }
  else if(id == 3){                              // restart
    if( m_pMC ) m_pMC->Stop();
  }
  else{                                          // reset go back and play
    hr = m_pMP->put_CurrentPosition(0);          // from start
    hr = m_pMC->Run();
  }
}
}
```

Listing 16.13. The other methods in the CMoviePlayer class. Constructor, destructor and method to use the Media Control interface for player control functions such as stop, play etc.

16.6   Capturing Stereo Pictures from a Pair of Video Cameras

So far, we haven't said much about how to acquire stereoscopic images. Chapter 10 discussed some special cameras and quite expensive HMDs with two built-in video cameras. In this section, we will develop a little program to let you use two inexpensive webcams for capturing stereoscopic images.

The application is based, to a considerable extent, on a combination of the code in Section 15.4.1, where we used a single camera to grab images, and some elements from the previous section on playing stereo movies. For this application, the implementation strategy is:

> Build two FilterGraphs to capture live video from different sources and render them into two memory frame buffers, for the left and right eyes respectively. Use an OpenGL stereo-enabled window for preview if desired (or two windows for individual left/right video). On command from the user, write a bitmap (or bitmaps) from the frame buffers to disk.

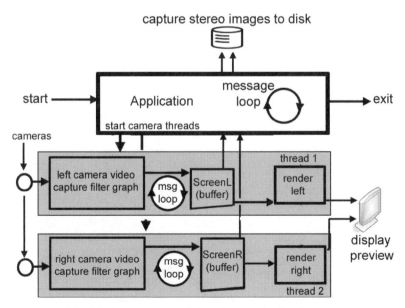

Figure 16.9. The block structure and data-flow paths for the stereo image capture application.

The concept is summarized in Figure 16.9, which illustrates how the previewed output from the stereo renderer is delivered to two OpenGL windows, one previewing the left camera's output and the other the right camera's output. The full program code offers the option to use stereo-enabled hardware. For now, though, we will examine the program that presents its output in two windows.

Note: If one uses two OpenGL windows in a single Windows application program, they should normally execute in separate threads. (We discussed threading in Section 15.1.1.) Consequently, we will write our application as a multi-threaded one.

```
struct
__declspec(uuid(\"{36F38352-809A-4236-A163-D5191543A3F6}"))CLSID_OpenGLRenderer;

class CRendererOpenGL : public CBaseVideoRenderer
{
public:   // Class constuctor and  destructor
    CRendererOpenGL(CMoviePlayer *pPres,
 // NEW VARIABLE FOR THREAD/CAMERA Identity
                    long id,
                    LPUNKNOWN pUnk,HRESULT *phr);
    ...      // same methods
private:  // member variables
 // identify whether rendering left or right camera
 // copy from "id"
        long m_instance;
    CGrabCamera * m_pCP;// replace  pointer to CMoviePlayer
    ....
};

CRendererOpenGL::CRendererOpenGL( // constructor
CGrabCamera *pPres,     // POINTER TO CAMERA OBJECT
           long *id,  // IDENTIFY LEFT OR RIGHT CAMERA
           LPUNKNOWN pUnk,
           HRESULT *phr )
:CBaseVideoRenderer(__uuidof(CLSID_OpenGLRenderer),
NAME("OpenGL Renderer"),pUnk,phr)
, m_pCP( pPres)
, m_instance(id){       // NEW !!
 ... // constructor body - same as movie player
}
```

Listing 16.14. Modification to the renderer filter for use in the *stereo image grab* program. This listing only highlights the changes for the class and its implementation. The omitted code is the same as in Listings 16.4 to 16.8.

```
HRESULT CRendererOpenGL::DoRenderSample( IMediaSample * pSample ){
... //
  if(m_instance == 1)S=ScreenL;   // RENDER INTO LEFT CAMERA BUFFER
  else              S=ScreenR;   // RENDER INTO RIGHT CAMERA BUFFER
...//
}

HRESULT CRendererOpenGL::SetMediaType(const CMediaType *pmt){
 if(m_instance == 1){
 // ALLOCATE SPACE FOR LEFT & RIGHT BUFFERS
   if(ScreenL != NULL)free(ScreenL); ScreenL=NULL;
   if((ScreenL = (unsigned char *)malloc(3*X*Y*2)) == NULL)return E_UNEXPECTED;
   }
 // BUFFER ALREADY ALLOCATED - ASSUMES LEFT CAMERA FOUND FIRST
  else ScreenR = ScreenL + 3*X*Y;
  ... //
  return hr;
}
```

Listing 16.14. (continued).

It makes sense to place the whole of the left and right capture tasks in different threads. The threads can write into different parts of the same frame buffer or into separate left and right frame buffers. It makes little difference.

Before looking at the code structure, we will quickly consider a minor change to the OpenGL renderer filter class. The changes are highlighted in Listing 16.14. In it, we see that the application class pointer is changed to reflect the different application name, CGrabCamera * m_pCP;. An extra member variable long m_instance is used to identify whether this instance of the filter is part of the graph capturing the left image or the graph capturing the right image. When it comes to copying the sample into ScreenL or ScreenR or allocating the screen pointers, identifier long m_instance shows the way.

To form the program's overall structure, we will make it as much like the movie player as possible. All of the work required to handle a single camera source (its DirectShow FilterGraph including the OpenGL rendering filter etc.) will be encapsulated in a C++ class, and an object of this class will be created to handle it. Listing 16.15 gives the CGrabCamera class and Listing 16.16 brings all the program's features together to show the flow of execution. Some remarks about a few of the statements in Listing 16.16 are appropriate:

```
class CGrabCamera {
public:
 CGrabCamera( );
 ~CGrabCamera();
 // render into the OpenGL window
 HRESULT Render(HWND,long, long);
 // choose the image format we wish to obtain from the camera
 HRESULT SetMediaFormat( GUID subtype);
 // window handle for window rendering camera preview
  HWND     m_hwnd;
 // left or right instance
 long      instance;
 // build the image capture graph
 HRESULT BuildCaptureGraph(long);
 void Cleanup(void);
private:
 // methods
 HRESULT CaptureVideo(IBaseFilter *pRenderer, long id);
 HRESULT FindCaptureDevice(IBaseFilter **ppSrcFilter, long *nfilters, long id);
 void CheckMovieStatus(void);

 CComPtr<ICaptureGraphBuilder2>  m_pCG;  // Helps to render capture graphs
 CComPtr<IGraphBuilder>          m_pGB;  // GraphBuilder interface
 CComPtr<IMediaControl>          m_pMC;  // Media Control interface
 CComPtr<IMediaEvent>            m_pME;  // Media Event interface
 ;
```

Listing 16.15. Class specification for the object which will acquire video frames from a camera and allow them to be previewed and captured to disk.

- *The* THREAD_DATA *structure.* A small structure is needed to pass more than one parameter to a thread function.

- CoInitializeEx(NULL,COINIT_MULTITHREADED). To use COM in an application, the correct initialization must be used.

- *The windows.* The program will create three windows. The main window will have a message handler function. The two image windows (created by each running thread) will use the same message handler code re-entrantly (so no local static variables are allowed).

- *The message-processing loop.* Since each thread is essentially a separate program executing concurrently they must have their own message processing loops.

- *Threading.* Different C/C++ compilers may create threads of execution in different ways. The _beginthread(ThreadFunctionName,

0, parameter) function is the preferred one in Visual C++. Its arguments carry the name of a function to be the entry point for the thread
and a parameter that will be passed on as the argument to the thread
function. The Win32 API has its own thread-creation function that
mirrors _beginthread very closely.

```
// structure to pass data
typedef struct tagTHREAD_DATA {
// to the thread processes
  HWND parent;  long id;  CGrabCamera *pP;
} THREAD_DATA;

HINSTANCE      hInstance     = 0;
// pointers to the camera objects
CGrabCamera* g_pCameraObject1=NULL,g_pCameraObject2=NULL;
// Left and Right Images
unsigned char *ScreenL=NULL,*ScreenR=NULL;
// Dimensions
long           X=0,Y=0;
HWND           hWndMain = NULL;

INT WINAPI WinMain( ... ){
// for multi-threaded COM
 CoInitializeEx(NULL,COINIT_MULTITHREADED);
// Create objects to look after
 g_pCameraObject1 = (CGrabCamera*)new CGrabCamera();
// left and right camera data.
 g_pCameraObject2 = (CGrabCamera*)new CGrabCamera();
 RegisterClassEx(...        // register two classed for Main and OpenGL windows
 hWndMain = CreateWindow(... // create main window
 THREAD_DATA p1,p2;         // declare two structures
// Fill structures to pass
 p1.id=1; p1.parent=hWndMain; p1.pP=g_pCameraObject1;
// to threads.
 p2.id=2; p2.parent=hWndMain; p2.pP=g_pCameraObject2;
// thread for left camera
 _beginthread(CameraThreadFunction,0,(void *)(&p1));
// thread for right camera
 _beginthread(CameraThreadFunction,0,(void *)(&p2));
 while(1){ ... /* message loop */ }
 .. // free screen and dynamic objects
// finished with COM & quit
 CoUninitialize(); return 0;
}
```

Listing 16.16. A skeleton outline of the stereoimage grabber program.

```
void CameraThreadFunction(void *arg){
 // This function is the main thread for looking after each camera.
 // It will know which thread (camrea) it is handling by the identifier
 // passed to it as part of the THREAD_DATA structure.
 HWND hwnd;
 THREAD_DATA *pp = (THREAD_DATA *)arg;
// create window to contain camera view
 hwnd = CreateWindow(CLASSNAME1.. .
// get ID of which thread we are
 pp->pP->m_hwnd=hwnd; pp->pP->instance=pp->id;
 // so that the window knows which thread this is
 SetWindowLongPtr(hwnd,GWLP_USERDATA,(LONG_PTR)pp);
// do most of the work
 hr = pp->pP->BuildCaptureGraph(id);
// refresh the preview window every 40ms
 SetTimer(hwnd, 101, 40 , NULL);
 while(1){ .. }  // thread message processing loop
                 // terminated by message from main program
// release Interfaces etc.
 pp->pP->Cleanup();
// closing down
 _endthread();
}

// render OpenGL
HRESULT CGrabCamera::Render(HWND hWnd, long id){
 .. //
// render the correct buffer
 if(id == 1 && ScreenL != NULL)
   glDrawPixels((GLsizei)X,(GLsizei)Y,GL_RGB,GL_UNSIGNED_BYTE,
   (GLvoid *)ScreenL);
 else if(ScreenR != NULL)
   glDrawPixels((GLsizei)X,(GLsizei)Y,GL_RGB,GL_UNSIGNED_BYTE,
   (GLvoid *)ScreenR);
 .. //
}

// main window handler
LRESULT WINAPI MsgProc(HWND hWnd,UINT msg,WPARAM wParam,LPARAM lParam){
 switch(msg){
  case WM_CHAR:
    if (wParam == VK_ESCAPE)PostMessage(hWnd, WM_CLOSE, 0, 0);
// Save the bitmap images
    else if(wParam == VK_SPACE)SaveImages();
   default:
    break;
 }
 return DefWindowProc( hWnd, msg, wParam, lParam );
}
```

Listing 16.16. (continued).

```
// thread window handler
LRESULT WINAPI GLProcessWndProc(HWND hWnd,UINT msg,WPARAM wParam,LPARAM lParam){
 switch(msg){
// fall through to next handler to render what the camera sees
  case WM_TIMER:
// render the camera's output - extract which thread we are.
  case WM_PAINT:
    THREAD_DATA *p=(THREAD_DATA *)GetWindowLongPtr( hWnd, GWLP_USERDATA);
    if(p && p->pP)p->pP->Render(hWnd,p->id);
  default:
    break;
 return DefWindowProc( hWnd, msg, wParam, lParam );
}
```

Listing 16.16. (continued).

- *Thread identifiers.* Each thread essentially uses the same code and the same message handler. So that the program can identify which thread is which (first or second), each thread is assigned an identity (ID) during its creation. The ID is passed to the thread function as part of the THREAD_DATA structure. Additionally, the preview window handler must know whether it is rendering left or right images, and so this structure is also passed to it using the small *user data* long word which every window makes available. This is written by:

  SetWindowLongPtr(hwnd,GWLP_USERDATA,(LONG_PTR)pp);

  It is read by the message-handling function with:

  GetWindowLongPtr(hWnd,GWLP_USERDATA);

- *Choosing cameras.* As part of the process of building the FilterGraphs, the thread that is created first chooses the first camera it can find. The second thread chooses the second camera it can find.

With the exception of these few points, the rest of the code in Listing 16.16 is *hopefully* self-explanatory. In the next section, we show how to put the FilterGraph together.

16.6.1 Building the FilterGraph to Capture Images

With one small exception, all the components of the FilterGraph that are needed to capture images from a *live* video source are present in the movie player example. The code in Listing 16.17 will build the FilterGraph. The most significant difference between this and earlier examples is the call to `FindCaptureDevice()`. The code of this function is designed to identify appropriate video sources. If id=1, it will return a pointer to the first suitable source, used for the left images. When id=2, the function returns a pointer

```
 // id=1 for left, id=2 for right
HRESULT CGrabCamera::BuildCaptureGraph(long id){
  HRESULT hr = S_OK;
  CComPtr<IBaseFilter> pRenderer;                  // pointer to renderer filter
 // pointer to video source (camera)
  CComPtr<IBaseFilter> pSrcFilter;
    // Create the filter graph and get its interface pointer
  hr = m_pGB.CoCreateInstance(CLSID_FilterGraph, NULL, CLSCTX_INPROC);
 // get the media-control interface pointer
  m_pGB.QueryInterface(&m_pMC);
 // get the media event interface pointer
  m_pGB.QueryInterface(&m_pME);

  // Create the custom renderer object (note: id)
  pRenderer=new CRendererOpenGL(this,id,NULL,&hr);
  hr = m_pGB->AddFilter(pRenderer, L"Stereo Renderer");
  // create the graph builder object and get its interface
  // (Note slightly different syntax)
  hr = CoCreateInstance (CLSID_CaptureGraphBuilder2,NULL CLSCTX_INPROC,
      IID_ICaptureGraphBuilder2, (void **) &(m_pCG.p));
  // Attach the existing filter graph to the capture graph
  // tell the graphbuilder about the graph
  hr = m_pCG->SetFiltergraph(m_pGB);
  // Use the system device enumerator and class enumerator to find
  // a live (camera) device.
  hr = FindCaptureDevice(&pSrcFilter,id);          // find the video source.
  // add this source to the graph
  hr = m_pGB->AddFilter(pSrcFilter, L"Video Capture");
  // Using the graph builder to render the stream buids the graph and adds any
  // necessary intermediate filters between source and renderer.
  hr = m_pCG->RenderStream (&PIN_CATEGORY_CAPTURE,&MEDIATYPE_Video,
  pSrcFilter,NULL,pRenderer);
  hr = m_pMC->Run();                                // run the graph
  return hr;
}
```

Listing 16.17. Constructing a FilterGraph to capture images from a video source using a custom renderer filter. Note the slightly different way that the ATL `CCComPtr` com pointer is used when creating the `Graph` and `GraphBuilder` objects.

to the second suitable source. Naturally, all function return values (the hr=
codes) need to be checked to see that everything worked as expected.

The last comment to make on this example is simply to say that a command given to the main window, via its menu for example, can initiate a process to write to disk the images in the ScreenL and ScreenR frame buffers. This could be in the form of Windows bitmaps or by using a compressed or lossy format, such as JPEG.

16.7 Making a Stereo Movie from Image Sequences

Many VR environments are built using 3D modeling software. Virtual tours are often made with computer animation packages too. But in most cases, the software packages are not really geared up for doing such things in stereo. This doesn't stop people using such software, because it is often possible to put two cameras in a scene. These mimic the viewer's eyes, and then the tour can be rendered from the left camera's point of view, followed by a rendering from the right camera's point of view. Under these circumstances, the usual way of working is to render sequences of images, one image per movie frame, one sequence for the right camera and one for the left.

To make use of such sequences, we will put together a short program that takes them and generates a left-over-right stereo movie AVI file suitable for playback using the player program described in Section 16.5. Stereo movie files (especially when they are written as uncompressed AVI files) will be large. The old *video for Windows (VfW)* technology which offers an API for all kinds of operations on AVI files is unsuitable for such large files. But, this is another application where DirectShow is ideal, because it can not only handle huge files, but it can include in the filter chain any compression technology that might be available on the host computer.

However, to write this utility, another type of DirectShow filter is needed, one that we haven't written before: a *source filter* this time. Luckily, the DirectShow SDK contains templates for all forms of filter, and we can model our program on one that comes close to meeting its needs. The \ProgramFiles\ dxsdk\Samples\C++\DirectShow\Filters\PushSource source filter project reads a few bitmap files and makes an AVI movie from them. Unfortunately, the SDK example extends beyond our needs into a fully compliant DirectShow source filter. Nevertheless, it shows what classes we must use in our

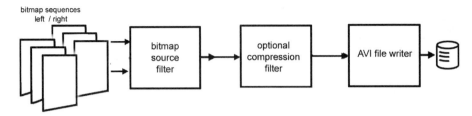

Figure 16.10. The FilterGraph for the AVI stereo movie compilation program.

source filter and how to insert the image files into the video stream. Once we have written the custom source filter to read the stereo image pairs, it fits into the FilterGraph concept that should be becoming familiar to you. Indeed, the graph for this application is not going to be significantly different from those we have already seen. Its details are set out in Figure 16.10.

In this example, we will start with a look at the overall application structure in Listing 16.18. The strategy is to make up a list of the file names for the bitmap image pairs. This we do by providing the first part (the root) of the name for the files. Then we add a sequence number at the end and put back the file name extension. Most 3D movie software uses this idea to write image sequences. For example, write each movie frame to an individual file: fileX001.bmp, fileX002.bmp etc. for frames 1, 2 etc.

In our code, we detect these sequences in function buildBMPfilelist() and store their names in the *FileList1[] and *FileList2[] character arrays. The rest of the code is mostly similar to what we have seen before. The new element concerns the source filter shown diagrammatically in Fig-

```
// define max number of frames (arbitrary)
#define MAX_FILES 8192
// number of files in Left and Right sequences
int nBitmapFiles1=0,nBitmapFiles2=0;
// list of filenames in left and right sequences
char *FileList1[MAX_FILES],*FileList2[MAX_FILES];

// DirectShow interfaces used for filter graph
IGraphBuilder          * g_pGraph   = NULL;
IBaseFilter            * g_pMux     = NULL;
ICaptureGraphBuilder2 * g_pCapture = NULL;
```

Listing 16.18. Building a movie from bitmaps—the application structure.

```
// Window handler
LRESULT CALLBACK WndMainProc(HWND hwnd,UINT msg,WPARAM wParam,LPARAM lParam){
    switch (msg){
        case WM_GRAPHNOTIFY:  HandleGraphEvent();  break;
// Stop capturing and release interfaces
        case WM_CLOSE:   CloseInterfaces(); break;
            break;
        case WM_DESTROY:        PostQuitMessage(0);  return 0;
    }
    return DefWindowProc (hwnd , message, wParam, lParam);
}

// main entry point
int PASCAL WinMain(HINSTANCE hInstance,HINSTANCE hIP, LPSTR lpCL, int nS){
 if(FAILED(CoInitializeEx(NULL, COINIT_APARTMENTTHREADED))) exit(1);
 RegisterClass( ..                       // usual Windows stuff
 ghApp = CreateWindow(...                 // usual Windows stuff
 buildBMPfilelist(hInstance,ghApp);       // build the list of files to be used
 BuildFilterGraph(hInstance);             // build and run the assembly filter graph
 while(1){ ... }                          // message loop
 CoUninitialize();
 return 0;
}

// When the graph has finished i.e. there are no more input files
HRESULT HandleGraphEvent(void){
// and event will be generated and the application will pause
 LONG evCode, evParam1, evParam2;
// this function monitors the event and sends the user a message
 HRESULT hr=S_OK;
// that the movie is built.
 if (!g_pME)return E_POINTER;
 while(SUCCEEDED(g_pME->GetEvent(&evCode,(LONG_PTR *)&evParam1,
                  (LONG_PTR *)&evParam2,0))){
   switch (evCode){
    case EC_COMPLETE: MessageBox(NULL,"Complete","Output",MB_OK);  break;
   }
   hr = g_pME->FreeEventParams(evCode, evParam1, evParam2);
 }
 return hr;
}
static void buildBMPfilelist(HINSTANCE hInstance,HWND hWnd){
   .. // Specify filename roots e.g.  "SeqRight" "SeqLeft"  then the functions
   .. // GetFilesInSequence will build up the list  with SeqRight001.bmp
   .. // SeqRight002.bmp etc. etc. and SeqLeft001.bmp SeqLeft002.bmp etc. etc.
   nBitmapFiles1=GetFilesInSequence(NameRoot1,FileList1);
   nBitmapFiles2=GetFilesInSequence(NameRoot2,FileList2);
 }
}
```

Listing 16.18. (continued).

```
HRESULT BuildFilterGraph(HINSTANCE hInstance){ //  make move from bitmaps
  HRESULT hr;
  CComPtr<IBaseFilter> pSource;                  // Our custom source filter
  hr = GetInterfaces();                          // get all the interface pointers
  hr = g_pCapture->SetFiltergraph(g_pGraph);
                                                 // create and put in our in place
                                                 // our source filter
  pSource = new CBitmapSourceFilter(min(nBitmapFiles1,nBitmapFiles2),NULL, &hr);
  hr = g_pGraph->AddFilter(pSource, L"BITMAP SOURCE");
  hr = g_pCapture->SetOutputFileName(
         &MEDIASUBTYPE_Avi,                      // specifies AVI for the target file
         GetMovieFilename()",                    // get output file name
         &g_pMux,                                // receives a pointer to the AVI mux
         NULL);
  hr = g_pCapture->RenderStream(0,0,pSource,0,g_pMux);
  hr = g_pMC->Run();                             // Start the build -  when it
  return S_OK;                                   // completes the graph will
                                                 // pause and a WM_GRAPHNOTIFY
                                                 // message will be set to parent
                                                 // window.

}
```

Listing 16.19. Building and executing the FilterGraph. Once finished, an event is sent to the application window's message handler, which it uses to pop up an alert message that the job is done and the program may be terminated.

ure 16.10. It is part of the FilterGraph that is constructed in the function BuildFilterGraph() and coded in Listing 16.19. We have omitted from our printed listings the details of how to select a compression filter. This is a minor point. How it fits into the code can be seen by examining the accompanying project files.

16.7.1   The Source Filter

A source filter is a little different from a renderer filter. The source filter pushes the video samples through the FilterGraph; it is in control. Following our discussion in Section 15.1 on DirectShow filters, the idea of *push* and *pull* in a FilterGraph and how filters connect through *pins* should not be a surprise. A source filter is written by deriving a class for the filter from a suitable base class and overriding one or more of its methods to obtain the specific behavior we want.

In the case of a source filter, we must derive from two classes: one from CSource for source filters and one from the source pin class CSourceStream. In our program, we call them CBitmapSourceFilter and CPushPinOnBSF, respectively. There is very little to do in the CSourceStream class, just the

```
class CPushPinOnBSF : public CSourceStream {
protected:
    // the following are member variables that are
    // standard for output pins on source filters
    CCritSec m_cSharedState;                 // Protects the filter's local data
    BOOL m_bZeroMemory;                      // Do we need to clear the buffer?
    CRefTime m_rtSampleTime;                 // The time stamp for each sample
    // The following class variables are application specific and we use them
    // when working with individually numbered bitmap files.
    int m_FramesWritten;                         // To track where we are in the file
    DWORD m_cbBitmapInfo;                    // Size of the bitmap headers
        HANDLE m_hFile;                      // Handles returned from CreateFile
    BOOL m_bFilesLoaded;                     // a bitmap has been loaded from
                                             // file correctly
    int m_iFrameNumber;                      // How many frames have been written
    int m_dFrames;                           // how many frames to write
    // the following are application specific and they define our bitmap data
    BITMAPINFO *m_pBmi1;                     // Pointer to the bitmap headers
    BYTE * m_pFile1;                         // Points to beginning of file buffers
        BYTE * m_pImage1;                    // Points to pixel bits of image 1
    BITMAPINFO *m_pBmi2;                     // Pointer to the bitmap headers
    BYTE * m_pFile2;                         // Points to beginning of file buffers
        BYTE * m_pImage2;                    // Points to pixel bits
public:
    // methods  constructor and destructor
    CPushPinOnBSF(int Nframes, HRESULT *phr, CSource *pFilter);
    ~CPushPinOnBSF();
    // These must be provided to ovrride base class
    HRESULT GetMediaType(CMediaType *pMediaType);
    HRESULT DecideBufferSize(IMemAllocator *pAlloc,
    ALLOCATOR_PROPERTIES *pRequest);
    HRESULT FillBuffer(IMediaSample *pSample);
    // Custom class specific function to load bitmap.
    void LoadNextBitmap(void);
};

class CBitmapSourceFilter: public CSource{// Our source filter class
public:
    CBitmapSourceFilter(int Nframes, IUnknown *pUnk, HRESULT *phr);
    ~CBitmapSourceFilter();
private:
    CPushPinOnBSF *m_pPin;                       // poiner to the filters output pin
public:
};
```

Listing 16.20. Filter classes: the source class and the filter pin class.

```
struct
__declspec(uuid("{4B429C51-76A7-442d-9D96-44C859639696}"))
CLSID_BitmapSourceFilter;

// Filter's class constructor and destructor
CBitmapSourceFilter::CBitmapSourceFilter(int Nframes, IUnknown *pUnk,
HRESULT *phr)
            : CSource(NAME("BitmapSourceFilter"), pUnk,
            __uuidof(CLSID_BitmapSourceFilter)){
// create the source filter's output pin
  m_pPin = new CPushPinOnBSF(Nframes, phr, this);
}
// delete the output pin
CBitmapSourceFilter::~CBitmapSourceFilter(){ delete m_pPin;}

// Filter's Pin class constructor
CPushPinOnBSF::CPushPinOnBSF(int nFrames, HRESULT *phr, CSource *pFilter)
// initialize base class
      : CSourceStream(NAME("Push Source BitmapSet"), phr, pFilter, L"Out"),
        m_FramesWritten(0),   // initialize - no frames written
        m_bZeroMemory(0),     // no buffer yet to clear
        m_iFrameNumber(0),    // no frames read
        m_rtDelta(FPS_25),    // PAL frame rate 25 frames per second
        m_bFilesLoaded(FALSE) // no files loaded
  // // Use constructor to load the first bitmap pair.
  .. //
  LoadNextBitmap();
  m_bFilesLoaded=TRUE;
  nFilesLoaded++;
}
CPushPinOnBSF::~CPushPinOnBSF(){ // filter's pin destructor
  // close any open files - free memory etc ...
}


void CPushPinOnBSF::LoadNextBitmap(void){   // load next bitmap in sequence
   .. // load next bitmap in sequence
   m_bFilesLoaded=TRUE;
}


HRESULT CPushPinOnBSF::GetMediaType(CMediaType *pMediaType){
    .. // See discussion in the text
    return S_OK;
}

HRESULT CPushPinOnBSF::DecideBufferSize(IMemAllocator *pAlloc,
ALLOCATOR_PROPERTIES *pRequest){
    .. // See discussion in the text
    return S_OK;
}
```

Listing 16.21. An overview of filter class and pin class methods.

declaration of a member variable of type `CPushPinOnBSF*` which is used
to store a pointer to the filter's output pin. Most of the work of the filter
takes place in the output pin object. This has to provide functions to tell
the downstream filters what type of media they can expect (in our case it is
24-bit RGB bitmaps) and how big the connecting buffer (between filters) has
to be. Most importantly, *it must have a method to fill the output buffer with the
bitmap pixels from the image sequence and provide correct timing information.*

Listing 16.20 shows the structure of the two classes that encapsulate the
filter's behavior. Listing 16.21 provides an overview of the class methods.
Listing 16.22 details the class method which pushes the bitmap image pixels
into the downstream filters.

For the overridden methods in the `CPushPinOnBSF` class, some brief
comments are warranted:

- `CPushPinOnBSF::GetMediaType()`. This method reports back on
  the format of the samples being fed out of the filter. In this context, a
  sample is a video frame. We want the samples to match the format of
  the input bitmaps (from the file sequences), so we must have a bitmap
  loaded before this method can execute. Since it is called only once,
  as the graph starts executing, we load the first bitmap in the class's
  constructor method.

- `CPushPinOnBSF::DecideBufferSize()`. This method reports how
  big the output sample buffer needs to be. For a video source filter,
  the output sample buffer will need to hold the pixel image. So for a
  24-bit RGB pixel format, this function must report back that it needs
  $w \times h \times 3$ bytes, where w and h are the width and height of the input
  bitmaps. (We are going to assume that all the bitmaps have the same
  width and height.)

- `CPushPinOnBSF::FillBuffer()`. This method does most of the work.
  It has two primary duties:

  1. To copy the image data from the left and right bitmaps into the
     pin's output buffer. It first copies the left image and then the
     right. Our program will assume that the bitmap's width must
     be a power of two, and we are going to use left-over-right stereo
     format. Consequently, these copies can be done as a block.

  2. To inform the filter what the reference time of the frame is. The
     reference time embedded in the AVI file allows a player to syn-

```
// This is where we insert the bitmaps into the video stream.
// FillBuffer is called once for every sample in the stream.
//
HRESULT CPushPinOnBSF::FillBuffer(IMediaSample *pSample){
 BYTE *pData;
 long cbData;
// this tells the filter to STOP (we have all frames)
 if(m_iFrameNumber== m_dFrames)return E_FAIL;
// first pair loaded in class constructor
 if(m_iFrameNumber > 0)LoadNextBitmap();
 if(!m_bFilesLoaded)return E_FAIL;
 CheckPointer(pSample, E_POINTER);
// to make sure no-other filter messes with our data
 CAutoLock cAutoLockShared(&m_cSharedState);
 // Access the output pin's data buffer
 pSample->GetPointer(&pData);
 cbData = pSample->GetSize();                  // how many bytes to copy
 //
 // Copy the pixel data from the bitmaps into the pins' output buffer - for
 memcpy(pData, m_pImage2,cbData);              // left image data
 pData += (DWORD)cbData;
 memcpy(pData, m_pImage1,cbData);              // right image data
 // calcuate the time in the movie at which this frame should appear
 REFERENCE_TIME rtStart = m_iFrameNumber * m_rtDelta;
 REFERENCE_TIME rtStop  = rtStart + m_rtDelta;
// set the time stamp of this frame
 pSample->SetTime(&rtStart, &rtStop);
// one more frame added
 m_iFrameNumber++;
// to make sure we get all the frames
 pSample->SetSyncPoint(TRUE);
 return S_OK;
}
```

Listing 16.22. Pushing the bitmap pixeldata into the AVI data stream for further filtering, compressing and ultimately archiving to disk.

chronize the movie with a real-time clock, dropping frames if
necessary. We must also ensure that the FilterGraph will *not*
ignore samples if we cannot read the bitmaps fast enough to
keep up in real time. This is done by calling the method
pSample->SetSyncPoint(TRUE).

When all the elements are brought together, the program will create stereo
movies from pairs of bitmap image files and write them into an AVI file using
any installed compression codec.

16.8 Summary

In this chapter, we have looked at the design and coding of a few Windows application programs for stereopsis. There are many other stereoscopic applications we could have considered. For example, we could add a soundtrack to the stereo movie or make a stereoscopic webcam [1]. Nevertheless, the programs we have explored here cover a significant range of components that, if assembled in other ways, should cover a lot of what one might want to do for one's own VR system.

Please do not forget that in the printed listings throughout this chapter, we have omitted statements (such as error checking) which, while vitally important, might have obfuscated the key structures we wanted to highlight.

Bibliography

[1] K. McMenemy and S. Ferguson. *Real-Time Stereoscopic Video Streaming. Dr. Dobb's Journal* 382 (2006) 18–22.

[2] T. Riemersma. "The FLIC File Format". http://www.compuphase.com/flic.htm, 2006.

[3] StereoGraphics Corporation. "Stereo3D Hardware Developer's Handbook". http://www.reald-corporate.com/scientific/developer_tools.asp, 2001.

[4] J. Siragusa et al. "General Purpose Stereoscopic Data Descriptor". http://www.vrex.com/developer/sterdesc.pdf, 1997.

17

Programming Input and Force Feedback

So far in Part II, everything that we have discussed has essentially been associated with the sense of sight. But, whilst it is probably true that the visual element is the most important component of any VR system, without the ability to interact with the virtual elements, the experience and sense of reality is diminished. We have seen in Section 4.2 that input to a VR system can come from a whole range of devices, not just the keyboard or the mouse. These nonstandard input devices are now indispensable components of any VR system. However, even equipping our VR systems with devices such as joysticks, two-handed game consoles or custom hardware such as automobile steering consoles, we don't come close to simulating real-world interaction. To be really realistic, our devices need to *kick back*; they need to resist when we push them and ideally stimulate our sense of touch and feeling. Again, looking back at Section 4.2, we saw that *haptic* devices can provide variable resistance when you try to push something and force feedback when it tries to push you, thus mediating our sense of touch within the VR.

In this chapter, we will look at how the concepts of nonstandard input, custom input and haptics can be driven in practice from a VR application program. Unfortunately for VR system designers on a budget, it is difficult to obtain any devices that, for touch and feel, come close to proving the equivalent of the fabulously realistic visuals that even a modestly priced graphics adapter can deliver. But in Section 17.5, we will offer some ideas as to how you can, with a modest degree of electronic circuit design skills, start some custom interactive experiments of your own.

From the software developer's perspective, there are four strategies open to us so that we can use other forms of input apart from the basic keyboard and mouse. These are:

1. DirectX, which was previously introduced in the context of real-time 3D graphics and interactive multimedia. Another component of the DirectX system, DirectInput, has been designed to meet the need that we seek to address in this chapter. We will look at how it works in Section 17.1.

2. Use a proprietary software development kit provided by the manufacturer of consoles and haptic devices. We shall comment specifically on one these in Section 17.3.1 because it attempts to do for haptics what OpenGL has done for graphics.

3. There is a class of software called *middleware* that delivers system independence by acting as a link between the hardware devices and the application software. The middleware software often follows a client-server model in which a server program communicates with the input or sensing hardware, formats the data into a device-independent structure and sends it to a client program. The client program receives the data, possibly from several servers, and presents it to the VR application, again in a device-independent manner. Often the servers run on independent host computers that are dedicated to data acquisition from a single device. Communication between client and server is usually done via a local area network. We explore middleware in a little more detail in Section 17.4.

4. Design your own custom interface and build your own hardware. This is not as crazy as it seems, because haptics and force feedback are still the subject of active R&D and there are no absolutes or universally agreed standards. We shall look at an outline of how you might do this and provide a simple but versatile software interface for custom input devices in Section 17.5.

We begin this chapter by describing the most familiar way (on a Windows PC) in which to program multi-faceted input and force feedback devices; that is, using DirectInput.

17.1 DirectInput

DirectInput is part of DirectX and it has been stable since the release of DirectX 8. It uses the same programming paradigm as the other components of DirectX; that is:

> Methods of COM interfaces provide the connection between application programs and the hardware drivers. An application program enumerates the attached input devices and selects the ones it wants to use. It determines the capabilities of the devices. It configures the devices by sending them parameters, and it reads the current state of a device (such as joystick button presses or handle orientation) by again calling an appropriate interface method.

DirectInput not only provides comprehensive interfaces for acquiring data from most types of input devices, it also has interfaces and methods for sending feedback information to the devices, to wrestle the user to the ground or give him the sensation of stroking the cat. We will give an example of using DirectInput to send force feedback in Section 17.2, but first we start with an example program to illustrate how two joystick devices might be used in a VR application requiring two-handed input. Since DirectInput supports a wide array of input devices, one could easily adapt the example to work with two mice or two steering-wheel consoles.

DirectInput follows the philosophy of running the data acquisition process in a separate thread. Therefore, an application program is able to obtain the state and settings of any input device, for example the mouse's position, by *polling* for them through a COM interface method.

In the remainder of this section, we will develop a small collection of functions that can be used by any application program to enable it to acquire (x, y, z) coordinate input from one, two or more joystick devices. The example will offer a set of global variables to hold input device data, and three functions to:

1. initialize DirectInput and find joysticks with `InitDirectInput(..)`;

2. release the DirectInput COM interfaces with `FreeDirectInput(..)`;

3. update the global variables by polling for the input device's position and button state with `UpdateInputState(..)`.

All the DirectInput code will be placed in a separate file for convenience.

A minimal joystick device will have two position movements, left to right (along an *x*-axis) and front to back (along a *y*-axis), along with a number of buttons. A different device may have a twist grip or slider to represent a *z*-axis rotation. It is the function of the DirectInput initialization stage to determine the capability of, and number of, installed joystick-like devices. Other devices, mice for example or mice with force feedback, should also be identified and configured. It is one of the tasks of a well designed DirectInput application program to be able to adapt to different levels of hardware capability.

17.1.1   Configuring DirectInput and Finding Devices

Since DirectInput uses COM to access its API, a *DirectInput* device object is created in a similar way to the Direct3D device.   In the function `InitDirectInput(..)`, the main task is to enumerate all joystick devices and pick the first two for use by the application. Device enumeration is standard practice in DirectX. It allows an application program to find out the capabilities of peripheral devices. In the case of joystick devices, enumeration is also used to provide information about the axes that it supports; for example, moving the stick to the left and right may be returned as an *x*-axis. Listing 17.1 sets up the framework for device initialization and enumeration, and Listing 17.2 tells DirectInput the name of the callback functions which permit the application to select the most appropriate joystick, or in this case the first two found.

During axis enumeration, the callback function takes the opportunity to set the *range* and *dead zone*. If, for example, we set the *x*-axis range to

```
// minimim header files for direct input
#include <windows.h>
#include <dinput.h>

// pointer to the DirectInput Object
LPDIRECTINPUT8       g_pDI              = NULL;
// Pointer to the first two joystick devices
LPDIRECTINPUTDEVICE8 g_pJoystick1       = NULL;
// found attached to the machine.
LPDIRECTINPUTDEVICE8 g_pJoystick2       = NULL;
```

Listing 17.1. The DirectInput COM interface is created, all attached joysticks are found and their properties determined.

$[-1000, 1000]$ then when our program asks for the current x-coordinate, DirectInput will report a value of -1000 if it is fully pushed over to the left. A dead zone value of 100 tells DirectX to report as zero any axis position determined to be less than 100. In theory, a dead zone should be unnecessary, but in practice the imperfections in cheap hardware can mean that the spring

```
HRESULT InitDirectInput(HWND hDlg, HINSTANCE g_hInst){
    HRESULT hr;
    // create the DirectInput object and obtain pointer to its interfaces
    hr = DirectInput8Create(g_hInst,DIRECTINPUT_VERSION,IID_IDirectInput8,
    (VOID**)&g_pDI,NULL);
    // Check all the joystick devices attached to the PC
    // a process called enumeration
    // we provide the callback function which assigns the global pointers.
    hr = g_pDI->EnumDevices(DI8DEVCLASS_GAMECTRL,EnumJoysticksCallback,
                            NULL,DIEDFL_ATTACHEDONLY);
    // do the things specifically for joystick 1 - set its data format to
    // the default
    hr = g_pJoystick1->SetDataFormat(&c_dfDIJoystick2);
    // our application program will have exclusive access
    // when it is running in the foreground
    hr = g_pJoystick1->SetCooperativeLevel(hDlg,DISCL_EXCLUSIVE|DISCL_FOREGROUND);
    // enumerate all axes for Joystick 1
    g_pJoystick1->EnumObjects(EnumAxesCallback,(VOID*)g_pJoystick1,DIDFT_AXIS);
    // enumerate to get any sliders
    g_pJoystick1->EnumObjects(EnumObjectsCallback,(VOID*)g_pJoystick1,DIDFT_ALL);
    // repeat for second joystick
    if( NULL != g_pJoystick2 ){ ... //
    }
    return S_OK;
}

HRESULT FreeDirectInput(){            // release the interfaces for the joysticks
    if( NULL != g_pJoystick1 )          {
// before releasing the joystick tell DI we don't neet it.
        g_pJoystick1->Unacquire();
        g_pJoystick1->Release();
        g_pJoystick1 = NULL;
    }
      // repeat here for second joystick ..
    ..//
    if( g_pDI ){ g_pDI->Release();  g_pDI = NULL; } // relase the DI interface
    return S_OK;
}
```

Listing 17.1. (continued).

```
// this is the function called to enumerate the joystick devices
BOOL CALLBACK EnumJoysticksCallback( const DIDEVICEINSTANCE* pdidInstance,
                                     VOID* pContext ){
HRESULT hr;
LPDIRECTINPUTDEVICE8 pJoystick;
// create the joystick device
hr = g_pDI->CreateDevice( pdidInstance->guidInstance,&pJoystick,NULL);
if( FAILED(hr) ) return DIENUM_CONTINUE;
nJoys++;    // we will stop when we have two
if(nJoys == 1)g_pJoystick1=pJoystick;
else          g_pJoystick2=pJoystick;
if(nJoys == 2)return DIENUM_STOP; // only need 2
return DIENUM_CONTINUE;
}
// this function will see if the joystick has more than 2 axes;
// some do, some don't
BOOL CALLBACK EnumAxesCallback( const DIDEVICEOBJECTINSTANCE* pdidoi,
VOID* pContext ){
DIPROPRANGE diprg; // a structure to pass data to a DI range property
DIPROPDWORD diprw; // a structure to pass data to a WORD size DI property
LPDIRECTINPUTDEVICE8 pJoystick = (LPDIRECTINPUTDEVICE8)pContext;
diprg.diph.dwSize       = sizeof(DIPROPRANGE);
diprg.diph.dwHeaderSize = sizeof(DIPROPHEADER);
diprg.diph.dwHow        = DIPH_BYOFFSET;
diprg.diph.dwObj        = pdidoi->dwOfs; // Specify the enumerated axis
diprg.lMin              = -1000;         // these are integer units and give
// the min and max range of the joystick
diprg.lMax              = +1000;
if( pdidoi->dwType & DIDFT_AXIS ){
  if(FAILED(pJoystick->SetProperty(DIPROP_RANGE,
  &diprg.diph)))return DIENUM_STOP;
// If this is an x, y, or z axis set the dead zone
  if(pdidoi->guidType == GUID_XAxis ||
    pdidoi->guidType == GUID_YAxis ||
    pdidoi->guidType == GUID_ZAxis){
    diprw.diph.dwSize       = sizeof(DIPROPDWORD);
    diprw.diph.dwHeaderSize = sizeof(DIPROPHEADER);
    diprw.diph.dwHow        = DIPH_BYOFFSET;
    diprw.diph.dwObj        = pdidoi->dwOfs;
    diprw.dwData            = 100;
    switch( pdidoi->dwOfs ){
      case DIJOFS_X:
        if(FAILED(pJoystick->SetProperty(DIPROP_DEADZONE,
        &diprw.diph )))return DIENUM_STOP;
        break;
```

Listing 17.2. For joystick devices, the number of available device axes has to be determined and configured so that values returned to the program when movement occurs along an axis fall within a certain range and have known properties.

```
// repeat for Y and Z axes
      }
    }
    return DIENUM_CONTINUE;
}
// this function checks for sliders on the joystick, sometimes used
// instead of twist axis
BOOL CALLBACK EnumObjectsCallback( const DIDEVICEOBJECTINSTANCE* pdidoi,
                                   VOID* pContext ){
 // Enumerate for other device attributes
    if (pdidoi->guidType == GUID_Slider){
 // apart from axes - we may wish to use a slider
      bSlider=TRUE;
    }
    return DIENUM_CONTINUE;
}
```

Listing 17.2. (continued).

in a joystick which is supposed to return it to dead center may actually return it to $+20$ the first time and -50 the second time. Thus, by setting up a dead zone, our application can ensure that if the user releases the handle, the joystick will always report an axis coordinate of zero.

17.1.2   Acquiring Data

Function `UpdateInputState(..)` (see Listing 17.3) is used to obtain the current state from two of the attached joystick devices. This information includes position, button presses etc. and requires a call to the `Poll()`

```
#define SCALE 0.0001f    // Scale the positions returned in the global
float x_rjoy_pos;        // variables so that the full range of motion
float y_rjoy_pos;        // on all axis is in the range [-1.000, +1.000]
float z_rjoy_pos;
float x_ljoy_pos;        // and the global variables for joystick2
float y_ljoy_pos;
float z_ljoy_pos;
long  nJoys=0;           // number of suitable joysticks found
```

Listing 17.3. Function to obtain the current axis positions and state of buttons pressed. Before we can obtain any information about a joystick, it must be acquired by the program. Otherwise, there could be a conflict with another application also wishing to use a joystick device.

```
HRESULT UpdateInputState( HWND hDlg ){
 HRESULT      hr;
 // joystick state information is provided in this strucure
 DIJOYSTATE2   js;
 if( g_pJoystick1 )          {          // do the first joystick
   hr = g_pJoystick1->Poll();  // see if we can access the joystick state
   if( FAILED(hr) ){          // if not - try to get the joystick back
     hr = g_pJoystick1->Acquire();
     while(hr == DIERR_INPUTLOST )hr = g_pJoystick1->Acquire();
     return S_OK;        // got it!
   }
   // GetDeviceState fills the state information structure
   if(FAILED(hr = g_pJoystick1->GetDeviceState(sizeof(DIJOYSTATE2)
   ,&js)))return hr;
   x_rjoy_pos=(float)(js.lX)*SCALE;  // scale it into range [-1.0 , 1.0]
   y_rjoy_pos=(float)(js.lY)*SCALE;  // This matches the integer range set
   z_rjoy_pos=(float)(js.lZ)*SCALE;  // during enumeration to [-1000,+1000].
   if(js.rgbButtons[0] & 0x80)..     // do someting with button 0
   if(js.rgbButtons[1] & 0x80)..     // button 1 etc.
 }
 // Repeat the code for JOYSTICK 2
 return S_OK;
}
```

Listing 17.3. (continued).

method on the joystick object followed by a call to GetDeviceState() on the same object. All the details are returned in a DIJOYSTATE2 structure (variable js). Structure members are scaled and assigned to global variables recording axis position etc. Note the use of the Acquire() method if the Poll() method returns an error code. An error will occur if another application has grabbed the use of the joystick. Before our application can again get data from the joystick, it must us the Acquire() method to claim exclusive use of it.

17.2   Force Feedback with DirectInput

Normally, one associates force feedback with highly specialized and expensive equipment, but it can be achieved in even modestly priced joystick devices because they are targeted at the game market and are in mass production. These devices cannot deliver the same accurate haptic response as the professional devices we discussed in Chapter 4, but they are able to provide a reasonable degree of user satisfaction. Since DirectInput was designed with the computer game developer in mind, it offers a useful set of functions for controlling force-feedback devices. In terms of the haptics and forces we discussed in Section 4.2, DirectInput can support constant forces, frictional forces, ramp forces and periodic forces.

```
// pointer to constant force effect
LPDIRECTINPUTEFFECT    g_pEffect1c        = NULL;
// friction force effect
LPDIRECTINPUTEFFECT    g_pEffect1f        = NULL;
// spring force effect
LPDIRECTINPUTEFFECT    g_pEffect1s        = NULL;
// constructed effect (an impulse)
LPDIRECTINPUTEFFECT    g_pEffect1poke     = NULL;
static BOOK bForce1=FALSE, bForce2=FALSE;


// enumerate - joysticks with force feedback - very similar to non FF joysticks.
static BOOL CALLBACK EnumFFDevicesCallback( const DIDEVICEINSTANCE* pInst,
                                   VOID* pContext ){
  LPDIRECTINPUTDEVICE8 pDevice;
  HRESULT            hr;
  hr = g_pDI->CreateDevice( pInst->guidInstance, &pDevice, NULL );
  if(FAILED(hr))return DIENUM_CONTINUE;
  nJoys++;
  if(nJoys == 1){g_pJoystick1=pDevice; bForce1=TRUE;}
  else          {g_pJoystick2=pDevice; bForce2=TRUE;}
  if(nJoys == 2)return DIENUM_STOP; // only need 2
  return DIENUM_CONTINUE;
}
// the following changes are needed to set up the force feedback joysticks
HRESULT InitDirectInput( HWND hDlg, HINSTANCE g_hInst){
  ..// enumerate for Force Feedback (FF) devices ONLY
  hr = g_pDI->EnumDevices(DI8DEVCLASS_GAMECTRL, EnumFFDevicesCallback,
                        NULL,DIEDFL_ATTACHEDONLY | DIEDFL_FORCEFEEDBACK);
  .. // if joystick 1 is found with FF create the effects for constant force,
     // friction and impulse.
  if(bForce1)CreateForceFeedbackEffects(&g_pEffect1c,&g_pEffect1f,
                             &g_pEffect1s,&g_pEffect1poke,
                             g_pJoystick1);
  ..// Do similar thing for second or more FF joystick
  return S_OK;
}
```

Listing 17.4. To find a joystick with force feedback, a device enumeration procedure similar to that for non-force-feedback devices is performed. The source code for CreateForceFeedbackEffects() is outlined in Listing 17.5.

```
static void CreateForceFeedbackEffects(LPDIRECTINPUTEFFECT *ppEffectC,
    LPDIRECTINPUTEFFECT *ppEffectF,LPDIRECTINPUTEFFECT *ppEffectS,
    LPDIRECTINPUTEFFECT *ppEffectPoke, LPDIRECTINPUTDEVICE8 pDevice){
DIEFFECT eff;              // effect parameter structure
DIPROPDWORD dipdw;
HRESULT     hr;
// set the effects to work in 2 dimensions (X Y)
DWORD           rgdwAxes[2]   = { DIJOFS_X, DIJOFS_Y };
// set default values for the parameters of the forces - these will be
// changes as necessary
// consult DirectInput Manual for details.
LONG            rglDirection[2] = { 0, 0 };
DICONSTANTFORCE cf            = { 0 };
DICONDITION     cd            = { 0, 10000,10000,0, 0, 0 };
DICONDITION     cd2[2]        = { 0, 10000,10000,0, 0, 0, 0, 10000, 10000,
                                     0, 0, 0};
dipdw.diph.dwSize       = sizeof(DIPROPDWORD);
dipdw.diph.dwHeaderSize = sizeof(DIPROPHEADER);
dipdw.diph.dwObj        = 0;
dipdw.diph.dwHow        = DIPH_DEVICE;
dipdw.dwData            = FALSE;
// tell force to automatically center the joystick on startup
if(FAILED(hr=pDevice->SetProperty(DIPROP_AUTOCENTER, &dipdw.diph)))return;
// the "eff" structure shows typical settings for a constant force.
ZeroMemory( &eff, sizeof(eff) );   // constant force effect
eff.dwSize              = sizeof(DIEFFECT);
eff.dwFlags             = DIEFF_CARTESIAN | DIEFF_OBJECTOFFSETS;
eff.dwDuration          = INFINITE;        // force continues indefinitely
eff.dwSamplePeriod      = 0;               // no time limit on force
eff.dwGain              = DI_FFNOMINALMAX; // maximum force
eff.dwTriggerButton     = DIEB_NOTRIGGER;  // force is NOT triggered
eff.dwTriggerRepeatInterval = 0;
eff.cAxes               = 2;               // number of axes to get force
eff.rgdwAxes            = rgdwAxes;        // which axes
eff.rglDirection        = rglDirection;    // direction of force
eff.lpEnvelope          = 0;               // no - envelope  constant force
eff.cbTypeSpecificParams = sizeof(DICONSTANTFORCE);
eff.lpvTypeSpecificParams = &cf;
eff.dwStartDelay        = 0;
// create the effect using the "eff"  effect structure - return
// pointer to effect
if(FAILED(hr=pDevice->CreateEffect(GUID_ConstantForce,&eff,ppEffectC,
NULL)))return;
..//
..// the other effects are set in a similar way.
  // Consult the full project code for details
return;
}
```

Listing 17.5. Create a constant force effect that can be played out to a joystick device.

The key to using force feedback in DirectInput is to think in terms of a *force effect*, such as an effect that simulates driving a car over a bumpy road. There are two types of effect, *supported effects* and *created effects*. A supported effect is a basic property of the input device, such as a constant resistance to pushing a joystick handle away from its center position. A created effect is a composite of the supported effects brought together to simulate a specific scenario, such as the bumpy road.

An application program which wishes to use force feedback begins by first enumerating the devices available to it. After enumeration of the supported effects, the program sets about building the created effects it intends to use. If no suitable hardware is available or the supported effects are insufficient for its needs, the program gives up. Once the effects are built, they can be used at any time. In the language of DirectInput, one says they are *played back* on the input device. Some effects can be modified just before playback (the direction of the force might be changed, for example). Some effects may be short lived (impulses, for example); others may go on indefinitely until actively canceled by the controlling application program.

Listing 17.4 shows the changes and additions to Listing 17.1 that are necessary in order to find and configure the force-feedback joysticks. Listing 17.5 shows how to create a constant force-feedback effect, and Listing 17.6 how to modify and apply different levels of force. Applying a constant force of 0 effectively leaves the joystick in a slack state.

Effects that are to last for a limited duration are created in the same way, and played back using the `Start(..)` and `Stop()` methods of the `DIRECTINPUTEFFECT` interface. Consult the accompanying project codes to see some typical examples.

DirectInput makes available many more force effect properties, and we have touched on only a few in these examples. The DirectInput [7] SDK provides full documentation on all the possible force configurations and attributes such as conditions, envelopes and offsets, gain etc. The SDK also comes with a small utility (called `fedit`) that lets a game developer design and test the effects and record them in a file that can then be loaded and played during the game. This is similar to DirectShow's GraphEdit program, which we looked at in Chapter 15.

Even thought DirectInput was designed to meet the needs of computer game developers, it does offer a consistent interface that could be used by more elaborate haptic devices. However, the basic force-feedback joysticks and their DirectInput control interface are limited, because they do not have a sense of location relative to a 3D frame of reference. To be truly able to

```
static INT              g_nXForce;
static INT              g_nYForce;
static INT              g_mForce;
static INT              nForce=0,nForce1=0,nForce2=0;

static HRESULT SetDeviceConstantForces(LPDIRECTINPUTEFFECT pEffect){
  // Modifying an effect is basically the same as creating a new one, except
  // you need only specify the parameters you are modifying
  LONG rglDirection[2] = { 0, 0 };
  DICONSTANTFORCE cf;
  // If two force feedback axis, then apply magnitude from both directions
  rglDirection[0] = g_nXForce;
  rglDirection[1] = g_nYForce;
  cf.lMagnitude   = g_mForce;
  DIEFFECT eff;
  ZeroMemory( &eff, sizeof(eff) );
  eff.dwSize              = sizeof(DIEFFECT);
  eff.dwFlags             = DIEFF_CARTESIAN | DIEFF_OBJECTOFFSETS;
  eff.cAxes               = 2;  // 2 axes
  eff.rglDirection        = rglDirection;
  eff.lpEnvelope          = 0;
  eff.cbTypeSpecificParams = sizeof(DICONSTANTFORCE);
  eff.lpvTypeSpecificParams = &cf;
  eff.dwStartDelay        = 0;
  // Now set the new parameters and start the effect immediately.
  return pEffect->SetParameters( &eff, DIEP_DIRECTION |
                                        DIEP_TYPESPECIFICPARAMS |
                                        DIEP_START );
}

SendDirectInputForce(int type){   //
  if(type == 0){ // stop constant force
    g_nXForce=0;
    g_nYForce=0;
    g_mForce=0;
    if(joyID == 1 && g_pEffect1c != NULL){
      SetDeviceConstantForces(g_pEffect1c);
      g_pEffect1c->Stop();
    }
  }
  else if(type == 1){ //start constant
    g_nXForce = 0;
    g_nYForce = -DI_FFNOMINALMAX;
    g_mForce = (DWORD)sqrt( (double)g_nXForce * (double)g_nXForce +
                            (double)g_nYForce * (double)g_nYForce);
    if(joyID == 1 && g_pEffect1c != NULL)SetDeviceConstantForces(g_pEffect1c);
  }
  else if(type ==  ...){ //

  }
}
```

Listing 17.6. Configuring and playing different types of forces in response to a simple integer command from the application program.

feel a 3D object, pick it up and move it around, one needs to be able to sense the 3D position of the point of contact with the object and apply reactionary forces in any direction. Chapter 4 described a few such devices, and we will now briefly look at the types of programming interfaces that have evolved to drive them.

17.3 Haptic Software Libraries and Tools

Haptic devices with multiple degrees of freedom and working under different operating conditions (desktop devices, or wearable devices, or custom devices) do not yet conform to a well-documented and universally agreed set of standard behaviors and software interface protocols. The things they do have in common are that they need to be able to operate in real time, move in three-dimensional space, apply forces in any direction and provide accurate feedback of their 3D position and orientation. The term *haptic rendering* (or *force rendering*) is used to describe the presentation of a scene that can be felt rather than one that can be seen. Haptics involves an additional complexity in that not only should the device present its user with a scene that she can feel, but it should also be able to allow its user to push, poke, prod and sculpt that scene. To put it another way, the device should not just oppose the movement of the user with a force to simulate a hard surface. It should be able to adapt to the movement of the user so as to simulate deformable or elastic material, or indeed go so far as to facilitate dynamics, friction and inertia as the user pushes a virtual object around the virtual world, lifts it or senses its weight. Imagine a virtual weight lifting contest.

We would not expect all these things to be doable by the haptic device hardware on its own, but in the future, just as today's 3D graphics cards can deliver stunningly realistic images to fool our eyes, we hope that a similar device will be available to fool our senses of touch and feel. In the meantime, the software that controls the haptic devices has a crucial role to play. Our software design goal should be programs that can, as far as possible, emulate the ideal behavior of a haptic device to completely facilitate our sense of *presence*. This is not a simple thing to achieve. As well as driving the haptic device, the application program has to:

- Store a description of the 3D scene in terms of surfaces that can be touched and felt.

- Calculate collisions between the virtual position of the end effector (the avatar) of the haptic device and the object's surfaces.

- Haptic devices are rarely operated blind. A 3D visualization of the scene with a indication of the location of the haptic point of contact must usually also be provided.

How these tasks are divided between an application program and a haptic rendering software development environment (SDK) is a major point of interest. In 3D graphics, the OpenGL and Direct3D software libraries take advantage of the GPU hardware to release CPU cycles for other tasks. It would be extremely useful to have equivalent processing power in a haptic device, say to carry out the collision detection between avatar and haptic scene, just as OpenGL provides a consistent graphics software API whether or not there is any hardware acceleration. Haptic device manufacturers do provide such libraries. SensAble Technologies, Inc.[1] is a major vendor of desktop haptic devices with its Phantom [15] range of products. It provides a software toolkit, the Ghost [14] SDK, which an application program can use to describe a haptic scene and interact with it in both a dynamic and a static way. SensAble has also introduced the OpenHaptics [16] toolkit, part of which is designed with the same philosophy underpinning OpenGL. It also takes advantage of OpenGL features such as calculating the visible surfaces in a scene which the haptics code can then use to render forces. Other vendors and research institutes have produced high-level haptic libraries and tools as well. H3D [17] is an open-source haptics system supported by SenseGraphics AB.[2] H3D is built on the X3D (see Section 5.3.1) standard for describing 3D Web environments, and since H3D is written to use XML, it is a good candidate for adding haptic descriptions of scenes for interaction over the Internet. HAL [2] is an extensible haptic library that can be used to drive the Phantom devices through the Ghost I/O driver.

However, the OpenHaptics system, with its tight integration with OpenGL, offers programmers the quickest route to add haptic interaction to their own desktop VR applications. We will take a closer look at the structure of OpenHaptics in Section 17.3.1.

There is one complication that programmers using haptics APIs need to be aware of when designing applications. The force-delivery mechanism in the haptic device uses servo motors that have to be commanded by a low-level software driver of some kind. Thus, there is always going to be a short lag between the user moving the haptic device to a position where he should feel the force and the actuation of that force. To the user, this gives the impression

---

[1] http://www.sensable.com/.

[2] http://www.sensegraphics.com/.

that the object is not as stiff as one would wish. More seriously, however, it can cause instability in the operation of the device. We briefly discussed this problem in Section 4.2.1. In practice, the haptic APIs use multi-threading, because it allows them to drive the servo motors from a high-priority thread that performs very little work other than to keep the servos correctly driven and report back the 3D location of the device's end effector within its working volume. The Ghost and OpenHaptics systems run what they call a *servo loop* at a rate of 1 kHz.

Some of the multi-threaded tasks a haptics application has to do might be hidden within a device driver in much the same way that DirectInput handles the hardware. However, a typical haptic application may need to run up to four concurrent threads:

1. A thread to render the scene visually. This runs at 30 Hz.

2. A fast force response and basic collision detection thread. This thread drives the servos and does the most simple of collision detection with a few bounding volumes such as spheres and planes. It typically runs at rates of between 300 Hz and 1 kHz.

3. A collision detection thread which calculates collisions between those parts of an object that are close to the avatar. This will be at a more detailed level than the collision detection in thread 2.

4. And finally, the simulation thread which performs such things as modeling surface properties, stiffness, elasticity, push and pull etc.

17.3.1 OpenHaptics

The OpenHaptics [16] system library from SensAble provides two software interfaces for Phantom devices. These are called the Haptics Device API (HDAPI) and Haptic Library API (HLAPI). The HDAPI is a low-level library which allows an application to initialize the haptic device, start a servo loop thread and perform some basic haptic commands, such as get the position of the end effector **p**, calculate a force as a function of **p** (for example, a coulombic force field) and pass that force to the servo loop. The HDAPI can execute synchronous calls, useful for querying the state of user operable push buttons, or asynchronous calls, such as rendering the coulombic force field.

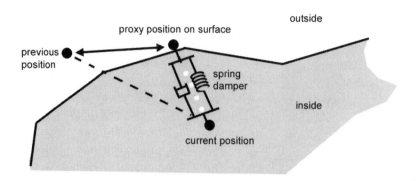

Figure 17.1. Using the proxy method for haptic rendering.

The HLAPI follows the OpenGL syntax and state machine model. It allows the shapes of haptic surfaces to be described by points, lines and triangular polygons, as well as sets material properties such as stiffness and friction. It offers the possibility of taking advantage of the feedback, depth rendering and the simultaneous 3D presentation which OpenGL offers. Indeed, there is such a close connection between being able to see something and touch something that the analogy with OpenGL could open the door to a much wider application of true haptic behavior in computer games, engineering design applications, or even artistic and sculpting programs.

OpenHaptics executes two threads, a servo thread at 1 kHz and a local collision detection thread which makes up simple local approximations to those shapes that lie close to the avatar. The collision detection thread passes this information to the servo thread to update the force delivered by the haptic device. OpenHaptics carries out proxy rendering as illustrated in Figure 17.1. The proxy follows the position of the haptic device, but it is constrained to lie outside all touchable surfaces. The haptic device, because of servo loop delays, can actually penetrate objects. The proxy will iterate from its previous position to a point on the surface closest to the haptic device's current position. (When the haptic device lies outside a touchable object, its position and the proxy position will coincide.) When the haptic device lies inside a touchable object, a restoring force using the spring damper force model will try to pull the haptic device back to the surface point where the proxy lies.

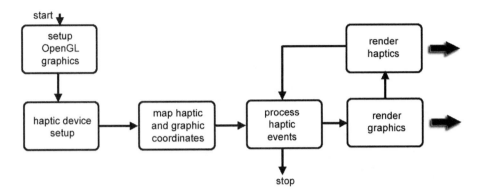

Figure 17.2. A flow diagram of a typical haptic program that uses the HLAPI. The *process events* block handles messages from the haptic device and will react to an *exit* message. The OpenGL render stage also renders a Z-depth buffer view from the avatar which may be used for collision detection.

OpenHaptics has three ways of generating forces:

1. *By shape*. The application program specifies geometric shapes which the haptics rendering thread uses to automatically compute the appropriate reaction forces to simulate touching the surface.

2. *By effect*. Global force effects such as viscosity and springs that are not related to any specific shape are defined by effect. This is an elaborate 3D version of the concept provided by DirectInput.

3. *By proxy*. The user can set a desired position and orientation for the haptic device's end effector, and forces will be generated by the device in order to move it towards the desired position.

A typical HLAPI application program has the structure shown in Figure 17.2. This figure depicts the close relationship between OpenGL and OpenHaptics. The graphics and haptics coordinate frames of reference are mapped onto each other with a transformation matrix. The haptics are rendered by using OpenGL's feedback or Z-buffer depth information to provide the object's surface location.

Full details of HLAPI programming with the Phantom devices and several useful example programs may be found in the OpenHaptics reference manual [16]. The full OpenHaptics software suite and drivers for Windows are included with Phantom device hardware.

Although the commercially available desktop haptic devices and force-feedback gloves are reducing in cost and becoming more available, you may still wish to design, build and use your own custom input devices. There are two possible approaches. First, a large body of knowledge from the open-source community can be tapped into through the use of *middleware*, or you may wish to work directly with a Windows PC. In the next two sections, we will explore the middleware concept and outline how a few simple electronic components can form an interface to the Windows PC through its universal serial bus (USB). In particular, we'll see how the USB can be accessed from an application program written in C without the need to write a device driver.

17.4   Middleware

In the context of VR, the term *middleware* relates to platform-independent software that acts as a bridge between the user interface hardware, which can range from joysticks to motion trackers, and the application programs that wish to use it. In this respect, it is not unlike DirectInput but in several respects it is much more powerful. Perhaps the most significant advantage is

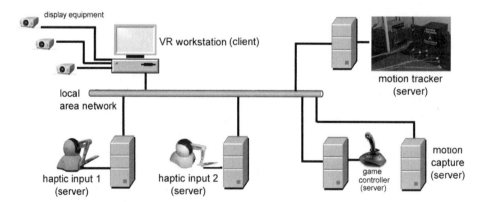

Figure 17.3. By using a middleware program library, the VR system can be distributed over a number of hosts. Each input device is controlled by its own dedicated controller. The input data is broadcast over a local area network and used as required by the VR software application.

that it allows the input devices to be physically remote from the VR application that uses the data. This effectively allows a VR system to operate in a distributed manner, in which input and display devices are connected to different hosts. Nevertheless, to the user, the system still appears to act as a single, tightly integrated entity. Figure 17.3 illustrates how this might work in practice.

Two mature, flexible and open source middleware tools that are easily built and extended are the Virtual-Reality Peripheral Network (VRPN) [20] and OpenTracker [9]. They both provide a degree of device abstraction and can be combined with a scene graph (see Section 5.2) visualization package to deliver powerful VR applications with only modest effort on the part of the developer. A helpful aspect of the architecture in these libraries is their use of the client-server model. This allows an individual computer to be dedicated to the task of data acquisition from a single device. Thus, the VR program itself receives its inputs in a device-independent fashion via a local area network (LAN). We will briefly offer comment on these two helpful pieces of software.

17.4.1   VRPN

The Virtual-Reality Peripheral Network system (see Taylor et al. [18]) was developed to address concerns that arise when trying to make use of many types of tracking devices and other resource-intensive input hardware. These concerns included:

- ease of physical access to devices, i.e., fewer wires lying around;

- very different hardware interfaces on devices that have very similar functions;

- devices may require complex configuration and set-up, so it is easier to leave them operating continuously; and

- device drivers may only be available on certain operating systems.

VRPN follows the client-server model in Figure 17.3 and has functions that can be built into the server program to manage the data acquisition from a large selection of popular hardware (see Section 4.3). For example, an Ascension Technology's Flock of Birds motion tracker can be connected to a VRPN server running under Linux via the /dev/tty0 serial device. The

VRPN server will multicast its data onto an Ethernet LAN which can then be picked up on a Windows PC running a VRPN client.

A good example application that demonstrates all these features is distributed with the OpenSceneGraph library [10]. The code for OpenScene-Graph's osgVRPNviewer project shows how to link directly to a VRPN server that reads its data from a spaceball device executing on the local host. Since VRPN and OpenSceneGraph are open-source applications, we suggest you download both of them and experiment with your own tracking and input devices.

17.4.2   OpenTracker

OpenTracker, introduced by Reitmayr and Schmakstieg [9], offers a library of C++ classes for supporting the addition of new tracking devices. However, from a practical point of view, it is probably more useful that it already has modules to support data acquisition from a wide range of currently available hardware: Flock of Birds, GPS devices and Intersense motion trackers are but a few examples. Like VRPN, it can operate in a distributed configuration such as that illustrated in Figure 17.3. If the device you wish to use for input in already known to OpenTracker then, to use it in your application, you simply have to write a short piece of client code and generate a configuration file. The configuration file is formatted in XML (see Section 5.3.1) and describes how the tracking data passes through the system from source to application. For example, tracking input can be taken from an ARToolKit (see Section 18.7) video tracking program and multicast onto an Ethernet network or, for testing purposes, reported on the local host's console.

17.5   Interfacing Custom VR Hardware with the Applications

In Section 17.1, we examined how VR applications running on Windows can make use of conventional input devices: the mouse, a joystick or some other sort of game console. But what if you need something special with more control, or you want to make your own human interface device? Is it going to be necessary to look into the inner workings of your computer's operating system in detail?

Fortunately not; for all the applications we can think of, there is an elegant but simple way to do this without resorting to the nightmare scenario: writing your own device driver for Windows or Linux. Of course, it is possible to do both—see [8] and [13]—but for all but the highest-speed interfaces, the universal serial bus (USB) will do the job. With a bit of clever programming, one can fool Windows into believing that it already has all the drivers it needs and is dealing with a joystick. Interfacing to a PC via USB could be the subject of several books in itself, so here we will only give an outline view of the hardware and explain how an application program written in C/C++ can read and write directly to a USB device.

17.5.1 A Brief Look at USB

The USB interface uses a serial protocol, and the hardware connections consist of four wires: +5V power and ground and two data lines called D+ and D−. The interface, depending on how it is configured, can be low, full or high speed. Low-speed transfers operate at 1.5 Mbits/s, full-speed at 12 Mbits/s and high-speed 480 Mbits/s. However, in addition to data, the bus has to carry control, status and error-checking signals, thus making the actual data rates lower than these values. Therefore, the maximum data rates are only 800 bytes per second for low speed, 1.2 megabits per second for full speed and 53 megabits per second for high speed. The USB Implementers Forum[3] provides a comprehensive set of documentation on USB and has many links to other useful resources.

There are four types of data transfers possible, each suited to different applications, which may or may not be available to low-speed devices. These transfers are control, bulk, interrupt and isochronous. Devices such as keyboards, mice and joysticks are low-speed devices, and use interrupt data transfer. The maximum possible transfer of data for this combination is 8 bytes per 10 milliseconds.

When a peripheral device is connected to a PC via USB, a process called enumeration takes place. During enumeration, the device's interface must send several pieces of information called *descriptors* to the PC. The descriptors are sets of data that completely describe the USB device's capabilities and how the device will be used. A class of USB devices falls into a category called human interface devices (HIDs), not surprisingly because these devices interact directly with people. Examples of HIDs are keyboards, mice and joysticks. Using the HID protocol, an application program can detect when someone moves a joystick, or the PC might send a force-feedback effect that the user can feel.

---

[3]http://www.usb.org/.

With many custom-made USB devices, the user has to create a driver for Windows. However, Windows has supported the HID class of USB devices since Windows 98. Therefore, when a USB HID is plugged into a PC running Windows, it will build a custom driver for the device. As a result, if we can construct a simple device that fools Windows into believing it is dealing with an HID, there will be no need to write custom device drivers and we can use DirectInput or write our own high-level language interface in only a few lines of code.

17.5.2 USB Interfacing

For anyone working on R&D in VR applications, it can be frustrating to find that what you want to do is just not possible with any commercially available equipment. This may be because it is just not viable for a manufacturer to go to the expense of bringing a product to market for a specialized application. Or it may be that until you convince the manufacturer that your idea works, they will not be interested. In these circumstances, one has only two alternatives, compromise on the hardware, for example use a force joystick where what you really want is a six-degrees-of-freedom haptic system with a working volume of four cubic meters, or build your own hardware. If one does decide to *DIY* it then the question is how to interface the device to the virtual world in the computer. In the early days of PC systems, developers turned to the serial and parallel I/O ports because they were easy to access and control from a high-level language program such as C. Nowadays, PC hardware has become much more complicated and the interfaces much more difficult to get at from an application program. Or have they?

The USB and HID interfaces offer a real alternative, since a number of programmable interrupt controllers (PICs) now offer a USB interface and have built in analog-to-digital (A/D) converters. This has put the software engineer back in control, because devices such as the Microchip 167865 PIC require only a few components and some embedded software to turn 6/8 analog voltages into a data stream which even a program executing without any special system privileges can access at a rate that is high enough to control the servo mechanism of a 3D haptic device. Figure 17.4 illustrates just such an interface built by Mack [5].

Books by Axelson [1] and Hyde [3] provide good starting points with guidelines for building interface devices for a PC using USB.

Once one has a way to measure and set several voltage levels in an external device under the control of an application program and at a high enough

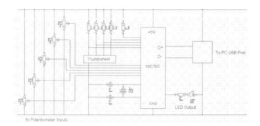 

Figure 17.4. A USB to analog voltage interface using a PIC. On the left is the circuit diagram of the device which converts the analog voltage on five potentiometers and one thumb wheel switch into a USB compatible data frame of nine bytes delivered to the PC at a rate of 1K bytes/s. No external power supply is necessary and the voltage scale of $[0, 5V]$ is converted to an unsigned byte value ($[0, 255]$).

speed, the places to which you can take your virtual world are only restricted by your imagination and the ability to construct your *human interface device design*. Apart from soldering a few electronic components, perhaps the hardest thing to do is write, develop and test the firmware that has to be written for the PIC. A good development system is essential. For example, the Proton+ Compiler [11] and some practical guidance may also be found in the PIC microcontroller project book [4].

At the core of the microcontroller code must lie routines that allow its USB interface to be enumerated correctly by Windows when it connects to the PC. This is done in a series of stages by passing back data structures called *descriptors*. Mack [5] gives a full program listing for PIC microcode that enumerates itself as a joystick to Windows. During enumeration, the PC requests descriptors concerning increasingly small elements of the USB device. The descriptors that have to be provided for a HID device so that it appears to Windows as a joystick are:

- *The device descriptor* has 14 fields occupying 18 bytes. The fields tell the PC which type of USB interface it provides, e.g., version 1.00. If a USB product is to be marketed commercially, a vendor ID (VID) must be allocated by the USB Implementers Forum. For hardware that will never be released commercially, a dummy value can be used. For example, if we assign 0x03E8 as the VID then an application program may look for it in order to detect our device (see Listing 17.7).

```
#include "setupapi.h"      // header file from Windows DDK
#include "hidsdi.h"        // header file from Windows DDK

// HIDs appear as normal Windows files
HANDLE h1=INVALID_HANDLE_VALUE,
// when open - use this handle to read/write.
       h2=INVALID_HANDLE_VALUE;
// are the reading threads running ?
BOOL   bLoop=FALSE;

// initialize TWO Human Interfaces Devices
BOOL InitHIDs(void){
 if((h1=FindUSBHID(0x04D8,0x0064,0x0001)) == INVALID_HANDLE_VALUE)return FALSE;
 if((h2=FindUSBHID(0x04D8,0x00C8,0x0001)) == INVALID_HANDLE_VALUE)return FALSE;
 _beginthread(UpdateHID1,0,(PVOID)NULL); // start a thread to handle each device
 _beginthread(UpdateHID2,0,(PVOID)NULL);
 return TRUE;
}

BOOL FreeHIDs(void){                          // release the devices
 bLoop=FALSE;  Sleep(100);                    // give the reading thread time to exit
 if(h1 != INVALID_HANDLE_VALUE)CloseHandle(h1);
 if(h2 != INVALID_HANDLE_VALUE)CloseHandle(h2);
 return TRUE;
}

HANDLE FindUSBHID(     // Find the USB device
 DWORD vendorID,       // We identify our HID joystick emulating devices by
 DWORD productID,      // looking for a unique vendor and product ID and version
                       // number.
 DWORD versionNumber){// These are set in the USB PIC's EEPROM.
 HANDLE deviceHandle = INVALID_HANDLE_VALUE;
// HIDs are identified by an interger in range 0 - N=number installed
 DWORD index = 0;
 HIDD_ATTRIBUTES deviceAttributes;
// Go through all possible HID devices until the one we want is found
// the code for function FindHID() is in the project code on the CD
 while((deviceHandle = FindHID(index++)) != INVALID_HANDLE_VALUE){
   // fill the attributes structure for this HID
   if (!HidD_GetAttributes(deviceHandle, &deviceAttributes))return
   INVALID_HANDLE_VALUE;
   if ((vendorID      == 0 || deviceAttributes.VendorID == vendorID) &&
       (productID      == 0 || deviceAttributes.ProductID == productID) &&
       (versionNumber == 0 || deviceAttributes.VersionNumber == versionNumber))
    return deviceHandle; // we have the one we want - return handle
    CloseHandle (deviceHandle);
  }
  return INVALID_HANDLE_VALUE;
}
```

Listing 17.7. Functions to initialize the USB device, acquire file handles to be used for reading data from the device, and start the thread functions that carry out the reading operation.

- *The configuration descriptor* has 8 fields occupying 9 bytes. This provides information on such things as whether the device is bus-powered. If the PC determines that the requested current is not available, it will not allow the device to be configured.
- *The interface descriptor* has 9 fields occupying 9 bytes. It tells the host that this device is an HID and to expect additional descriptors to provide details of the HID and how it presents its reports. (For example, in the case of a joystick, the report describes how the nine-byte data frame is formatted to represent the x-, y-, z-coordinates, button presses etc.)

Field	Value	Description	Attribute
Usage page	0x05	Generic Desktop	0x01
Usage	0x09	Joystick	0x04
Collection	0xA1	Application	0x01
Usage page	0x05	Simulation controls	0x02
Usage	0x09	Throttle	0xBB
Logical Minimum	0x15	0	0x00
Logical Maximum	0x26	255	0x00FF
Physical Minimum	0x35	0	0x00
Physical Maximum	0x46	255	0x00FF
Report size	0x75	Number of bits = 8	0x08
Report count	0x95	Number of above usages = 1	0x01
Input	0x81	Data, variable, absolute	0x02
Usage page	0x05	Generic Desktop	0x01
Usage	0x09	Pointer	0x01
Collection	0xA1	Physical	0x00
Usage	0x09	x-axis	0x30
Usage	0x09	y-axis	0x31
Usage	0x09	z-axis	0x32
Usage	0x09	Slider	0x36
Usage	0x09	RZ	0x35
Report count	0x95	Number of above usages = 5	0x05
Input	0x81	Data, variable, absolute	0x02
End collection	0xC0	—	—
End collection	0xC0	—	—

Table 17.1. The report descriptor of a USB HID that is interpreted as emanating from a three-axis joystick. The value parameter tells the enumerating host what is being defined and the Attribute parameter tells the host what it is being defined as. For example, a descriptor field value of 0x09 tells the host that a joystick axis is being defined and an attribute of 0x30 says it is an x-axis.

- *The endpoint descriptor* has 6 fields occupying 7 bytes. It specifies how the USB device is to be driven; for example, by the interrupt mechanism with a polling interval of 10 ms that allows the PC to receive position updates 100 times per second.

- *The HID descriptor* has 7 fields occupying 9 bytes. It informs the PC of the number of report descriptors and to which USB standard it is working, e.g., USB 1.

- *The report descriptor* is the one that specifies the format of the data being sent by the USB device; for example, 6 bytes of data to the PC. There is no *correct* way of writing this report, as any one of several variations could produce a similar result. To emulate a common joystick device, for example, the descriptor would specify a single byte of data for each of the six inputs in the following order: throttle setting, x-axis position, y-axis position, z-axis position, slider setting and rotational position about the z-axis. Table 17.1 shows how the fields in a descriptor table would be set in order to emulate such a joystick. When attached to a PC, the game controller's dialog box would announce the presence of the device and show its properties, as illustrated in Figure 17.5. Such a device could then use DirectInput to read the joystick axis positions and use them as coordinates in an absolute 3D frame of reference for the end effector of an experimental haptic device, or an experimental 3D location sensor or indeed any device that can interface to an analog voltage in the range $[0, 5V]$.

A full description of the descriptor tables for HIDs is given in the HID usage tables [19].

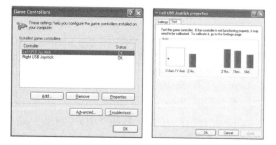

Figure 17.5. A USB device configured with the device descriptor given in Table 17.1 will appear to Windows like a joystick device and can be calibrated as such and used by DirectInput application programs in the same way that a joystick would.

Not only does the embedded microcontroller program have to provide correct enumeration of the USB descriptors for the device it is emulating, it must also configure its own A/D converters and move the data from the input registers to the RAM locations where the USB interface expects them to be. This will differ from PIC to PIC, but typically the PIC program will provide an interrupt handler to service an interrupt from the USB interface. In response to the interrupt, either enumeration or data transfer from the internal buffer to PC will take place. The PIC program will also go through an initialization procedure to load the USB descriptors, configure its A/D converters for input and then enter an infinite loop to alternate between reading from the A/D input and writing to the USB data transfer buffer. Examples of PIC program design may be found in [4] and specifics of the Microchip 167865 from its datasheet [6].

17.5.3 Accessing the USB HID Directly from C/C++

Using the hardware described in the previous section as an illustration, it is perfectly possible to read the emulated joystick status and position via the DirectInput API. Indeed, we could read two, three or more attached joysticks using DirectInput. Nevertheless, it may be that a custom interface which can pick up the nine-byte data stream directly is more useful. This can be done without having to resort to the trouble of writing a full Windows device driver.

Any HID USB interface devices can be accessed directly from an application program written in C/C++ by simply using the Win32 API functions `ReadFile(...)` and `WriteFile(...)`.

Listings 17.7 and 17.8 show how this is done for the pseudo-joystick device discussed in Section 17.5.2. We will need a few components from the Windows device driver development kit (DDK). Specifically, the header files `hidusage.h`, `hidpi.h` and `setupapi.h` need to be included in our application code, since functions they define are called during initialization. They themselves require the header file `pshpack4.h` from the DDK, and the application must be linked with the DDK stub libraries `hid.lib` and `setupapi.lib`. We only need a few functions from these, but it is easer to find the header and stub libraries than to make them up yourself. Microsoft used to make the DDK freely available for download, but that has been withdrawn. Nevertheless it is easy to find the necessary files with a Web search.

Reading from the device is done in a separate thread function (see Listing 17.8), so that it can execute at a high priority, just as the haptic servo

threads do for controlling force rendering. How the device is initialized and
data acquisition threads are started is shown in Listing 17.7. The application
program searches for two devices with appropriate product IDs (these match
the ones set in the device descriptors as part of the PIC firmware). Function FindUSBHID(..) searches all the installed HIDs looking for the desired
vendor and product ID. Function FindHID(int index) (see the full code
listings on CD for the source code of this function) calls into the Windows
DDK to obtain a file handle for the first or second HID device installed on

```
extern float x_rjoy_pos;    // variables to hold the x-y-z joystick position
extern float y_rjoy_pos;
extern float z_rjoy_pos;
extern BOOL  bLoop;         // set to TRUE elsewhere to terminate thread

static void UpdateHID1(PVOID pparams){ // haptic servo - loop
 int i,n;
 BYTE f[9];                 // the USB HID delivers up to a 9 byte frame
 float x1,y1,z1;
 // external flag to terminate thread
 bLoop=TRUE;
 // wait in continuous loop till thread terminates
 while(bLoop){
 // Read the 7 bytes of data from the USB - HID frame. Put this
   for(i=0;i<4;i++){
 // in a loop to discard any frames that arrived since last read.
    ReadFile(h1,f,9,&n,NULL);
 // n=number of bytes read if we got less than seven we've got the lot
    if(n < 7)break;
   }
   x1=((float)(f[2]))/255.0;  // x joystick position is in byte[2] of USB frame
   y1=((float)(f[3]))/255.0;
   z1=((float)(f[4]))/255.0;
   x_rjoy_pos=(2.0*x1-1.0)     // scale joystick values to [-1, +1] range
   y_rjoy_pos=(2.0*y1-1.0);
   z_rjoy_pos=(2.0*z1-1.0);
   Sleep(20);                 // put the thread to sleep for 20 ms
 }
 _endthread();
 return;
}
```

Listing 17.8. The data acquisition function. The USB devices appear as standard
conventional Windows files, thus a ReadFile(h1,...) statement will retrieve
up to nine bytes from the latest USB frame obtained by Windows from the device.
Executing the read command four times purges any buffered frames. Doing this
ensures that we get the latest position data.

the computer, as defined by the argument index. Once the last device has been checked, the function returns the code INVALID_HANDLE_VALUE. To get the vendor and product ID, the DDK function HidD_GetAttributes() is called to fill out the fields in a HIDD_ATTRIBUTES structure. The structure members contain the information we have been looking for.

17.6   Summary

In this chapter, we have looked at a varied assortment of programming methods for acquiring human input from sources other than the traditional keyboard and mouse. A sense of touch, two-handed input and force feedback are all vital elements for VR, whether it be on the desktop or in a large-scale environmental simulation suite. In the next and final chapter, we shall seek to bring together the elements of graphics, multimedia and a mix of input methods in some example projects.

Bibliography

[1] J. Axelson. *USB Complete: Everything You Need to Develop Custom USB Peripherals,* Third Edition. Madison, WI: Lakeview Research, 2005.

[2] HAL. "Haptic Library". http://edm.uhasselt.be/software/hal/.

[3] J. Hyde. *USB Design by Example: A Practical Guide to Building I/O Devices,* Second Edition. Portland, OR: Intel Press, 2001.

[4] J. Iovine. *PIC Microcontroller Project Book: For PIC Basic and PIC Basic Pro Compilers,* Second Edition. Boston, MA: McGraw-Hill, 2004.

[5] I. Mack. *PhD Differentiation Report.* School of Electrical Engineering, Queen's University Belfast, 2005. http://www.ee.qub.ac.uk/graphics/reports/mack_report_2005.pdf.

[6] Microchip Technology Inc. "DS41124C". http://ww1.microchip.com/downloads/en/devicedoc/41124c.pdf, 2000.

[7] Microsoft Corporation. "DirectX SDK". http://msdn.microsoft.com/directx/sdk/, 2006.

[8] W. Oney. *Programming the Microsoft Windows Driver Model,* Second Edition. Redmond, WA: Microsoft Press, 2002.

[9] "OpenTracker". http://studierstube.icg.tu-graz.ac.at/opentracker/, 2006.

[10] OSG Community. "OpenSceneGraph". http://www.openscenegraph.org, 2007.

[11] Proton+. *PIC 2.1.5.3 User Manual.* Cambridge, UK: Crownhill Associates, 2004.

[12] G. Reitmayr and D. Schmalstieg. "An Open Software Architecture for Virtual Reality Interaction". In *Proceedings of the ACM Symposium on Virtual Reality Software and Technology*, pp. 47–54. New York: ACM Press, 2001.

[13] A. Rubini and J. Corbet. *Linux Device Drivers,* Second Edition. Sebastopol, CA: O'Reilly, 2001.

[14] SensAble Technologies, Inc. "GHOST SDK Support". http://www.sensable.com/support-ghost-sdk.htm.

[15] SensAble Technologies, Inc. "Haptic Devices". http://sensable.com/products-haptic-devices.htm.

[16] SensAble Technologies, Inc. "OpenHaptics Toolkit". http://www.sensable.com/support-openhaptics.htm.

[17] SenseGraphics AB. "H3D API". http://www.h3d.org/, 2006.

[18] R. Taylor et al. "VRPN: A Device-Independent, Network-Transparent VR Peripheral System". In *Proceedings of the ACM Symposium on Virtual Reality Software and Technology*, pp. 55–61. New York: ACM Press, 2001.

[19] USB Implementers Forum. "USB HID Usage Tables Version 1.11". http://www.usb.org/developers/devclass_docs/Hut1_11.pdf, 2001.

[20] "Virtual Reality Peripheral Network". http://www.cs.unc.edu/Research/vrpn/, 2007.

18 Building on the Basics, Some Projects in VR

In this, the final chapter, we thought it would be useful to show you how the code and examples in Chapters 13 through 17 can be rearranged and combined to build some additional utility programs. We will also discuss how all the elements of desktop VR can be made to work together and the issues that this raises in terms of program architecture and performance. We won't be printing code listings in this chapter for two main reasons. First, the book would be too long, and second, we would only be repeating (more or less) a lot of the code you have already seen.

All the code, project files, help and *readme* comments can be found on the accompanying CD. We have written our programs using Microsoft's Visual Studio 2003.NET and Visual C++ Version 6. Although Visual Studio 2005 is the latest version of this tool, we chose to use 2003 .NET and VC 6 for our implementation because VS 2005 can easily read its predecessor version's project files but the converse is not true.

So, enough said about where to find the codes for our projects; let's have a look at them in detail.

18.1 Video Editing

We start with video editing. If you are compiling some video content for your VR environment, you will need to be able to mix, match and collate some disparate sources.

On the face of it, a video-editing application should be a very substantial project. There are many commercial applications that have had years of work devoted to their development. If all you want is a good video editing tool then the code we include for our little project is not for you. However, if you have followed our examples in Chapter 15 on using DirectShow and you want to build some editing capability into an application in your VR program portfolio, you could not do better than use the *DirectShow Editing Services* (DES) that are part of DirectShow. Microsoft's Movie Maker program uses this technology.

18.1.1 DirectShow Editing Services

The concept underlying DES is quite straightforward. In only a few lines of code, it is possible to plug together functions that join, cut, mix, dub or resize a collection of AVI files. You will find the programs for several of these ideas on the CD. Hopefully, the comments in the code should make the logic easy to follow, once you appreciate the philosophy and jargon behind DES. We won't cover every aspect of DES here; for that, consult Pesce [8] and the DirectShow SDK [7]. Instead, we shall concentrate on a couple of key features.

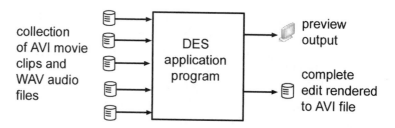

Figure 18.1. A DES application program selects a number of input movies and sound files. It arranges all of them, or sections (clips) from them, in any order and mixes them together over time. The application can present a preview of the arrangement and compose an output movie file containing the arrangement and with a specified resolution and format.

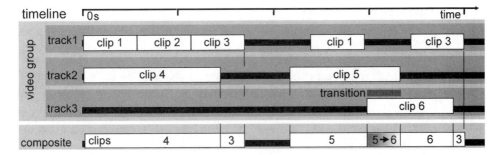

Figure 18.2. DES video editing. Several video clips, each in an AVI file are combined into a single video stream. The clips are placed on a track and starting times defined. The final video is produced by superimposing track 3 on track 2 on track 1. In any gaps between the clips on track 3 the composite of tracks 1 and 2 will appear. A transition interval can be used to mix or wipe between tracks.

Figures 18.1 and 18.2 illustrate everything we need to know about DES. A few COM interfaces provide all the editing functionality one could wish for. DES is based on the concept of the *timeline*. A DES application program lays out the video clips along this timeline by providing a time for the clip to start and a time for it to finish. Clips are concatenated into tracks, and if there are any time gaps between clips, a blank section is included in the track. Clips cannot overlap in time within a track. The tracks are mixed together to form the video *composition*[1] using a layer model and another DES element called a *transition*. Video compositions are joined with a parallel structure for the audio tracks to form the final movie.

The layer-mixing model works similar to the way images are layered in a photo-editing program. A first track is laid down against the timeline (gaps between clips appear as blanks), and this becomes the current composition. As the second track is processed, it overwrites the current composition in those intervals where they overlap in time unless a *transition* is used to set a time interval during which the incoming track gradually replaces the current composition. When this process is complete, we have a new and updated current composition. This process is repeated until all the tracks have been processed. Figure 18.2 illustrates the mixing process and the DES structure.

---

[1]DES allows a hierarchical combination of compositions, but this facility is something we don't need in a basic video editor.

18.1.2   Recompression

Whilst the DES can be used to change the compression codec used in an AVI file, it is a bit of an overkill. It is easier to build a simple DirectShow FilterGraph to do the same thing. All one needs are a source filter, the compression filter of choice and a file-writing filter. The smart rendering capability of the `GraphBuilder` object will put in place any necessary decrypting filters. To select the compression filter, our application creates a selection dialog box populated with the *friendly names* of all the installed compression filters, which we can determine through enumeration. Then a call to the media control interface's `Run()` method will run the graph. We have seen all the elements for this project before, with the exception of telling the FilterGraph that it must not drop frames if the compression cannot be accomplished in real time. This is done by disengaging the filter synchronization from the FilterGraph's internal clock with a call to the `SetSyncSource(NULL)` method on the `IMediaFilter` interface.

18.2   Stereo Movie and Picture Output for a Head-Mounted Display

For the most part, we rely on the fact that nearly all graphics adapters have two video outputs and that the Windows desktop can be configured to span across the two outputs. Stereo-enabled HMDs, like that illustrated in Figure 10.1, have separate VGA inputs for left and right eyes, and these can be connected to the pair of outputs on the video card. To generate the stereoscopic effect, no special hardware is required; we simply ensure that images and movies are presented in the-left-beside right format illustrated in Figure 16.4.

Thus, when an image or movie is shown in a full-screen window, its left side will appear in one of the video outputs and its right side in the other. As a result, any movie or image display program that scales the output to fill the desktop will do the job.

18.3   Video Processing

This project provides a framework for an application to apply all kinds of 3D processing effects to a video file. It is not as comprehensive as the video editor in Section 18.1, but the essence of it could be formed into a DirectShow

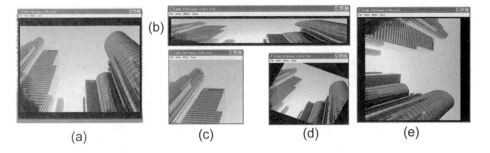

Figure 18.3. The video editing project allows the AVI clip in (a) to be resized, as in (b) with any aspect ratio, or a section of it cut out and rendered (c), or some image process applied to it, such as dynamic or fixed rotations (d) and (e).

filter for inclusion in a DES application. The idea behind the application is to use DirectShow to decode an AVI video file and render it into an OpenGL texture. We showed how to do this in Section 15.6. This mesh can be animated, rotated, rolled up—almost any morphing effect could be created as the movie is playing in real time. By capturing the movie as it is rendered in the onscreen window, a new movie is generated and stored in an AVI file. To capture the movie from the renderer window; the program uses a FilterGraph with the same structure we used in Chapter 15 to generate an AVI movie from a sequence of bitmaps.

Taking advantage of OpenGL's texturing means that the processed movie will be antialiased and small sections of the movie's frame could also be magnified. In addition, if we call on the GPU fragment processor (through the use of OpenGL 2 functions), we will able to apply a wide range of computationally intensive imaging processes in real time. Figure 18.3 illustrates a few features of this project.

The comments in the project code will guide you through the application.

18.4   Chroma-Keying

Chroma-key is the well-known effect used in TV news programs and weather forecasts where the presenter appears in front of a background video or computer-generated display. A video signal from a camera is used as one source. The speaker stands/sits in front of a blue or green screen, and the electronic effects box mixes any parts of the picture where the blue/green background is visible with a second video.

We can put together a program to produce the chroma-key effect using a webcam and an AVI movie. DirectShow provides all the software components needed, and we've covered most of the other detail in Chapter 15 already. It is simply a matter of building a suitable FilterGraph. If you have two webcams, you can even do it in stereo. We will base most of our application on the code from Section 15.4 concerning live video capture into an OpenGL texture. The program has three threads of execution:

1. A thread to run the parent window and the control panel dialog box.

2. A thread to run the camera video capture FilterGraph (the output is written to an intermediate memory buffer) and OpenGL (OGL) texture renderer. The OGL renderer paints the texture from the intermediate buffer onto a mesh rectangle that has been scaled to exactly fill a child window of the application. Before rendering the texture, the chroma-key effect is generated by mixing the video frame data in the memory buffer with the movie frame data obtained in the third thread.

3. A thread to run a FilterGraph which plays an AVI movie file using our custom renderer from Chapter 15. The custom renderer filter sends the movie frame images to a second memory buffer.

The structure of the application's FilterGraph is illustrated in Figure 18.4, and the chroma-key effect is demonstrated in Figure 18.5. The important code that carries out the mixing is in the CopyMediaSample(..) function. It operates on the red, green and blue color components and uses some user-defined thresholds to define at what levels the chroma-key kicks in.

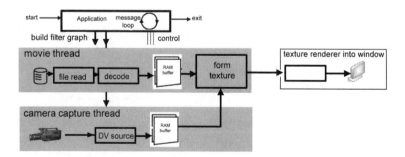

Figure 18.4. The FilterGraph for the chroma-key project. The video RAM buffer content is combined with the AVI movie buffer content by chroma-keying on a blue or green key color in the video buffer.

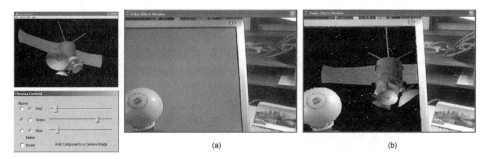

Figure 18.5. Chroma-key in action. At the top left is a frame from a movie. The dialog below sets the color thresholds. At (a), the camera image shows a desktop (a yellow ball sits in front of a screen showing a green color). When the chroma-key is turned on, at (b), parts of the green screen that are visible are replaced with the frames from the AVI movie.

Generating a chroma-key effect using a blue or green background is not the only way one might achieve a similar result. An alternative (and one you might care to try for yourselves) is to *key on differences*. The idea behind keying on differences is:

> Set up the webcam, take a photo of the background (no need for a constant color) and store this background. Now let someone or something move into the scene. By comparing the pixel colors from the captured photo with the color from the new scene, we can determine whether a value is part of the background or not. When operating in chroma-key mode, the program replaces anything it thinks belonged to part of the original background with the material we wish to key in as the new background.

Another possible alternative is to key on a color specified by its hue, saturation and value (HSV). The HSV model provides a more natural way to select color, and it may be easier to choose a key color using this model than with the RGB components. Or, the application could offer a *color picker* facility where the user points to a pixel in the scene and that color (or a close variant of it) is chosen as the key color.

In practice, of course, few things work exactly as one would expect. Unless the lighting conditions are good (almost up to studio conditions), the cameras have adequate color performance and the background color is highly monochromatic, it can be very difficult to get the key color to match the background as we would wish.

18.5   A Multi-Projector VR Theater System

In Chapter 4, we discussed building a basic immersive VR environment by projecting overlapping images, movies and real-time interactive 3D content onto curved surfaces. A design based on two projectors can be driven by one PC system, because most graphics adapters have two video outputs on which one can display different parts of the desktop.[2]

The project described in this section offers a simple solution for the case where we need to drive a four-projector set-up such as that illustrated in Figure 4.13. The controlling software provides a collection of programs to correct for the distortions that arise when a conventional projector system throws its output onto a cylindrical screen, as in Figure 4.11. The project will also offer some code and suggestions for how to link and control two PCs using an Ethernet network so that the four outputs can be generated by using two stereo-ready graphics adapters, without any requirement for special-purpose hardware.

The basic idea underlying the project is to divide the image or movie into four sections. The two central sections are displayed by a master computer,

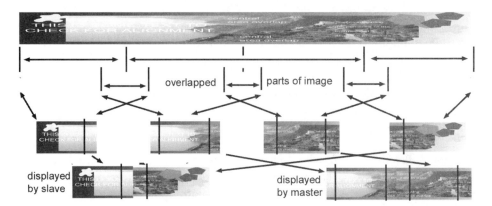

Figure 18.6. Dividing up an image for display in four pieces. Overlaps are used so that the projected pieces can be blended. The master machine drives the two projectors displaying the central area. The slave machine drives the two projectors showing the extreme left and extreme right image segments.

---

[2]We will do this by setting up one wide desktop, rendering the output to go to one of the projectors from the left side of the desktop and the output for the other projector from the right half of the desktop.

and the two peripheral sections are displayed by a slave computer under the remote control of the master. This arrangement is described in Figure 18.6. As we have noted before, it is necessary to arrange some overlap between the images. This complication necessitates sending more than one half of the image to each of the two pairs of projectors.

We don't have space here to go into a line-by-line discussion of how the code works. As you might guess, however, we have already encountered most of the coding concepts we need to use[3] in previous chapters. It is now only a matter of bringing it together and adapting the code we need. By knowing the overall design of the project software, you should be able to follow the fine detail from the documentation on the CD and comments in the code.

18.5.1   Hardware and Software Design

Since we are going to be driving four projectors, we will need two PCs, each with a dual-output graphics adapter. The PCs will be connected together through their Ethernet [2] interfaces. You can join the PCs in back-to-back fashion or use a switch/hub or any other network route. In our program, we assume that the PCs are part of a private network and have designated private IP addresses [5] of 192.168.1.1 and 192.168.1.2. The two machines will operate in a client-server model with one of them designated *the server* and the other *the client*. On the server machine, a program termed the *master* will execute continuously, and on the client machine another program termed the *slave* will do the work. It is the job of the master program to load the images (or movies), display their central portion and serve the peripheral part out to the slave. It is the job of the slave to acquire the peripheral part of the image (or movie) from the master and display it[4].

Whilst the same principles apply to presenting images, movies or interactive 3D content across the multi-projector panorama, we will first discuss the display of wide-screen stereoscopic images. The software design is given in block outline in Figure 18.7. In Section 18.5.3, we will comment on how the same principles may be used to present panoramic movies and 3D graphics.

---

[3]Actually, we have not discussed Windows sockets and TCP/IP programming before but will indicate the basic principles in this section.

[4]We don't want to complicate the issue here, but on the CD you will find a couple of other utilities that let us control both the master and the slave program remotely over the Internet. You will also see that the architecture of the master program is such that it can be configured to farm out the display to two slave programs on different client machines. It could even be easily adapted to serve out the display among four, eight or more machines.

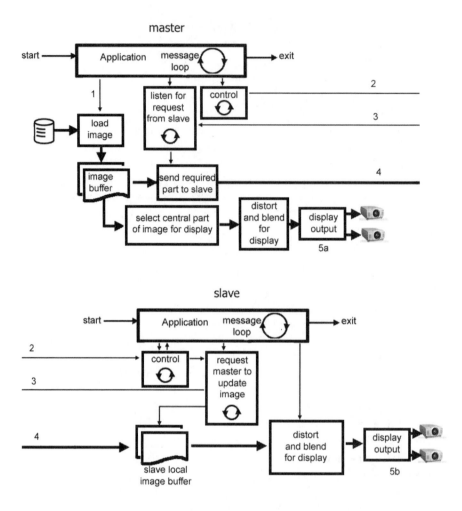

Figure 18.7. A block outline of the software design for the master/slave quad-projector display. The steps labeled 1 to 5 illustrate the sequence of events which occur when an image (or sequences of images are displayed.)

The logical execution of the system proceeds as follows:

- Master and slave applications both enter window message-processing loops and await commands. The master application starts two additional communications threads (using Windows sockets) to handle communications. One thread will handle requests from the slave to send the part of the image that the slave is to display. The second thread will handle control messages to be sent to the client.

The slave also starts two communications threads. One will retrieve the image from the master; the other waits for commands from the master and posts messages into the message loop to execute those commands.

- When the master is instructed to display an image, it reads and decodes the pixel data into an RAM buffer (stereoscopic images are stored in left-over-right format). This is labeled 1 in Figure 18.7.

- The master sends a control signal to the slave to tell it that a new image is ready to be grabbed (Step 2).

- The slave tells the master to dispatch the part of the image it needs to form the peripheral edges of the display (Step 3). The slaves local buffer is filled (Step 4).

- The master and slave apply the distortion and blending to their respective parts of the image and send the output to their pair of display projectors (Steps 5a and 5b).

18.5.2   Distortion and Blending

Distortion and blending are configured manually using a test image and grid. The parts of the image, as shown in Figure 18.6, are rendered as a texture on a quadrilateral mesh. By interactively moving the corners of the mesh and controlling the bending along the edges, the shape of the projection is altered to facilitate some overlap between adjacent projections and correct for a nonflat screen. The application program and these features are partially shown in action in Figure 4.11. The system only has to be configured once. Parameters to control blending and overlap between adjacent projections are recorded in a file. This data is used to automatically configure the projections each time the programs start.

Displaying images and movies as objects painted with an OpenGL texture works very well. It may be used for a stereoscopic display and can easily cope with the refresh rates required for video output. Rendering the output of a 3D graphics program requires an additional step. Since we must render an OpenGL-textured mesh to form the output, we cannot render a 3D scene into the same output buffers. Thus, we do not render the scene into either the front or back frame buffers; instead, we render it into an auxiliary buffer.

(Most graphics adapters that support OpenGL have at least two auxiliary buffers.) The content of the auxiliary buffer is then used as the source of a texture that will be painted onto the mesh object. This may involve an extra step, but it does not usually slow down the application, because none of these steps involve the host processor. Everything happens within the graphics adapter.

18.5.3 Synchronization

Synchronization, command passing and data distribution between the master and slave programs are accomplished using TCP/IP[5] over Ethernet. Our application uses the Windows sockets[6] programming interface, and because it is multi-threaded, we don't have to worry too much about the blocking action that used to beset Windows sockets programming. The master and slave programs open listening sockets on high-numbered ports and wait for data or commands to be sent from the other side.

We use our own simple protocol for sending commands by encoding the action into an eight-byte chunk; for example, to initiate a data transfer. The slave's control thread waits in an infinite loop using the `recv(...)` function, which only returns once it gets eight bytes of data from the server which were sent using the `send(..)` function. The slave acts upon the command by using `PostMessage(...)` to place a standard Windows `WM_COMMAND` message into the message loop. When the slave's control thread has posted the message, it goes back to sleep, again waiting on the `recv(...)` function for the next command. All the other master/slave conversations follow this model. When requesting that an image be sent by the master to the slave, the eight-byte data chunk tells the slave how much data to expect and the two programs cooperate to send the whole image. The TCP protocol guarantees that no data is lost in the transmission.

At first glance, the code for all these client-server conversations look quite complex, but that is only because of the care that needs to be taken to check for errors and unexpected events such as the connections failing during transmission. In fact, we really only need to use four functions to send a message from one program to another: `socket(..)` creates the socket, `connect()`

---

[5]Transmission Control Protocol/Internet Protocol is a standard used for the communication between computers, in particular for transmitting data over networks for all Internet-connected machines.

[6]Windows sockets (Winsock) is a specification that defines how Windows network software should access network services, particularly TCP/IP.

connects through the socket with a listening program, `send(...)` sends the data and `closesocket(..)` terminates the process.

The receiving code is a little more complex because it usually runs in a separate thread, but again it basically consists of creating a socket, binding the listener to that socket with `bind(..)`, accepting connections on that socket with `accept(..)` and receiving the data with `recv(..)`.

Along with the main programs, you will find an example of a small pair of message sending and receiving programs. This illustrates in a minimal way how we use the Windows sockets for communication over TCP/IP.

So far, we have concentrated on displaying image sequences using two PCs and four video outputs. To extend this idea to similar applications that present panoramic movies and 3D content poses some difficulties:

- Movies must present their images at rates of 30 fps. Even the smallest delay between the presentation of the central and peripheral parts of the frame would be quite intolerable. Wide-screen images require a high pixel count, too; for example, a stereoscopic movie is likely to require a frame size of 3200×1200 pixels, at the minimum. The movie player project we include with the book restricts the display to a two-output stereoscopic wide-screen resolution of 1600×1200, so it requires only a single host. To go all the way to a four output movie player, it would be necessary to render two movies, one for the center and one for the periphery. Each would need to be stored on its own PC, and the master and slave control signals designed to keep the players in synchronism.

- In a four-output real-time 3D design, the communications channel would need to pass the numerical descriptions of the 3D scene and image maps from master to slave. Under most conditions, this data distribution would only need to be done in an initialization stage. This transfer might involve a considerable quantity of data, but during normal usage (with interaction or scripted action, for example), a much smaller amount of data would need to be transferred—carrying such details as viewpoint location, orientation, lighting conditions and object locations. Our example project code restricts itself to using a single host and projecting onto a curved screen with two projectors.

In this section, we have highlighted some key features of a suite of core programs to deliver a multi-projector cave-type immersive virtual environment. Larger commercial systems that can cost in excess of $100k will typically implement these functions in custom hardware. However, the common PC's processing and GPU capability now allow anyone with the right software to experiment with VR caves at a fraction of that cost. More details of how this project works are included in the commentary on the CD.

18.6   Using Image-Processing and Computer-Vision Libraries

In Section 8.4, the OpenCV and Vision SDK software libraries were introduced. These two libraries can be easily integrated into programs that need special purpose functionality such as overlays and removing perspective distortion. Since the subjects of image processing and computer vision could each fill a book, we are only going to provide a framework for using OpenCV, typically for processing moving images.

Our examples come in the form of a special inline DirectShow filter and two short applications programs that use the filter. The programs comprise:

- *A filter.* As part of the OpenCV distribution [6], a DirectShow filter called ProxyTrans provides a method to apply any of the OpenCV image processing functions. Unfortunately, it didn't work for us, and so we provide our own *simpler* code based on a minor modification of the EZRGB24 sample that ships as part of DirectX. By adding one additional interface method to the EZRGB24 project, our filter allows any program which inserts it into a FilterGraph datastream to define a callback function. The callback function is passed as a pointer to an OpenCV image structure representing the image samples. Once we have access to every image sample passing through the FilterGraph, it is possible to apply any OpenCV process to it. Please consult the *readme* file for details of how the component is registered with the operating system. A registered DirectShow filter may be used by any program (and in the `graphedit` utility).

- *AVI processor.* This example shows how to use our filter in a scenario where we wish to process a movie stored in an AVI file. It previews the action of the image process and then writes it into another AVI file.

With the exception of illustrating how to instantiate our filter component and use the callback function, all the elements of the program have been discussed in earlier chapters. The same applies to our next program.

- *Camera-processor.* This example illustrates how to apply OpenCV image processing functions to a real-time video source.

18.7   Augmented Reality (AR)

In Chapter 8, we discussed virtual-reality videoconferencing, where the idea is to map video images onto virtual screens. This is one form of an exciting series of recent developments going broadly under the heading of AR. We can put together a small project by using the ARToolKit [1] to illustrate another form of AR: that is, placing virtual objects into real-world scenes.

The ARToolKit is a group of functions that allows us to easily give the illusion that virtual elements actually exist in the real environment around us. This can be done stereoscopically, but with a simple webcam and a cheap set of video glasses (such as the Olympus Eye-Trek that are normally used for watching TV or DVDs), we can achieve the same effect (see Figure 18.8).

In this project, we will adapt one of the ARToolKit's simple examples so that a 3D mesh model may be placed *virtually* into the real world we see before us, as for example in Figure 18.9.

Figure 18.8. The combination of a basic webcam stuck to a simple pair of video classes forms a low cost form of head mounted display suitable for the augmented reality project.

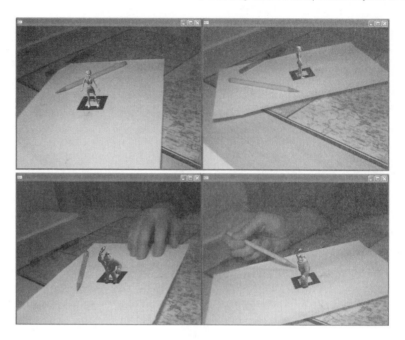

Figure 18.9. Placing miniature mesh figures onto a *real* desktop illustrates the capability of an AR head-mounted display.

18.7.1   The ARToolKit

At the core of the ARToolKit is the action of placing the 3D object into the field of view of a camera so that it looks as if the object were attached to a marker in the real world. This is best illustrated with an example. First we see the scene with the square marker indicated on the white sheet of paper in Figure 18.10(a).

The software in the ARToolKit produces a transformation matrix that, when applied to an OpenGL scene, makes it look as if the object were centered on the marker and lying in the plane of the marker. (The marker is actually a black outline with the word "Hiro" inside; this detail is not clear in Figures 18.9 and 18.10). The nonsymmetric text in the marker allows a definite frame of reference to be determined. When we render an OpenGL mesh, it will appear to sit on top of the marker (see Figure 18.10(b)). If we move the piece of paper with the marker around (and/or move the camera) then the model will appear to follow the marker, as in Figure 18.10(c).

Figure 18.10. In the ARToolKit, a marker in the camera's field of view (a) is used to define a plane in the real world on which we can place a synthetic object so that it looks like it is resting on the marker (b). Moving the marker causes the object to follow it and so gives the illusion that the synthetic object is actually resting on the marker.

In the examples shown, the 3D mesh model is centered 50 mm above the center of the marker so the marker itself is partially visible and the object appears to hover over it. As the marker is moved, the object moves with it. If the marker is moved away from the camera, it gets smaller and so does the object. The ARToolKit software has a lot of clever work to do in order to make this simple looking thing possible. Fortunately, in practice, it is quite easy to use the ARToolKit provided one understands the coordinate systems.

In Figure 18.11, all the coordinate systems in use are shown. The AR-ToolKit uses two coordinate frames of reference, one based at the camera center (ARC) and one based at the marker center (ARM). OpenGL is used for rendering, and so all 3D mesh drawings take place in their own frame of reference (OGL). The marker's coordinates coincide exactly with the OpenGL coordinate system, so to draw something normal to the plane of the marker, one increases the z-coordinate in OpenGL. The modeling software (OpenFX or 3DStudio) use a mathematical coordinate system where z is considered vertical as opposed to OGL's, in which the y-coordinate is vertical. So, when rendering a mesh model, one must remember to swap the z-value to y and the y-value to $-z$. This important property, and a reminder of the coordinate systems we need to use, is depicted in Figure 18.11.

Because the AR marker and OpenGL coordinate systems are the same, all one has to do to render a 3D mesh model so that it looks as if it is stuck to the marker is draw its polygons using OpenGL coordinates. So, for example, to render a line running vertically for 10 units from the center of the maker relative to the plane of the maker, one uses the OpenGL library function calls:

```
glBegin(GL_LINES);
glVertex3f(0.0,0.0, 0.0);
glVertex3f(0.0,0.0,10.0);
glEnd();
```

Note: to obtain an orientation, such as that depicted for the plane in Figure 18.10, when we are loading the mesh from an OFX or 3DS file, we must perform an additional rotation of 90 degrees about the *x*-axis. This is because if we just carried out the swap of coordinates as indicated above (*z* to *y* etc.) the model (of the plane etc.) would appear to point down—towards the marker.

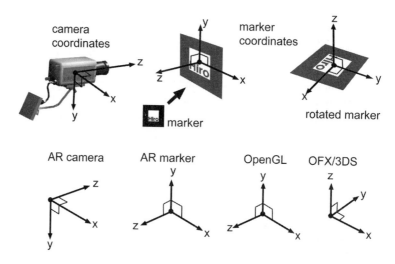

Figure 18.11. Coordinate frames of reference used in the ARToolKit and the 3D modeling software packages.

18.7.2   How the ARToolKit Works

As we discussed in Chapter 8, the transformation (the homography) that maps a plane in which the marker lies (and with it the marker's coordinate frame of reference) must be obtained by identifying at least the four points at the corners of the marker and in their correct order. Once these points have been identified, the transformation from camera frame of reference to marker frame of reference may be determined. Since the camera's frame of reference is considered the global frame of reference, any 3D model can be made to behave as if it were lying in the marker's frame of reference by applying the transformation to it.

In the examples covered in Chapter 8, identification of the points around the square marker was done manually. In the ARToolKit, this process is automated by carrying out an automatic pattern recognition to identify the four corners of the marker. However, before this can be done, the camera must be calibrated. Calibration of the camera is necessary to determine any distortion in the lens as well as the field of view and aspect ratio. The camera parameters are stored in a file "data/camera_para.dat" in the example code. Correcting for distortion is necessary because it helps in the recognition of the marker. By knowing the field of view, aspect ratio and the marker's real-world size, one can also get an approximate value for the distance it is away from camera.

The process of recognizing the marker and hence determining its corner positions in the 2D camera's image requires some knowledge of the pattern in the marker. This information is stored in a file; for the basic ARToolKit pattern, this file is "data/patt.hiro".[7] The process of recognition is only explained briefly in the ARToolKit documentation, but it suggests that there are several stages. These are:

1. Turn the image into a binary bitmap, i.e., every pixel is represented by a binary state—black or white. A threshold intensity level is set. Above this threshold the pixel value is 1; below it the pixel value is 0.

2. Using the binary image, the ARToolKit attempts to identify a set of lines in the image. (Some of these will hopefully be identified as the edges round the marker.)

3. Try to find the places where these lines meet to form the corners of the possible image of the marker.

4. Read the pattern from the marker's pattern file and, using it as a template, attempt to match it to the region bounded by the possible corners and edges of the putative marker.

5. When a match is detected with some degree of confidence, the homography that maps the marker's and camera's frames of reference may be determined by using an equation analogous to Equation (8.4).

---

[7]It is possible to build this pattern information for other "custom" markers; you should consult the ARToolKit documentation for details on how to do this. It is also possible to instruct the ARToolKit to look for several different markers in the same image; again, consult the documentation for information on how to do this.

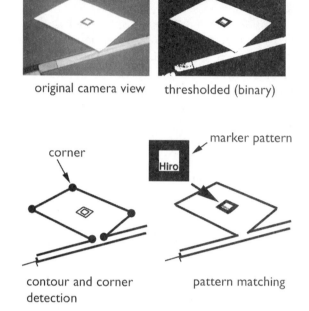

Figure 18.12. The steps performed by the ARToolKit as it tracks a marker in the camera's field of view and determines the mapping homology between the plane in which the marker lies and the camera's projection plane.

This process is illustrated in Figure 18.12, where the matched rectangle contains the four corner locations that can be used to determined the transformation from the camera to marker frames of reference.

Full implementation details can be found from following the comments in the code and the project's commentary on the CD.

18.8   Virtual Sculpting in 3D

To finish off this chapter, we thought we'd put together a small project for a desktop VR system that brings together some of the concepts we have been developing in the last few chapters. Actually, it is quite a large project It is a desktop program that brings together input from two joysticks that will control a couple of virtual tools that we can use to push and pull a 3D polygonal mesh in order to sculpt it into different shapes. The display will be stereoscopically enabled and force feedback is simulated via the joysticks. Mathematical models will be applied to the deformation of the mesh (as we push and pull it) to make it look like it is made out of elastic material.

As has been the case with all the other projects proposed in this chapter, we are not going to examine the code line by line; it's far too long for that. But, after reading this section and the previous chapters, you should be able to follow the logic of the flow of execution in the program using the commentary and comments within the code and *most importantly* make changes, for example to experiment with the mathematical model of deformation.

Figure 18.13 illustrates how our project looks and behaves. In (a). a few shapes made up from polygon meshes will be formed into a new shape by pushing and pulling at them with the tools. In (b), after a bit of prodding, the shapes have been roughly deformed. In (c), part of the mesh is visible behind some of the shapes. Bounding boxes surround the shapes in the scene. By testing intersections between bounding boxes, the speed of the collision detection functions is increased considerably. *These simple shapes are loaded from 3D mesh files, so it is possible to use almost any shape of mesh model.*

Now, we will briefly highlight some key points in the design of the program.

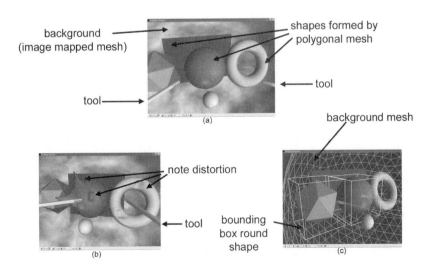

Figure 18.13. The sculpting program in action.

18.8.1   The Idea

One of the key points about an application like this is that it must operate in real time, so there are always going to be compromises that will have to be made in designing it. The program has to run continuously, and it should repeat the following tasks as quickly as possible:

1. Read the position and orientation of the joystick devices that are controlling the sculpting implements/tools.

2. Check for collision between the tools and the elastic models.

3. Since the tools are regarded as solid and the models as deformable, if they collide (and where they overlap), move the vertices in the model out of the way. Use a mathematical algorithm to determine how other parts of the mesh model are deformed so as to simulate the elastic behavior.

4. Send a signal back to the joysticks that represent the tools. This signal will generate a force for the user to feel. On simple joysticks, all that we can really expect to do is simulate a bit of a kick.

5. Render a 3D view of the scene, i.e., the elastic models and sculpting tools.

The project's program code must do its best to follow the logic of these steps. There are a number of functions to render the scene. These are called in response to a `WM_PAINT` message. They act almost independently from the rest of the code. When the program needs to indicate a change in the scene, typically as a result of moving the tools or the math model of elasticity moving some of the vertices in the mesh, a `WM_PAINT` message is dispatched.

The main action of the program is driven by a `WM_TIMER` message that calls function `RefreshScene()` in which all these steps take place. Timing is critical, and so the rate set for this timer is a very important design parameter. If it is too slow, we will observe an unpleasant lag between our action in the real world and the implements in the virtual sculptor. On the other hand, if it is too fast, there will be insufficient time for the collision detection routines to do their work or the mathematical elasticity simulation to come up with an answer.

Choosing the timer interval is one of a number of critically important decisions that have to be made in designing a program like this. The others relate to which values to use for the parameters in the mathematical model of elastic behavior. There is a good theoretical basis for the choice of many parameter values in the mathematical model, but often in practice we have to compromise because of the speed and power of the computer's processor. In our code, the timer setting and mathematical model parameters we quote have evolved after experimentation, so you should feel free to see what effect changing them has on your own system. In order to do this, we will give a brief review of the theory underlying the core algorithm simulating elastic deformation.

18.8.2   The Mathematical Model

When devising a method to simulate some phenomenon using a computer, the best way to start is to see if anyone has proposed a mathematical model for it (or for a very similar problem) before. In this case, there are many interesting models around. Our problem is analogous to the fascinating subject of—the computer animation of cloth and similar materials [4]. Some mathematical models for the dynamic behavior of cloth are very accurate, and they give realistic results. Unfortunately, these rarely work in real time. Other models may be less precise but still give acceptable-looking results, and they do work well in real time. The model that we choose to use is one of a spring-mass particle system, as illustrated in Figure 18.14.

In Figure 18.14(a), a 3D shape is represented by a mesh network of interconnecting springs along each edge in the shape. The vertices are represented by particles at points such as **p**. In Figure 18.14(b), pulling **p** away from its original position establishes restoring forces in the springs that would tend to re-

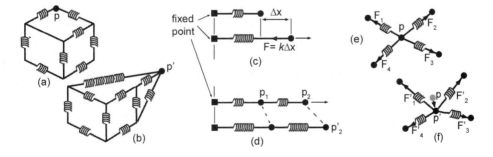

Figure 18.14. A spring-mass particle system model for elastic deformation.

turn the mesh to its original shape when we release the vertex at point $\mathbf{p}'$. In a simple one-dimensional spring (Figure 18.14(c)), the restoring force is proportional to the extension in the length of the spring. In Figure 18.14(d), when two or more springs are connected, all the nonfixed points such as $\mathbf{p}_1$ may move. All new positions must be obtained simultaneously; in this 1D case an analytic solution of the math model may be obtained. In Figure 18.14(e), if the springs forms a complex 3D network then the effect of a distortion must be obtained using a numerical solution technique. In Figure 18.14(f), when the point $\mathbf{p}$ is moved to $\mathbf{p}'$, the forces in all attached springs will change and the shape of the network will adjust to a new state of equilibrium.

This model fits very well with our polygonal representation of shapes. The edges in our *mesh* model correspond to the little springs connected into a 3D network by joining the vertices together. We think of the original configuration of the model as being its rest state, and as we pull one or more of the vertices, the whole network of springs adjusts to take on a new shape, just as they would if we really built the model out of real springs and pulled or pushed on one of them.

The behavior of this network of springs is modeled mathematically by using the well-known laws of Hooke and Newton. Hooke tells us that the restoring force $\mathbf{F}$ in a spring is proportional to the extension in the spring $\Delta\mathbf{p}$. That is, $\mathbf{F} = k\Delta\mathbf{p}$, where k is the constant of proportionality. Newton tells us that the acceleration a of a particle (in our case a vertex in our mesh) with a mass m is proportional to the force applied to it. That is, $\mathbf{F} = m\mathbf{a}$. We also know that acceleration is the second differential of the distance moved, so this equation can be rewritten as $\mathbf{F} = m\,\ddot{\mathbf{p}}$. This differential equation would be relatively easy to solve for a single vertex and one spring, but in a mesh with hundreds of vertices interconnected by springs in three dimensions, we have to use a computer to solve it and also use a numerical approximation to the modeling equations.

Several numerical techniques can be used to solve the equations that simulate how the vertices respond to an applied disturbance (House and Breen [4] discuss this in depth). One technique that is fast and simple and has been used by physicists for solving dynamic particle problems has also been shown by computer game developers to be particularly good for our type of problem. It is called *Verlet integration* [9], and whilst it allows us to avoid the complexity of carrying out an *implicit* solution for the equations, we can also still avoid some of the instabilities that occur in the simpler *explicit* solution techniques. If you want to know all about the numerical solution of such problems and all the challenges this poses, a good place to start is with [3].

Combining the laws of Hooke and Newton, we obtain a second-order differential equation in 3D:

$$\mathbf{a} = \ddot{\mathbf{p}}_i = \frac{\mathbf{F}}{m} = \frac{k}{m} \sum_{j=1}^{n} \Delta(\mathbf{p}_j - \mathbf{p}_i), \tag{18.1}$$

which we must solve at every vertex in the mesh. The term $\sum_{j=1}^{n} \Delta(\mathbf{p}_j - \mathbf{p}_i)$ sums the change in the length of each of the n edges that are connected to vertex i. We must discretize Equation (18.1) by replacing the second order differential $\ddot{\mathbf{p}}_i$ with an approximation. The key feature of Verlet integration recognizes that the first order differential of p, that is $\dot{\mathbf{p}}_i$, is a velocity, and that if we are given initial values for position and velocity, we can write an iterative pair of equations which may be used to determine how a vertex moves in a short time interval Δt. If $\mathbf{p}$ is the current position of a vertex and $\mathbf{v}$ is its current velocity then the updated position $\mathbf{p}'$ and updated velocity $\mathbf{v}'$ are given by

$$\begin{aligned} \mathbf{v}' &= \mathbf{v} + \mathbf{a}\Delta t, \\ \mathbf{p}' &= \mathbf{p} + \mathbf{v}'\Delta t. \end{aligned}$$

Normally, the velocity is not calculated explicitly and is approximated by a finite difference expression for the differential. If the previous position is given by $\bar{\mathbf{p}}$ then $\mathbf{v} = \frac{\mathbf{p} - \bar{\mathbf{p}}}{\Delta t}$. Thus we can rewrite our expression to compute $\mathbf{p}'$ as

$$\mathbf{p}' = 2\mathbf{p} - \bar{\mathbf{p}} + \mathbf{a}\Delta t^2.$$

And of course, we need to update our previous position to our current position; that is, $\bar{\mathbf{p}} = \mathbf{p}$. By substituting for $\mathbf{a}$ as given in Equation (18.1), we have a remarkably simple iterative algorithm that we can use to solve for the transient behavior of a mesh. With the right choice of parameters, it offers a visually reasonable approximation to the behavior of a squashy/elastic/deformable object.

We can improve slightly on the spring-mass model of Equation (18.1) by adding in the effect of a *damper* in parallel with the spring. This helps to make the object appear less elastic. It is easy to model this effect mathematically

```
// something has changed - may be set in many places
static bChange=FALSE;

// key function - called in response to Windows timer
static void RefreshScene(void){
// indicates if we are already busy !!!!!!
static BOOL bBusy=FALSE;
// indicates if collision detected between L & R tools
static BOOL bToolsHit=FALSE;
// acquire the latest data from the joysticks
UpdateInputState(glWndFrame);
// Move the virtual sculpting tools to mimic the
UpdateRightObject();
// actions of the joysticks.
UpdateLeftObject();
// can't do anything if we not finished last call
if(bBusy)return;
// flag that we are busy
bBusy=TRUE;
RelaxDeformation(bChange);
// If something has changes e.g. a tool
if(bChange){
// has moved, then apply the algorithm.

// get new positions of tools
  UpdateObjectTransformations(FREE);
// check to see if tools hit each other
  if(CheckToolIntersections()){
// If they hit and we have force feedback
    if(!bToolsHit){
// devices send them a force forcing tools
      bToolsHit=TRUE; SendForce(4,0);
// apart.
    }
// don't let tools do the physically impossible
    RestoreOldToolPositions();
// do it again!
    UpdateObjectTransformations(FREE);
  }
// if the tools are now apart - switch of repulsion
  else if(bToolsHit){
    bToolsHit=FALSE; SendForce(0,1); SendForce(0,2);
  }
```

Listing 18.1. The sculpt program's most vital actions are dispatched from function RefreshScene(), which has to do the data acquisition, all the collision detection and simulation of elastic deformation within the available time interval (about 30 to 50 times per second).

```
    // do the collision detection of tool into objects
      CheckIntersection();
    // apply the spring mass deformation model
      UpdateDeformation();
    }
    if(IsWindow(glWnd))InvalidateRect(glWnd, NULL, FALSE);
    bChange=FALSE;
    bBusy=FALSE;
    return;
}
```

Listing 18.1. (continued).

because the damper provides a force which is proportional to the velocity of a vertex: $\mathbf{F}_d = k_d(\mathbf{p} - \bar{\mathbf{p}})/\Delta t$. This force is added to the spring restoration force in Equation (18.1).

The stability and type of behavior of the simulation is governed by the choice of values for the coefficients Δt, m, k and k_d. In the code, we have used experimentally determined values that work quite well. As we mentioned before the values we use will govern the stability and convergence of our simulation but the study of this is well outside the scope of this book.

18.8.3 Implementation

The polygonal mesh models we will use for the elastic surfaces and the sculpting tools are loaded as part of the program initialization. Rectangular bounding boxes are built around them. These bounding boxes are used so that collision detection can be done as quickly as possible. That is, a rectangular box around a sculpting tool is checked for intersection with a rectangular box around each of the elastic objects. If the boxes intersect then every polygon in a tool is checked for intersection against every polygon in the object.

As mentioned before, the key steps in the program are initiated from function `RefreshScene()`, which is shown in outline in Listing 18.1. The input for the two instruments is obtained from two joysticks by using DirectInput. The mesh models are rendered in real time using OpenGL, which allows for stereoscopic display.

It only remains to sum up by saying that this has been quite a complex project in terms of what it set out to achieve. But apart from the mathematical model for deformation, all of the other elements arise just by the application of common sense, and most of the coding details, e.g., OpenGL/DirectInput, have been described before.

18.9   Summary

Well, this brings us to the end of our projects chapter and indeed to the end of the book itself. There are many more project ideas we might have followed up, but as we hope you've seen from these few examples, most of the elements that such projects would be built on are not that different from those we explored in detail in Part II. We hope that our book has shown you some new things, excited your imagination to dream up new ways of using VR and provided some clues and pieces of computer code to help you do just that. Have fun!

Bibliography

[1]   *The Augmented Reality Toolkit.* http://www.hitl.washington.edu/artoolkit/.

[2]   M. Donahoo and K. Calvert. *TCP/IP Sockets in C: Practical Guide for Programmers.* San Fransisco, CA: Morgan Kaufmann, 2000.

[3]   C. Gerald. *Applied Numerical Analysis.* Reading, MA: Addison Wesley, 1980.

[4]   D. House and D. Breen (Editors). *Cloth Modeling and Animation.* Natick, MA: A. K. Peters, 2000.

[5]   C. Hunt. *TCP/IP Network Administration,* Second Edition. Sebastopol, CA: O'Reilly, 1998.

[6]   Intel Corporation. *Open Source Computer Vision Library.* http://www.intel.com/technology/computing/opencv/, 2006

[7]   Microsoft Corporation. "DirectShow SDK". http://msdn.microsoft.com/directx/sdk/, 2006.

[8]   M. Pesce. *Programming Microsoft DirectShow for Digital Video and Television.* Redmond, WA: Microsoft Press, 2003.

[9]   L. Verlet. "Computer Experiments on Classical Fluids". *Physical Review* 159 (1967) 98–103.

Appendices

A   Rotation with Quaternions

It is perfectly feasible to define the orientation of an object in 3D space using only three numbers, such as the Euler angles $(\varphi, \vartheta, \alpha)$. However, when we make the jump to animation, these three values are no longer adequate to represent the orientation of the object at a given frame and how it rotates from that orientation to a new orientation at the next frame. Indeed, we can represent this change by a 3×3 transformation matrix, but a nine-element matrix is difficult to manipulate, and there is an easier way to achieve the same results with only four numbers! That's where quaternion mathematics gets involved.

In 1843, Hamilton [1] developed the mathematics of the quaternion as part of his attempt to find a generalization of the complex number. The significance of the quaternion for computer graphics and animation was first recognized by Shoemake [3]. Since then, it has become the standard way of implementing angular interpolation.

In essence, the advantage of using quaternions in 3D computer graphics work is that that interpolation between two orientations $(\varphi_0, \vartheta_0, \alpha_0)$ and $(\varphi_1, \vartheta_1, \alpha_1)$ when expressed in their quaternion form is easily done utilizing the shortest path between both of them. We recall that *linear* interpolation between two position vectors gives a straight line, and in Cartesian geometry a straight line is the shortest path between two points. When dealing with 3D rotations, however, the shortest distance is no longer a straight line but rather a curve. Quaternion mathematics allows us to interpolate between two orientations along this shortest-distance curve. How is this possible? Essentially, a unit quaternion represents an axis of rotation, and at the same time, an amount of rotation about that axis.

To make use of quaternions in VR, it is unnecessary to explore the details of the quaternion. We are interested in using quaternions to help us achieve angular interpolation between given orientations, and this requires that we can interpolate between quaternions and switch back and forward between equivalent representations of orientation, i.e., Euler angles, quaternions and rotation matrices. In this appendix we hope to scratch the surface of this fascinating subject and prove how useful they are in 3D animation, computer graphics and VR.

A.1   The Quaternion

A quaternion q is an ordered pair $(w, \mathbf{v})$ of a scalar and three-dimensional vector $\mathbf{v}$ with components (x, y, z). Like vectors, a *unit length* or normalized quaternion must satisfy

$$w^2 + x^2 + y^2 + z^2 = 1.$$

Quaternions have their own algebra with rules for addition and multiplication. Addition is straightforward: add the scalar components together and add the vector components together.

Multiplication is more interesting. Given two quaternions q_1 and q_2 the product is the quaternion

$$q_1 q_2 = (w, \mathbf{v}) = (w_1 w_2 - \mathbf{v_1} \cdot \mathbf{v_2}, w_1 \mathbf{v_2} + w_2 \mathbf{v_1} + \mathbf{v_1} \times \mathbf{v_2}).$$

A conjugate quaternion to q is defined as $\bar{q} = (w, -\mathbf{v})$. The magnitude of a quaternion q is determined by multiplying it with its conjugate quaternion; that is

$$q\bar{q} = |q|^2 = (w^2 - \mathbf{v} \cdot \mathbf{v}, -w\mathbf{v} + w\mathbf{v} + \mathbf{v} \times \mathbf{v}).$$

Since the cross-product of any vector with itself is 1, this expressions simplifies to

$$q\bar{q} = |q|^2 = w^2 + |\mathbf{v}|^2.$$

Note that if q is of unit magnitude, its inverse q^{-1} equals its conjugate, $q^{-1} = \bar{q}$ and $q\bar{q} = 1$.

A.2   Quaternions and Rotation

We have seen in Section 6.6.3 that the action of rotating a vector $\mathbf{r}$ from one orientation to another may be expressed in terms of the application of a transformation matrix R which transforms $\mathbf{r}$ to $\mathbf{r}' = R\mathbf{r}$, which has a new orientation. The matrix R is independent of $\mathbf{r}$ and will perform the same (*relative*) rotation on any other vector.

One can write R in terms of the Euler angles, i.e., as a function $R(\varphi, \vartheta, \alpha)$. However, the same rotation can also be achieved by specifying R in terms of a unit vector $\hat{\mathbf{n}}$ and a single angle γ. That is, $\mathbf{r}$ is transformed into $\mathbf{r}'$ by rotating it round $\hat{\mathbf{n}}$ through γ. The angle γ is positive when the rotation takes place in a clockwise direction when viewed along $\hat{\mathbf{n}}$ from its base. The two equivalent rotations may be written as

$$\mathbf{r}' = R(\varphi, \vartheta, \alpha)\mathbf{r};$$
$$\mathbf{r}' = R'(\gamma, \hat{\mathbf{n}})\mathbf{r}.$$

At first sight, it might seem difficult to appreciate that the same transformation can be achieved by specifying a single rotation round one axis as opposed to three rotations round three orthogonal axes. It is also quite difficult to imagine how $(\gamma, \hat{\mathbf{n}})$ might be calculated, given the more naturally intuitive and easier to specify Euler angles $(\varphi, \vartheta, \alpha)$. However, there is a need for methods to switch from one representation to another.

> In a number of important situations, it is necessary to use the $(\gamma, \hat{\mathbf{n}})$ representation. For example, the Virtual Reality Modeling Language (VRML) [2] uses the $(\gamma, \hat{\mathbf{n}})$ specification to define the orientation adopted by an object in a virtual world.

Watt and Watt [4, p. 359] derive an expression that gives some insight into the significance of the $(\gamma, \hat{\mathbf{n}})$ specification of a rotational transform by determining $R\mathbf{r}$ in terms of $(\gamma, \hat{\mathbf{n}})$:

$$\mathbf{r}' = R\mathbf{r} = \cos \gamma \mathbf{r} + (1 - \cos \gamma)(\hat{\mathbf{n}} \cdot \mathbf{r})\hat{\mathbf{n}} + (\sin \gamma)\hat{\mathbf{n}} \times \mathbf{r}. \qquad (A.1)$$

This expression is the vital link between rotational transformations and the use of quaternions to represent them. To see this consider two quaternions:

1. $p = (0, \mathbf{r})$, a quaternion formed by setting its scalar part to zero and its vector part to $\mathbf{r}$ (the vector we wish to transform).

2. $q = (w, \mathbf{v})$, an arbitrary quaternion with unit magnitude: $q\bar{q} = \bar{q}q^{-1} = 1$.

The product qpq^{-1} gives the quaternion

$$qpq^{-1} = (0, (w^2 - (\mathbf{v} \cdot \mathbf{v})\mathbf{r} + 2(\mathbf{v} \cdot \mathbf{r})\mathbf{v} + 2w(\mathbf{v} \times \mathbf{r}))). \qquad (A.2)$$

Since q ($q = (w, \mathbf{v})$) is an arbitrary quaternion of unit magnitude, there is no loss of generality by substituting $\cos\left(\frac{\gamma}{2}\right)$ for w and $\sin\left(\frac{\gamma}{2}\right)\hat{\mathbf{n}}$ for $\mathbf{v}$. This results in the following:

$$q = \left(\cos\left(\frac{\gamma}{2}\right), \sin\left(\frac{\gamma}{2}\right)\hat{\mathbf{n}}\right). \qquad (A.3)$$

Now, utilizing this new formulation for q, we can rewrite Equation (A.2). After some simplification it can be seen that the vector part of qpq^{-1} is identical, term for term, with the rotation expressed by Equation (A.1).

Therefore, if we express a rotation in matrix form $R(\gamma, \hat{\mathbf{n}})$, its action on a vector $\mathbf{r}$ is equivalent to the following four steps:

1. Promote the vector $\mathbf{r}$ to the quaternion $p = (0, \mathbf{r})$.

2. Express matrix $R(\gamma, \hat{\mathbf{n}})$ as the quaternion $q = (\cos\frac{\gamma}{2}, \sin\frac{\gamma}{2}\hat{\mathbf{n}})$.

3. Evaluate the quaternion product $p' = qpq^{-1}$.

4. Extract the transformed vector $\mathbf{r}'$ from the vector component of $p' = (0, \mathbf{r}')$. (Note that the scalar component of operations such as these will always be zero.)

This isn't the end of our story of quaternions because, just as rotational transformations in matrix form may be combined into a single matrix, rotations in quaternion form may be combined into a single quaternion by multiplying their individual quaternion representations together. Let's look at a simple example to show this point. Consider two rotations R_1 and R_2 which are represented by quaternions q_1 and q_2 respectively.

By applying q_1 to the point p, we will end up at p'. By further applying q_2 to this new point, we will be rotated to p''; that is,

$$
\begin{aligned}
p' &= q_1 p q_1^{-1}; \\
p'' &= q_2 p' q_2^{-1}; \\
p'' &= q_2(q_1 p q_1^{-1})q_2^{-1}; \\
p'' &= (q_2 q_1)p(q_1^{-1} q_2^{-1}).
\end{aligned}
$$

To simplify the expression, we can represent q_1q_2 as the composite matrix q_c. And since quaternions satisfy $(q_2q_1)^{-1} = q_1^{-1}q_2^{-1}$, we can write $(q_1^{-1}q_2^{-1}) = (q_2q_1)^{-1}$. This simplifies our expression to

$$p'' = q_c p q_c^{-1}.$$

A.2.1 Converting Euler Angles to a Quaternion

From the Euler angles $(\varphi, \vartheta, \alpha)$, a quaternion that *encapsulates the same information* is constructed by writing quaternions for rotation of φ about the z-axis, ϑ about the y-axis and α about the x-axis:

$$q_x = \left[\cos\frac{\alpha}{2}, \sin\frac{\alpha}{2}, 0, 0\right],$$

$$q_y = \left[\cos\frac{\vartheta}{2}, 0, \sin\frac{\vartheta}{2}, 0\right],$$

$$q_z = \left[\cos\frac{\varphi}{2}, 0, 0, \sin\frac{\varphi}{2}\right],$$

and then multiplying them together as $q = q_z q_y q_x$. The components of q are

$$w = \cos\frac{\alpha}{2}\cos\frac{\vartheta}{2}\cos\frac{\varphi}{2} + \sin\frac{\alpha}{2}\sin\frac{\vartheta}{2}\sin\frac{\varphi}{2},$$

$$x = \sin\frac{\alpha}{2}\cos\frac{\vartheta}{2}\cos\frac{\varphi}{2} - \cos\frac{\alpha}{2}\sin\frac{\vartheta}{2}\sin\frac{\varphi}{2},$$

$$y = \cos\frac{\alpha}{2}\sin\frac{\vartheta}{2}\cos\frac{\varphi}{2} + \sin\frac{\alpha}{2}\cos\frac{\vartheta}{2}\sin\frac{\varphi}{2},$$

$$z = \cos\frac{\alpha}{2}\cos\frac{\vartheta}{2}\sin\frac{\varphi}{2} - \sin\frac{\alpha}{2}\sin\frac{\vartheta}{2}\cos\frac{\varphi}{2}.$$

A.2.2 Converting a Quaternion to a Matrix

In Section 6.6.3, it was shown that a rotational transformation in matrix form could be applied to a position vector to pivot it into a new orientation $\mathbf{p}' = T_\vartheta \mathbf{p}$. A quaternion contains rotational information, but it cannot be directly applied to a position vector in the same way that a matrix can. Therefore, it is useful to have a method of expressing the rotational information in the quaternion directly as a matrix, which in turn can be used to rotate position vectors.

For a normalized quaternion $[w, x, y, z]$, the corresponding 4×4 matrix is

$$
\begin{bmatrix}
1 - 2y^2 - 2z^2 & 2xy + 2wz & 2xz - 2wy & 0 \\
2xy - 2wz & 1 - 2x^2 - 2z^2 & 2yz + 2wx & 0 \\
2xz + 2wy & 2yz - 2wx & 1 - 2x^2 - 2y^2 & 0 \\
0 & 0 & 0 & 1
\end{bmatrix}.
$$

It is important to note that a 4×4 matrix can encapsulate positional transformations as well as rotational ones. A unit quaternion only describes pure rotations. So, when quaternions are combined, the complex rotation they represent is with respect to axes passing through the coordinate origin $(0, 0, 0)$.

A.3   Converting a Matrix to a Quaternion

If the rotational matrix is given by

$$
M =
\begin{bmatrix}
a_{00} & a_{01} & a_{02} & 0 \\
a_{10} & a_{11} & a_{12} & 0 \\
a_{20} & a_{21} & a_{22} & 0 \\
0 & 0 & 0 & 1
\end{bmatrix}
$$

and the quaternion q by

$$
q = [w, x, y, z],
$$

Shoemake's [3] algorithm in Figure A.1 obtains q given M.

A.4   Converting a Quaternion to Euler Angles

To convert the quaternion to the equivalent Euler angles, first convert the quaternion to an equivalent matrix, and then use the matrix-to-Euler angle conversion algorithm. Sadly, matrix-to Euler-angle conversion is unavoidably ill-defined because the calculations involve inverse trigonometric functions. To make the conversion, use the algorithm shown in Figure A.2, which converts the matrix M with elements a_{ij} to Euler angles $(\varphi, \vartheta, \alpha)$.

The angles $(\varphi, \vartheta, \varphi)$ lie in the interval $[-\pi, \pi]$, but they can be biased to $[0, 2\pi]$ or some other suitable range if required.

$$w = \tfrac{1}{4}(1 + a_{00} + a_{11} + a_{22})$$

if $w > \varepsilon$ {

$\qquad w = \sqrt{w}$

$\qquad w_4 = \dfrac{1}{4w}$

$\qquad x = w_4(a_{12} - a_{21})$

$\qquad y = w_4(a_{20} - a_{02})$

$\qquad z = w_4(a_{01} - a_{10})$

}

else {

$\qquad w = 0$

$\qquad x = -\tfrac{1}{2}(a_{11} + a_{22})$

$\qquad$ if $x > \varepsilon$ {

$\qquad\qquad x = \sqrt{x}$

$\qquad\qquad x_2 = \dfrac{1}{2x}$

$\qquad\qquad y = x_2 a_{01}$

$\qquad\qquad z = x_2 a_{02}$

$\qquad$ }

$\qquad$ else {

$\qquad\qquad x = 0$

$\qquad\qquad y = \tfrac{1}{2}(1 - a_{22})$

$\qquad\qquad$ if $y > \varepsilon$ {

$\qquad\qquad\qquad y = \sqrt{y}$

$\qquad\qquad\qquad z = \dfrac{a_{12}}{2y}$

$\qquad\qquad$ }

$\qquad\qquad$ else {

$\qquad\qquad\qquad y = 0$

$\qquad\qquad\qquad z = 1$

$\qquad\qquad$ }

$\qquad$ }

}

Figure A.1. An algorithm for the conversion of rotational transformation matrix M with coefficients $a_{i,j}$ to quaternion q with coefficients (w, x, y, z). The parameter ε is the machine precision of zero. A reasonable choice would be 10^{-6} for floating-point calculations. Note: only elements of M that contribute to rotation are considered in the algorithm.

$$\sin \vartheta = -a_{02}$$
$$\cos \vartheta = \sqrt{1 - \sin^2 \vartheta}$$
if $|\cos \vartheta| < \varepsilon$ {

   It is not possible to distinguish heading
   from pitch and the convention that
   φ *is* 0 *is assumed, thus:*
   $$\sin \alpha = -a_{21}$$
   $$\cos \alpha = a_{11}$$
   $$\sin \varphi = 0$$
   $$\cos \varphi = 1$$
}
else {

   $$\sin \alpha = \frac{a_{12}}{\cos \vartheta}$$
   $$\cos \alpha = \frac{a_{22}}{\cos \vartheta}$$
   $$\sin \varphi = \frac{a_{01}}{\cos \vartheta}$$
   $$\cos \varphi = \frac{a_{00}}{\cos \vartheta}$$
}
$$\alpha = ATAN2(\sin \alpha, \cos \alpha)$$
$$\vartheta = ATAN2(\sin \vartheta, \cos \vartheta)$$
$$\varphi = ATAN2(\sin \varphi, \cos \varphi)$$

Figure A.2. Conversion from rotational transformation matrix to the equivalent Euler angles of rotation.

A.5 Interpolating Quaternions

As we mentioned at the start of this appendix, the beauty of quaternions is that they allow us to interpolate from one orientation to the next along the shortest rotation path. Let's now look at how this is possible.

A clue to the answer comes from the concept of latitude and longitude. Latitude and longitude are angular directions from the center of the earth to a point on its surface whose position is desired. Thus a pair of values (*latitude, longitude*) represents a point on the earth's surface. To go from one place to another by the shortest route, one follows a *great circle*, illustrated in Figure A.3. The great circle is the line of intersection between a sphere and a plane that passes through the center of the sphere. Intercontinental flight paths for aircraft follow great circles; for example, the flight path between London and Tokyo passes close to the North Pole. We can therefore say that just as the shortest distance between two points in a Cartesian frame of reference is by a straight line; the shortest distance between two (*latitude*, *longitude*) coordinates is along a *path* following a great circle. The (*latitude*, *longitude*) coordinates at intermediate points on the *great circle* are determined by interpolation, in this case by *spherical interpolation*.

Quaternions are used for this interpolation. We think of the end points of the path being specified by quaternions, q_0 and q_1. From these, a quaternion q_i is interpolated for any point on the great circle joining q_0 to q_1. Conversion of q_i back to (*latitude*, *longitude*) allows the path to be plotted. In terms of angular interpolation, we may think of latitude and longitude as simply two of the Euler angles. When extended to the full set $(\varphi, \vartheta, \alpha)$, a smooth interpolation along the equivalent of a great circle is the result.

Using the concept of moving along a great circle as a guide to angular interpolation, a *spherical interpolation function*, *slerp*(), may be derived. The form of this function that works for interpolating between quaternions $q_0 = [w_0, x_0, y_0, z_0]$ and $q_1 = [w_1, x_1, y_1, z_1]$ is given in [3] as

$$slerp(\mu, q_0, q_1) = \frac{\sin(1 - \mu)\vartheta}{\sin \vartheta} q_0 + \frac{\sin \mu\vartheta}{\sin \vartheta} q_1,$$

where μ, the interpolation parameter, takes values in the range $[0, 1]$. The angle ϑ is obtained from $\cos \vartheta = q_0 \cdot q_1 = w_0 w_1 + x_0 x_1 + y_0 y_1 + z_0 z_1$.

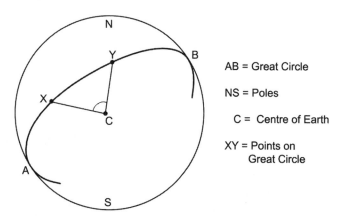

Figure A.3. A great circle gives a path of shortest distance between two points on the surface of a sphere. The arc between points X and Y is the shortest path.

$$\sigma = w_0 w_1 + x_0 x_1 + y_0 y_1 + z_0 z_1$$

if $\sigma > 1$ then normalize q_0 and q_1 by
dividing the components of q_0 and q_1 by σ

$$\vartheta = \cos^{-1}(\sigma)$$

if $|\vartheta| < \varepsilon$ {

$$\beta_0 = 1 - \mu$$
$$\beta_1 = \mu$$

}

else {

$$\beta_0 = \frac{\sin(1 - \mu)\vartheta}{\sin \vartheta}$$
$$\beta_1 = \frac{\sin \mu\vartheta}{\sin \vartheta}$$

}

$$w_i = \beta_0 w_0 + \beta_1 w_1$$
$$x_i = \beta_0 z_0 + \beta_1 x_1$$
$$y_i = \beta_0 y_0 + \beta_1 y_1$$
$$z_i = \beta_0 z_0 + \beta_1 z_1$$

Figure A.4. Algorithmic implementation of the *slerp()* function for the interpolated quaternion $q_i = [w_i, x_i, y_i, z_i]$.

Figure A.4 gives an algorithmic implementation of the *slerp()* function for the interpolated quaternion $q_i = [w_i, x_i, y_i, z_i]$ which avoids the problem of division by zero when ϑ is close to zero.

Bibliography

[1] W. R. Hamilton. "On Quaternions: Or on a New System of Imaginaries in Algebra". *Philosophical Magazine* 25 (1844) 10–14.

[2] J. Hartman, J. Wernecke and Silicon Graphics, Inc. *The VRML 2.0 Handbook*. Reading, MA: Addison-Wesley Professional, 1996.

[3] K. Shoemake. "Animating Rotation with Quaternion Curves". *Proc. SIG-GRAPH '85, Computer Graphics* 19:3 (1985) 245–254.

[4] A. Watt and M. Watt. *Advanced Animation and Rendering Techniques: Theory and Practice*. Reading, MA: Addison-Wesley Professional, 1992.

B The Generalized Inverse

This appendix gives a definition for the inverse of a nonsquare matrix.

The definition of the inverse of a matrix A is a matrix A^{-1} such that if $|A| \neq 0$ then A and A^{-1} satisfy

$$AA^{-1} = A^{-1}A = I.$$

This definition of the inverse only applies to square matrices. In the case of any matrix A, it is possible to define a generalized inverse A^- which satisfies

$$AA^-A = A. \tag{B.1}$$

If A is square and $|A| \neq 0$ then $A^- = A^{-1}$. Multiplying both sides of Equation (B.1) by the transpose of A gives us

$$AA^-(AA^T) = AA^T.$$

AA^T is a square matrix, and therefore if $|AA^T| \neq 0$, we can find its *conventional* inverse $(AA^T)^{-1}$ and write

$$AA^-(AA^T)(AA^T)^{-1} = AA^T(AA^T)^{-1}, \text{ or}$$
$$AA^- = AA^T(AA^T)^{-1}.$$

The implication here is that for any (not necessarily square) matrix, if we need its inverse, we may use its (generalized) inverse A^- given by

$$A^- = A^T(AA^T)^{-1}. \tag{B.2}$$

This is exactly the expression that we need in order to invert the n × m Jacobian matrix.

For more information on generalized inverses and their properties and limitations, consult [1]. Specifically, the two points most important for IK are the existence of an inverse for AA^T and the fact that normally we have more unknowns than equations (i.e., $m > n$). The practical implication of $m > n$ is that the articulation can attain its goal in more than one configuration.

Bibliography

[1] T. Boullion and P. Odell. *Generalized Inverse Matrices.* New York: John Wiley and Sons, 1971.

Aligning Two Images in a Panoramic Mosaic

C

An algorithm that attempts to align two images in a rotational panorama must attempt to determine the parameters h_{ij} given by Equation (8.4) in the homography $\mathbf{X}' = H\mathbf{X}$. These parameters transform pixels at image coordinate (x, y) in the existing panorama I_c so that they overlap the incoming image I_1, at location (x', y') (as given by Equations (8.2) and (8.3)) with the minimum difference — to be precise, with the minimum least squared error (LSE) between all pixels:

$$\varepsilon = \sum_{i=all\ pixels} \left[I_1(x_i', y_i') - I_c(x_i, y_i) \right]^2.$$

In the following argument, we follow the work of Szeliski and Shum [2].

To register the image $I_1(\mathbf{x}')$ with $I_c(\mathbf{x})$ where $\mathbf{x}' = H\mathbf{x}$, we first compute an approximation to I_1, i.e.,

$$\widetilde{I}_1(\mathbf{x}) = I_1(H\mathbf{x}),$$

and then find a deformation of $\widetilde{I}_1(\mathbf{x})$ which brings it into closer alignment with $I_c(\mathbf{x})$ by updating H.

The algorithm is iterative. It starts with a guess for H, which might be the null transform. It then gradually updates homography H, which we can write as

$$H \leftarrow (I + D)H, \qquad (C.1)$$

with

$$D = \begin{bmatrix} d_0 & d_1 & d_2 \\ d_3 & d_4 & d_5 \\ d_6 & d_7 & d_8 \end{bmatrix},$$

by obtaining a D that reduces ε. This process is repeated until ε is minimized. The LSE algorithm we follow is detailed in [1], which also provides suitable computer code.

Recognizing that H has only eight independent coefficients, we may set $d_8 = 1$. Also, finding the value of $I_1(\mathbf{x}')$ where $\mathbf{x}' = (I + D)H\mathbf{x}$ is equivalent to finding the value of $\widetilde{I}_1(\mathbf{x}'')$ using

$$\mathbf{x}'' = (I + D)\mathbf{x}. \qquad (C.2)$$

Therefore

$$x'' = \frac{(1 + d_0)x + d_1 y + d_2}{d_6 x + d_7 y + (1 + d_8)}, \qquad (C.3)$$

and

$$y'' = \frac{d_3 x + (1 + d_4)y + d_5}{d_6 x + d_7 y + (1 + d_8)}. \qquad (C.4)$$

Letting $\mathbf{d} = D\mathbf{x}$ in $\mathbf{x}'' = (I + D)\mathbf{x}$, we can write:

$$\mathbf{x}'' = \mathbf{x} + \mathbf{d},$$

and therefore it is possible to express the error between the aligned images I_1 and I_c as a function of the vector $\mathbf{d} = (d_0, d_1, d_2, d_3, d_4, d_5, d_6, d_7, d_8)$:

$$\varepsilon(\mathbf{d}) = \sum_i \left[\widetilde{I}_1(\mathbf{x}''_i) - I_c(\mathbf{x}_i) \right]^2 = \sum_i \left[\widetilde{I}_1(\mathbf{x}_i + \mathbf{d}) - I_c(\mathbf{x}_i) \right]^2. \qquad (C.5)$$

The term $\widetilde{I}_1(\mathbf{x}_i + \mathbf{d})$ in Equation (C.5) may be approximated as a truncated Taylor series expansion.

For a function I of one variable $I(d)$, the first two terms in the Taylor series expansion made about a point x are

$$I(x + d) = I(x) + \left. \frac{dI}{dd} \right|_x d + \cdots .$$

When d is a function of another variable, say x'', the expansion becomes

$$I(x + d) = I(x) + \frac{dI}{dx''} \frac{dx''}{dd}\bigg|_x d + \cdots .$$

When both d and x are vectors, the expansion becomes

$$I(\mathbf{x} + \mathbf{d}) = I(\mathbf{x}) + \nabla I(\mathbf{x}) \frac{\partial \mathbf{x}''}{\partial \mathbf{d}}\bigg|_{\mathbf{x}} \mathbf{d} + \cdots . \tag{C.6}$$

In the case of the expansion of $\tilde{I}_1(\mathbf{x}''_i)$, in terms of $\mathbf{d}$ about $\mathbf{x}_i$,

$$\tilde{I}_1(\mathbf{x}''_i) = \tilde{I}_1(\mathbf{x}_i + \mathbf{d}) = \tilde{I}_1(\mathbf{x}_i) + \nabla \tilde{I}_1(\mathbf{x}_i) \frac{\partial \mathbf{x}''}{\partial \mathbf{d}}\bigg|_{\mathbf{x}_i} \mathbf{d} + \cdots , \tag{C.7}$$

where

$$\nabla \tilde{I}_1(\mathbf{x}_i) = \left[\frac{\partial \tilde{I}_1}{\partial x}, \frac{\partial \tilde{I}_1}{\partial y} \right]_{\mathbf{x}_i}$$

and

$$\frac{\partial \mathbf{x}''}{\partial \mathbf{d}} = \left[\begin{array}{ccccccc} \frac{\partial x''}{\partial d_0} & \cdot & \cdot & \cdot & \cdot & \cdot & \frac{\partial x''}{\partial d_8} \\ \frac{\partial y''}{\partial d_0} & \cdot & \cdot & \cdot & \cdot & \cdot & \frac{\partial y''}{\partial d_8} \end{array} \right] .$$

Then at $\mathbf{x}_i$, one obtains from Equations (C.3) and (C.4),

$$\frac{\partial \mathbf{x}''}{\partial \mathbf{d}}\bigg|_{\mathbf{x}_i} = \left[\begin{array}{ccccccccc} x & y & 1 & 0 & 0 & 0 & -x^2 & -xy & -x \\ 0 & 0 & 0 & x & y & 1 & -xy & -y^2 & -y \end{array} \right] , \tag{C.8}$$

defining

$$\mathbf{g}_i^T = \left[\frac{\partial \tilde{I}_1}{\partial x}, \frac{\partial \tilde{I}_1}{\partial y} \right]_{\mathbf{x}_i}$$

and

$$J_i^T = \left[\begin{array}{ccccccccc} x & y & 1 & 0 & 0 & 0 & -x^2 & -xy & -x \\ 0 & 0 & 0 & x & y & 1 & -xy & -y^2 & -y \end{array} \right] .$$

Equation (C.7) may be written as

$$\tilde{I}_1(\mathbf{x}_i + \mathbf{d}) = \tilde{I}_1(\mathbf{x}_i) + \mathbf{g}_i^T J_i^T \mathbf{d} + \cdots . \tag{C.9}$$

Truncating the approximation in Equation (C.9) after first-order terms and substituting it into Equation (C.5) gives

$$\varepsilon(\mathbf{d}) \approx \sum_i \left[\mathbf{g}_i^T J_i^T \mathbf{d} + (\tilde{I}_1(\mathbf{x}_i) - I_c(\mathbf{x}_i)) \right]^2, \qquad (C.10)$$

where $\tilde{I}_1(\mathbf{x}_i) - I_c(\mathbf{x}_i)$ is the intensity error in pixel i.

This is a conventional linear least squares error minimization problem using the normal equations where we wish to minimize ε with respect to each of the elements in $\mathbf{d}$ (d_0, d_1 etc.). This requires writing nine simultaneous equations,

$$\frac{\partial \varepsilon(\mathbf{d})}{\partial d_j} = 0 \quad j = 0, \ldots, 8$$

and solving for the d_i.

To see how to proceed, we observe that the term $\mathbf{g}_i^T J_i^T$ in Equation (C.10) is a 1×9 matrix; let us call this $A_i = [a_{i0} \ldots a_{i8}]$. $(\tilde{I}_1(\mathbf{x}_i) - I_c(\mathbf{x}_i))$ is a scalar, so let us call this e_i. With these substitutions in Equation (C.10), we need to find

$$\frac{\partial \left(\sum_i (A_i \mathbf{d} + e_i)^2 \right)}{\partial d_j} = 0 \quad j = 0, \ldots, 8.$$

Writing explicitly the case where $j = 0$,

$$\frac{\partial \left(\sum_i (a_{i0}d_0 + a_{i1}d_1 + \cdots + a_{i8}d_8 + e_i)^2 \right)}{\partial d_0} = 0$$

and differentiating gives

$$\sum_i (a_{i0}d_0 + a_{i1}d_1 + \cdots + a_{i8}d_8 + e_i) a_{i0} = 0$$

or by rearranging

$$\sum_i a_{i0}a_{i0}d_0 + a_{i0}a_{i1}d_1 + \cdots + a_{i0}a_{i8}d_8 = -\sum_i e_i a_{i0}.$$

Bringing in the other cases $j = 1, 2 \cdots 8$ gives us nine equations in the nine d_i unknowns:

$$\sum_i a_{i0}a_{i0}d_0 + a_{i0}a_{i1}d_1 + \cdots + a_{i0}a_{i8}d_8 = -\sum_i e_i a_{i0},$$

$$\sum_i a_{i1}a_{i0}d_0 + a_{i1}a_{i1}d_1 + \cdots + a_{i1}a_{i8}d_8 = -\sum_i e_i a_{i1},$$

$$\cdots = \cdots$$

$$\sum_i a_{i8}a_{i0}d_0 + a_{i8}a_{i1}d_1 + \cdots + a_{i8}a_{i8}d_8 = -\sum_i e_i a_{i8}.$$

These equations can be written in matrix form:

$$\sum_i A_i^T A_i \mathbf{d} = -\sum_i e_i A_i^T,$$

and if we resubstitute $\mathbf{g}_i^T J_i^T$ for A_i and $J_i \mathbf{g}_i$ for A_i^T, we arrive at the normal equations

$$\left(\sum_i J_i \mathbf{g}_i \mathbf{g}_i^T J_i^T\right)\mathbf{d} = -\left(\sum_i e_i J_i \mathbf{g}_i\right). \tag{C.11}$$

These represent a set of linear equations of the form

$$A\mathbf{d} = \mathbf{b},$$

with

$$A = \sum_i J_i \mathbf{g}_i \mathbf{g}_i^T J_i^T \quad \text{and} \quad \mathbf{b} = -\sum_i e_i J_i \mathbf{g}_i,$$

which upon solution yields the nine coefficients of $\mathbf{d}$. Hence, the updated homography in Equation (C.1) may be determined. With a new H known, the iterative algorithm may proceed.

The solution of Equation (C.11) may be obtained by a method such as Cholesky decomposition [1] and the whole algorithm works well, provided the alignment between the two images is only off by a few pixels. Following the argument in [2], it is a good idea to use a modified form of the minimization procedure requiring only three parameters. This corresponds to the case in which each image is acquired by making a pure rotation of the camera, i.e., there is no translational movement or change of focal length.

If a point $\mathbf{p}$ (specified in world coordinates (p_x, p_y, p_z)) in the scene projects to image location $\mathbf{x}$ (i.e., homogeneous coordinates $(x, y, 1)$) in image I_c then after the camera and the world coordinate frame of reference have been rotated slightly, it will appear in image I_i to lie at $\mathbf{p}''$ where

$$\mathbf{p}'' = R\mathbf{p}, \qquad (C.12)$$

where R is a 3×3 rotation matrix. For a pinhole camera of focal length f, we have $\mathbf{x} = F\mathbf{p}$ and $\mathbf{x}'' = F\mathbf{p}''$ where

$$F = \begin{bmatrix} f & 0 & 0 \\ 0 & f & 0 \\ 0 & 0 & 1 \end{bmatrix}.$$

If we know $\mathbf{x}$ then we can write (up to an arbitrary scale factor) $\mathbf{p} = F^{-1}\mathbf{x}$. Using Equation (C.12), one obtains the homography

$$\mathbf{x}'' = FRF^{-1}\mathbf{x}. \qquad (C.13)$$

The rotation R may be specified in a number of different ways, one of which defines an axis of rotation and an angle of rotation around that axis. Appendix A gave an expression for this in Equation (A.1). In the notation of this appendix where we wish to rotate a point $\mathbf{p}$ around an axis $\hat{\mathbf{n}} = (n_x, n_y, n_z)$ by an angle ϑ, it becomes

$$\mathbf{p}'' = R\mathbf{p} = \cos\vartheta\,\mathbf{p} + (1 - \cos\vartheta)(\hat{\mathbf{n}} \cdot \mathbf{p})\hat{\mathbf{n}} + (\sin\vartheta)\hat{\mathbf{n}} \times \mathbf{p}. \qquad (C.14)$$

Since we are matching images that are only mildly out of alignment (by a few pixels), the rotation of the camera which took the images must have been small (say by an angle $\Delta\vartheta$). Using the small angle approximation $(\sin(\Delta\vartheta) \approx \Delta\vartheta,\ \cos(\Delta\vartheta) \approx 1)$ allows Equation (C.14) to be simplified as

$$\mathbf{p}'' = \mathbf{p} + \Delta\vartheta(\hat{\mathbf{n}} \times \mathbf{p}). \qquad (C.15)$$

In our argument, it is more convenient to write the vector $\times$ product in matrix form; thus, Equation (C.15) is

$$\mathbf{p}'' = \mathbf{p} + W\mathbf{p}$$

or

$$\mathbf{p}'' = (I + W)\mathbf{p}, \qquad (C.16)$$

where

$$W = \begin{bmatrix} 0 & -w_z & w_y \\ w_z & 0 & -w_x \\ -w_y & w_x & 0 \end{bmatrix} = \begin{bmatrix} 0 & -n_z\Delta\vartheta & -n_y\Delta\vartheta \\ n_z\Delta\vartheta & 0 & -n_x\Delta\vartheta \\ -n_y\Delta\vartheta & n_x\Delta\vartheta & 0 \end{bmatrix}.$$

Comparing Equations (C.16) and (C.12), one can see that for small rotations $R = I + W$, and so Equation (C.13) becomes

$$\mathbf{x}'' = F(I + W)F^{-1}\mathbf{x} = (I + FWF^{-1})\mathbf{x}. \qquad (C.17)$$

Notice that Equation (C.17) is exactly the same as Equation (C.2) if we let $D = FWF^{-1}$, and so the same argument and algorithm of Equations (C.2) through (C.11) may be applied to determine the matrix of parameters W. However, in the purely rotational case, W contains only three unknowns w_x, w_y and w_z. This makes the solution of the linear Equations (C.11) much more stable. A couple of minor modifications to the argument are required in the case of the homography $H \leftarrow (I + D)H$ with D given by $D = FWF^{-1}$.

The vector $\mathbf{d}$ becomes $\mathbf{d} = (0, -w_z, fw_z, w_z, 0, -fw_x, -\frac{w_y}{f}, \frac{w_x}{f}, 0)$. The Taylor expansion in Equation (C.6) must account for the fact that $\mathbf{d}$ is really only a function of three unknowns $\mathbf{d} \equiv \mathbf{w}$ with $\mathbf{w} = (w_x, w_y, w_z)$. Thus it becomes

$$I(\mathbf{x} + \mathbf{w}) = I(\mathbf{x}) + \nabla I(\mathbf{x}) \frac{\partial \mathbf{x}''}{\partial \mathbf{d}} \frac{\partial \mathbf{d}}{\partial \mathbf{w}}\bigg|_{\mathbf{x}} \mathbf{w} + \cdots, \qquad (C.18)$$

where $\partial \mathbf{d}/\partial \mathbf{w}$ is the 9×3 matrix:

$$\frac{\partial \mathbf{d}}{\partial \mathbf{w}} = \begin{bmatrix} \frac{\partial d_0}{\partial w_x} & \frac{\partial d_0}{\partial w_y} & \frac{\partial d_0}{\partial w_z} \\ \frac{\partial d_1}{\partial w_x} & \frac{\partial d_1}{\partial w_y} & \frac{\partial d_1}{\partial w_z} \\ \frac{\partial d_2}{\partial w_x} & \frac{\partial d_2}{\partial w_y} & \frac{\partial d_2}{\partial w_z} \\ \cdots & \cdots & \cdots \\ \frac{\partial d_8}{\partial w_x} & \frac{\partial d_8}{\partial w_y} & \frac{\partial d_8}{\partial w_z} \end{bmatrix} ; \quad \text{i.e.,} \quad \frac{\partial \mathbf{d}}{\partial \mathbf{w}} = \begin{bmatrix} 0 & 0 & 0 \\ 0 & 0 & -1 \\ 0 & 0 & f \\ 0 & 0 & 1 \\ 0 & 0 & 0 \\ -f & 0 & 0 \\ 0 & -1/f & 0 \\ 1/f & 0 & 0 \\ 0 & 0 & 0 \end{bmatrix}.$$

Thus, with $\partial \mathbf{x}''/\partial \mathbf{d}$ given by Equation (C.8), we have

$$J_i^T = \frac{\partial \mathbf{x}''}{\partial \mathbf{d}} \frac{\partial \mathbf{d}}{\partial \mathbf{w}}\bigg|_{\mathbf{x}_i} = \begin{bmatrix} -\frac{xy}{f} & f + \frac{x^2}{f} & -y \\ -f - \frac{y^2}{f} & \frac{xy}{f} & x \end{bmatrix}.$$

If the camera's focal length is not known, it may be estimated from one or more of the transformations H obtained using the eight-parameter algorithm. When H corresponds to a rotation only, we have seen that

$$H = \begin{bmatrix} h_0 & h_1 & h_2 \\ h_3 & h_4 & h_5 \\ h_6 & h_7 & 1 \end{bmatrix} = FRF^{-1} \sim \begin{bmatrix} r_{00} & r_{01} & r_{02}f \\ r_{10} & r_{11} & r_{12}f \\ r_{20}/f & r_{21}/f & r_{22} \end{bmatrix}.$$

The symbol $\sim$ indicates equal up to scale, and so, for example, $r_{02} = h_2/f$. We don't show it here, but for H to satisfy the rotation condition, the norm of either the first two rows or the first two columns of R must be the same. And at the same time, the first two rows or columns must also be orthogonal. Using this requirement allows us to determine the camera focal length f with

$$f = \begin{cases} \sqrt{\dfrac{h_2 h_5}{h_0 h_3 + h_1 h_4}} & \text{if} \quad h_0 h_3 \neq h_1 h_4, \\[2ex] \sqrt{\dfrac{h_5{}^2 - h_2{}^2}{h_0^2 + h_1^2 - h_3^2 + h_4^2}} & \text{if} \quad h_0{}^2 + h_1{}^2 \neq h_3{}^2 + h_4{}^2. \end{cases}$$

Bibliography

[1] W. Press et al. (editors). *Numerical Recipes in C++: The Art of Scientific Computing,* Second Edition. Cambridge, UK: Cambridge University Press, 2002.

[2] H. Shum and R. Szeliski. "Panoramic Image Mosaics". Technical Report MSR-TR-97-23, Microsoft Research, 1997.

D A Minimal Windows Template Program

A Windows program is made up of two main elements: the source code and the resources that the compiled code uses as it executes. Resources are things such as icons, little graphics to represent the mouse pointer, bitmaps to appear in toolbars and a description of the dialog boxes. This information is gathered together in a `.RC` file which is a simple text file that is normally only viewed by using an integrate development environment such as Microsoft's Developer Studio. The resources are compiled in their own right (into a `.RES` file) and then this and the object modules that result from the compilation of the C/C++ source codes are combined with any libraries to form the executable image of the application program. A Windows application will not only need to be linked with the standard C libraries but also a number of libraries that provide the graphical user interface (GUI) and system components.

A large proportion of the Window operating system is provided in dynamically linked libraries (DLLs) that are loaded and used only when the program is executing. `SHELL32.DLL`, `GDI32.DLL` and `USER32.DLL` are examples of these components. When linking a program that needs to call a system function in one of these, it is necessary to bring in a dummy, or *stub*, static library (shell32.lib, gdi32.lib etc.) that contains just enough code to prevent an error message of the form "function not found" being generated and which will point at the correct location of the function in the DLL once the application program gets up and running. The majority of MFC application programs also implement the class library in a DLL: MFC41.DLL, MFC71.DLL etc.

```
// resource identifiers - also included in application
#include "resource.h"
#include "windows.h"
// Icon - define the icon identifier and file from which the are to be compiled
IDI_ICON1                ICON    DISCARDABLE      "ico1.ico"
// BITMAP - same as incos for small pictures (typically used for toolbars
IDB_BMP1                 BITMAP  MOVEABLE PURE    "bitmap1.bmp"
// CURSOR - application specific mouse cursors
IDC_HAND                 CURSOR  DISCARDABLE      "finger.cur"
// Menu - the program's main menu
IDC_MENU MENU DISCARDABLE
BEGIN
    POPUP "&File"
    BEGIN
        MENUITEM "E&xit",                IDM_EXIT
    END
END
// Dialog  - every program has an ABOUT dialog box
IDD_ABOUTBOX DIALOG DISCARDABLE  22, 17, 230, 75
STYLE DS_MODALFRAME | WS_CAPTION | WS_SYSMENU
CAPTION "About"
FONT 8, "System"
BEGIN
    ICON             IDI_ICON1,IDC_MYICON,14,9,16,16
    LTEXT            "Template Version 1.0",IDC_STATIC,49,10,119,8,SS_NOPREFIX
    DEFPUSHBUTTON    "OK",IDOK,195,6,30,11,WS_GROUP
END
// String Table - all text strings can be stored as
// resources for internationalization
STRINGTABLE DISCARDABLE
BEGIN
    IDS_APP_TITLE        "Skeleton Windows Application"
END
```

Listing D.1. The text of a Windows resource file before compilation. It shows how resources are identified, a small menu is specified and a dialog box described.

Name △	Size	Type	Modified
[h] resource.h	1 KB	C Header file	05/11/2005 10:03
[c] StdAfx.c	1 KB	C Source file	05/11/2005 10:02
[h] StdAfx.h	1 KB	C Header file	05/11/2005 10:03
[c] main.c	5 KB	C Source file	05/11/2005 10:03
[h] main.h	1 KB	C Header file	05/11/2005 10:03
main.rc	2 KB	Resource Template	05/11/2005 10:06
icon1.ico	1 KB	Icon	05/10/1995 17:55
FINGER.cur	1 KB	Cursor	25/09/1999 18:44
icon2.ICO	1 KB	Icon	14/10/1995 03:49
bitmap1.bmp	2 KB	Bitmap Image	23/10/1999 16:46

Figure D.1. A typical collection of the files required to build a minimal Windows program. Icons, mouse cursors and small bitmaps are all used to *beautify* the appearance of the application program to the user.

> DLLs are both a very clever idea and one of the biggest curses to the Windows programmer. See Section 12.4 for more on DLLs.

It can be quite interesting to sneak a look at a few lines from a resource file (see Listing D.1) and also to see a list of the files that need to be brought together in the creation of even the smallest of Windows applications (see Figure D.1).

All the work for the basic template is done by the code in the C source file main.c and its co-named header file. The files StdAfx.h and Stdafx.h are used to help speed up the build process by precompiling the rather large system header files and can usually be left alone. In main.c, there are two principal functions: the program's entry point and a function to process and act on messages sent to the application by Windows. *Handling messages is the key concept in Windows programming.* Messages are sent to the program's handler function whenever the user selects a command from the menu bar or moves the mouse across the window. Setting up the shape, size and behavior of the application's user interface and interacting with Windows is done by calling functions in the application programmer interface (API). All these functions exist within one of the system DLLs (mentioned earlier). Function prototypes for the API functions, useful constants and the stub libraries are all provided and documented in the SDK.

For our minimal application, the template must exhibit the key features of message loop, definition of window class and message handling function. It will have an entry point in the conventional C sense, the WinMain() function.

The `WinMain()` function performs the following three tasks:

1. Register a window class (not to be confused with a C++ class). Each window class has an associated function that acts upon the messages dispatched to any window of that class. There is no practical limit to the number of windows for each class. One window is designated as the main window.

2. Create one or more visible windows of that class and go into a loop to check for and dispatch *messages* to the appropriate window function.

3. When a *quit* message is received by the program's main window, the loop is broken and the program exits.

Cutting things down to the absolute minimum, our Windows template application can be summarized in two listings. Listing D.2 shows the main function and how it creates the windows class, defines the message handling function and the message loop. The skeleton outline of the message-processing function appears in Listing D.3.

The message-processing function, `WndProc(...)`, in Listing D.3 has the task of acting on any messages that the program wants to respond to. There are many messages that could be acted upon. If no action is to be taken for a particular message, it must be passed to the default system handler `DefWindowProc(..)`. A switch/case construct is typically used to respond to messages which are given appropriately named identifiers in the `windows.h` header file. Arguments passed to `WndProc` provide a *handle* (which

```
// The windows SDK header file
#include <windows.h>
// C library - other headers can go here
#include <stdlib.h>
// The application specific header file
#include "resource.h"
// used to identity resources symbolically

// Prototype of message processing function
LRESULT CALLBACK WndProc(HWND,UINT,WPARAM,LPARAM);
// a name to identify the program's type of window
char szWindowClass[]="frameworkclass1";
```

Listing D.2. The `WinMain()` function.

```
// All windows programs use this as the
int APIENTRY WinMain(HINSTANCE hInstance,
// entry point the parameters give access
                    HINSTANCE hPrevInstance,
// to the command line and a HANDLE to the
                    LPSTR     lpCmdLine,
// application to enable accesss to resources.
                    int       nCmdShow) {
WNDCLASSEX wcex;
HWND hWnd;
MSG  msg;
// fill members of the class structure to define the style of the window
wcex.cbSize         = sizeof(WNDCLASSEX);
wcex.style          = CS_HREDRAW | CS_VREDRAW;
wcex.lpfnWndProc    = (WNDPROC)WndProc;            // POINTER TO MESSAGE HANDLER
wcex.cbClsExtra     = 0;
wcex.cbWndExtra     = 0;
wcex.hInstance      = hInstance;
wcex.hIcon          = LoadIcon(hInstance, (LPCTSTR)IDI_ICON1);
// standard arrow mouse pointer
wcex.hCursor        = LoadCursor(NULL, IDC_ARROW);
// standard background
wcex.hbrBackground = (HBRUSH)(COLOR_WINDOW+1);
// menu ID from resource file
wcex.lpszMenuName  = (LPCSTR)IDC_MENU;
// the unique window ID
wcex.lpszClassName = szWindowClass;
wcex.hIconSm        = LoadIcon(wcex.hInstance, (LPCTSTR)IDI_ICON2);
// tell Windows about our window
if(!RegisterClass(&wcex)) return 1;               // quit if fails
// create the program's main window
hWnd = CreateWindow(szWindowClass,"Application Window", WS_OVERLAPPEDWINDOW,
        CW_USEDEFAULT, 0, CW_USEDEFAULT, 0, NULL, NULL, hInstance, NULL);
if (!hWnd) return FALSE;
ShowWindow(hWnd, nCmdShow);                        // make window visible
UpdateWindow(hWnd);                               // make sure it is drawn
while(GetMessage(&msg,NULL,0,0)){                 // MAIN MESSAGE LOOP
  if (!TranslateAccelerator(msg.hwnd,NULL,&msg)){ // any keyboard
        TranslateMessage(&msg);                   // Send any messages
        DispatchMessage(&msg);                    // to correct [lace
  }
}
return msg.wParam;
}
```

Listing D.2. (continued).

```
LRESULT CALLBACK WndProc(HWND hWnd, UINT message, WPARAM wParam, LPARAM lParam){
// local variables to identify the menu command
 int wmId, wmEvent;
// required to handle the WM_PAINT message
 PAINTSTRUCT ps;
// allows us to draw into the client area of window
 HDC hdc;
// act on the message
 switch (message)   {
    case WM_COMMAND:
// the low 16 bits of wParam give the menu command
       wmId    = LOWORD(wParam);
       wmEvent = HIWORD(wParam);
       switch (wmId){
// process the menu command selected by the user
         case IDM_EXIT:
// tell windows to send us back a WM_DESTROY message
            DestroyWindow(hWnd);
            break;
         default:
            return DefWindowProc(hWnd, message, wParam, lParam);
         }
         break;
    case WM_PAINT:
// tells Windows "WE" are going to draw in client area
       hdc = BeginPaint(hWnd, &ps);
// which gives us a "handle" to allow us to do so!

// Just for fun!
       DrawText(hdc, "Hello", strlen("Hello"), &rt, DT_CENTER);
// paired with BeginPaint
       EndPaint(hWnd, &ps);
       break;
    case WM_DESTROY:
// send a message to break out of loop in WinMain
       PostQuitMessage(0);
       break;
    default:
// default message handling
       return DefWindowProc(hWnd, message, wParam, lParam);
 }
 return 0;
}
```

Listing D.3. The message-processing function.

is a bit like a pointer) to identify the window, an integer to identify the message and two 32-bit data words holding information for the message. The message data could be a command identifier, or in the case of a user-defined message, a pointer to a major data structure in memory.

Our minimal template acts upon three of the most commonly handled messages:

1. WM_COMMAND. This message typically arises from a program's menu or a toolbar. The low-order 16 bits of the WPARAM data word hold the integer identifier of the menu command selected. For example, in the resources script file in Listing D.1, the EXIT command is given an identifier of IDM_EXIT. This is defined as the integer 1001 in the header file resource.h. Since resource.h is included in both resource and source files, the same symbolic representation for the exit command may be used in both places.

2. WM_PAINT. This is probably the most significant message from the point of view of any graphical application program. Anything that is to appear inside the window (in its so called *client area*) must be drawn in response to this message. So, the OpenGL, the DirectShow and the Direct3D examples will all do their rendering in response to this message. To be able to draw in the client area,[1] Windows gives the program a handle (pointer) to what is called a *device context* (HDC hDC;) which is needed by most drawing routines, including the 3D ones. The simple example in Listing D.3 draws the word *Hello* in response to the WM_PAINT message.

3. WM_DESTROY. This is the message sent to the window when the *close button* is clicked. If this is a program's main window, it usually is a signal that the program should terminate. PostQuitMessage(0) sends the NULL message, which will break out of the message-processing loop (Listing D.2) when it is processed for dispatch.

---

[1] The client area is the part of the window within the border, below any menu and title bar. Applications draw their output in the client area, in response to a WM_PAINT message.

An MFC Template Program

E

In programs that use the Microsoft Foundation Class (MFC) library, the elements of message loop, window class and message processing function described in Appendix D are hidden inside the library classes. The application programmer is thus mainly concerned with only having to write code to respond to the messages of interest to them. This is done in short handler functions without the need for the *big switch/case* statement. This programming model is somewhat similar to that used in Visual Basic. For example, one might want to process mouse button clicks and/or respond to a menu command. These functions are formed by deriving application-specific C++ classes from the base MFC classes.

The MFC is based on the object-oriented programming model exemplified by the C++ language. Each program is represented by an object, called an *application object*. This has a *document* object associated with it and one or more *view* objects to manage the image of the document shown to the user of the program. In turn, these objects are instances of a class derived from an MFC base class.

The relationship amongst the key objects is illustrated in Figure E.1. Our framework code will show how these classes are used in practice.

It is up to the programmer to decide what information is recorded about the document and how that is processed. The MFC base classes help process

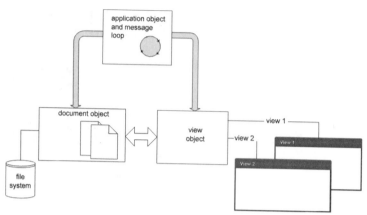

Figure E.1. The key elements in an application program using the MFC.

the document, for example reading or writing it to/from disk files. It is also the responsibility of the application program to draw in the view object's window(s) those aspect of the document it wishes to display. In the context of the MFC, a document can be as diverse as a text file, a JPEG picture or a 3D polygonal CAD model.

The program's application, document and view objects are instances of classes derived from the MFC base classes CWinApp, CDocument and CView respectively. There are over 100 classes in the MFC. Figure E.2 illustrates the hierarchical relationship amongst the ones related to drawing, documents and views.

Perhaps the main advantage of the MFC is that it provides a number of detailed user interface objects, for example, dockable toolbars, tooltips and status information, with virtually no extra work on the part of the programmer. However, the documentation for the MFC is extensive and it takes quite a while to become familiar with the facilities offered by it. The complexity of the MFC has reached such a stage that it is more or less essential to use an integrated development environment such as that provided by Visual C++ to help you get a simple application up and running. The so-called Application and Class wizards, which are nothing more than sophisticated code-generating tools, make the construction of a Windows application relatively painless. For more information on the use of the MFC and Visual C++ programming in general, consult one of the texts [1] or [2]. When you use the Class wizard to create a template framework that will be suitable for programs involving OpenGL, Direct3D or DirectShow, the developer tools will put the

Partial list of MFC classes (as of version 7)

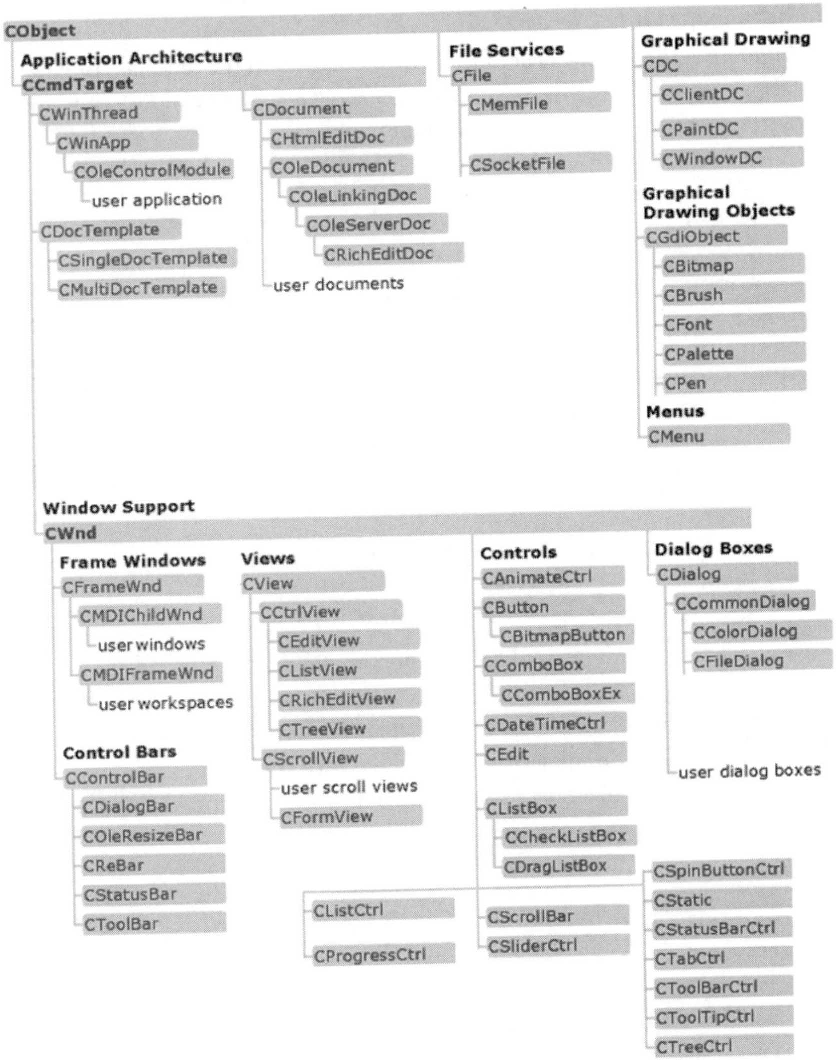

CObject

Application Architecture
CCmdTarget
- CWinThread
  - CWinApp
    - COleControlModule
    - user application
- CDocTemplate
  - CSingleDocTemplate
  - CMultiDocTemplate

- CDocument
  - CHtmlEditDoc
  - COleDocument
    - COleLinkingDoc
      - COleServerDoc
      - CRichEditDoc
  - user documents

File Services
- CFile
  - CMemFile
  - CSocketFile

Graphical Drawing
- CDC
  - CClientDC
  - CPaintDC
  - CWindowDC

Graphical Drawing Objects
- CGdiObject
  - CBitmap
  - CBrush
  - CFont
  - CPalette
  - CPen

Menus
- CMenu

Window Support
CWnd

Frame Windows
- CFrameWnd
  - CMDIChildWnd
    - user windows
  - CMDIFrameWnd
    - user workspaces

Control Bars
- CControlBar
  - CDialogBar
  - COleResizeBar
  - CReBar
  - CStatusBar
  - CToolBar

Views
- CView
  - CCtrlView
    - CEditView
    - CListView
    - CRichEditView
    - CTreeView
  - CScrollView
    - user scroll views
    - CFormView

- CListCtrl
- CProgressCtrl

Controls
- CAnimateCtrl
- CButton
  - CBitmapButton
- CComboBox
  - CComboBoxEx
- CDateTimeCtrl
- CEdit

- CListBox
  - CCheckListBox
  - CDragListBox

- CScrollBar
- CSliderCtrl

Dialog Boxes
- CDialog
  - CCommonDialog
    - CColorDialog
    - CFileDialog

  - user dialog boxes

- CSpinButtonCtrl
- CStatic
- CStatusBarCtrl
- CTabCtrl
- CToolBarCtrl
- CToolTipCtrl
- CTreeCtrl

Figure E.2. The MFC class hierarchy as it relates to documents and views. Many classes have been omitted.

important classes into separate files, each with their own header to define the derived class. The names of the files are derived from the application name (in this example, it is *FW*).

Figure E.3, depicting a list of the files and classes in the template program, contains a wealth of detail. As you can see, the *Class View* window contains a lot of information. It lists four important classes:

1. Class CFwApp. This is the main application class. It is derived from the MFC class CWinApp and the whole application is based on the static creation of an object of this class. Here is the line of code from fw.cpp that handles everything:

```
///////////////////// The one and only CFwApp object!
CFwApp theApp;  // create static instance of application object
```

The class specification is in file fw.h and its implementation is in file fw.cpp.

2. Class CFwDoc. The application's document class derives all the benefit of the functions from the CDocument class and its predecessors. Notably, this includes file selection dialogs and error handling. The framework application must provide the body of a function to read and write the document to a file of some sort. This is done by providing the body of a *virtual* C++ function called:

```
void CFwDoc::Serialize(CArchive& ar){
   if (ar.IsStoring()) { /* TO-DO: add storing code here */ }
   else                 { /* TO-DO: add loading code here */ }
}
```

Of course, one needs to add to the CFwDoc class appropriate members to represent the contents of the document. It might be a pointer to an array of vertices and triangular polygons from a 3D rendering project, for example. The class specification is in file fwDoc.h and its implementation in file fwDoc.cpp.

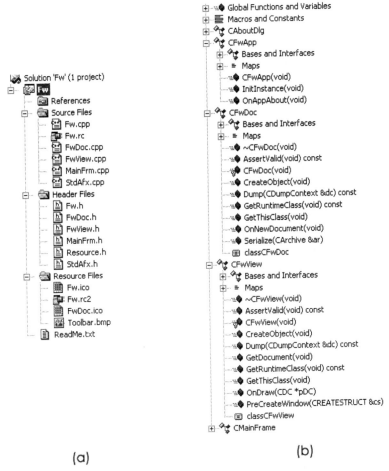

(a)   (b)

Figure E.3. The file view (a) and the class view (b) of a *basic* MFC template application.

3. Class CFwView. This class represents the client area of the program's window. Of prime importance in this class is the function[1] that mimics the WM_PAINT message in a conventional Windows program. In this class, it is the virtual function OnDraw(..) that gets called when a WM_PAINT message is received by the window. In the MFC, functions that begin with On... are designated as message handler functions. In Visual C++, the wizard that helps you build the program will also

---

[1]In object-oriented programming, the functions in a class are also known as its *methods*. We will use the terms function and method interchangeably.

helpfully add the code for most message handlers. Here is the code for this one:

```
void CFwView::OnDraw(CDC* pDC){
  // TO-DO: add draw code for drawing in the client area here
}
```

pDC is a pointer to a class which wraps up as a member variable (m_hDC) the device context handle (the HDC) we discussed earlier when talking about the WM_PAINT message in Appendix D. The class specification is in file fwView.h and its implementation is in file fwView.cpp.

4. Class CMainFrm. As you might guess from the naming convention, this class is mostly independent of the program. An object of this class is responsible for providing the frame that appears around the view controlled by the CFwView class. It also handles the application menu (passing commands on to either the view or document classes), the toolbar and maximize/minimize and close buttons. The class specification is in file MainFrm.h and its implementation is in file MainFrm.cpp.

There are two other features of the MFC that must be mentioned:

1. *Message maps.* If you look at the source code files for the template program (Listings E.2, E.4 or E.6), near the beginning you will see the following lines of code:

```
BEGIN_MESSAGE_MAP(CFwDoc, CDocument)
  //{{AFX_MSG_MAP(CFwDoc)
  // DO NOT EDIT what you see in these blocks of generated code!
  //}}AFX_MSG_MAP
END_MESSAGE_MAP()
```

It shows macro functions which set up the link between window messages including menu or toolbar commands and the C++ code that the programmer writes to handle the message. So, for example, in our framework application we want to respond to commands from the menu that displays an *about* dialog, create a new document or open an existing one. To achieve this, the MESSAGE_MAP would be augmented with another macro making the link between the menu command identifier and the handler function as follows:

```
BEGIN_MESSAGE_MAP(CFwApp, CWinApp)
    //{{AFX_MSG_MAP(CFwApp)
      ON_COMMAND(ID_APP_ABOUT, OnAppAbout)
    //}}AFX_MSG_MAP
    // Standard file based document commands
    ON_COMMAND(ID_FILE_NEW, CWinApp::OnFileNew)
    ON_COMMAND(ID_FILE_OPEN, CWinApp::OnFileOpen)
END_MESSAGE_MAP()
```

The rather bizarre-looking comments such as //}}AFX_MSG_MAP are inserted in the code by the helper wizard in Microsoft Visual C++ to allow it to add, and later remove, message mappings and other features without the programmer having to do it all herself. So, from the message map, one can say that if the program's user selects the File->Open command from the menu, a WM_COMMAND message will be generated, its wParam decoded to a value of ID_FILE_OPEN and the CWinApp class function called. (All of this happens in the background, deep inside the code of the class library, and therefore it does not clutter up the application with pages of repeated code or a hugely bloated switch/case construct.)

2. *DoDataExchange.* One of the most tedious jobs in building a large and richly featured Windows program in C/C++ is writing all the code to manage the user interface through dialog boxes and dialog box controls, getting information back and forward from edit controls, handling selection in drop down and combo boxes etc., and then verifying it etc. This is one area where the MFC offers a major advantage because it builds into the classes associated with dialog controls a mechanism to automatically transfer and validate the control values, copying back and forward and converting from the text strings in text boxes etc. to/from the program's variables. This mechanism is called *DoDataExchange.* Our template code does not use this, but if you examine the template code at the bottom of the fw.cpp file (on the CD) where the "about" dialog class is declared, you will see the following lines of code:

```
//{{AFX_VIRTUAL(CAboutDlg)
protected:
virtual void DoDataExchange(CDataExchange* pDX);     // DDX/DDV support
//}}AFX_VIRTUAL
```

```
class CFwApp : public CWinApp{   // The Application Class is derived
public:                          // from the MFC class CWinApp.
    CFwApp();
    // Overrides
    // ClassWizard generated virtual function overrides
    //{{AFX_VIRTUAL(CFwApp)
public:
    virtual BOOL InitInstance();// A function that will be specified for set-up
    //}}AFX_VIRTUAL
    // Implementation
    //{{AFX_MSG(CFwApp)
    afx_msg void OnAppAbout();
    //    NOTE - the ClassWizard will add and remove member functions here.
    //    DO NOT EDIT what you see in these blocks of generated code !
    //}}AFX_MSG
    // MACRO to allow message processing by class functions
    DECLARE_MESSAGE_MAP()
};
```

Listing E.1. The application class.

The effect of this is to specify a function that the mechanism of automatically exchanging and validating data in dialog boxes would use if there was any data to exchange. Just to illustrate the point: had the *about* dialog contained an edit control (for entering text) and the CAboutDlg class contained a Cstring member to hold the text, then including the following code for the DoDataExchange function:

```
void CAboutDlg::DoDataExchange(CDataExchange* pDX) {
    CDialog::DoDataExchange(pDX);
    DDX_Control(pDX, IDC_EDIT1, m_EditText);
}
```

will result in the text contents of the edit control being copied to the m_EditText member variable of the CAboutDlg class when the dialog is dismissed by the user.

To sum up: the complete MFC framework (see Listings E.1 and E.2) presents the key features of the application class, Listings E.3 and E.4 show the document class and Listings E.5 and E.6 show the view class.

```
#include "stdafx.h"
#include "Fw.h"
#include "MainFrm.h"
#include "FwDoc.h"
#include "FwView.h"

CFwApp theApp;                                      // The one and only CFwApp
                                                    // object

BEGIN_MESSAGE_MAP(CFwApp, CWinApp)                  // message map for application
    //{{AFX_MSG_MAP(CFwApp)
    ON_COMMAND(ID_APP_ABOUT, OnAppAbout)
        // NOTE - the ClassWizard will add and remove mapping macros here.
        //    DO NOT EDIT what you see in these blocks of generated code!

    //}}AFX_MSG_MAP
    ON_COMMAND(ID_FILE_NEW, CWinApp::OnFileNew)   // handle the NEW command here
    ON_COMMAND(ID_FILE_OPEN, CWinApp::OnFileOpen)// handle Open command here
END_MESSAGE_MAP()

CFwApp::CFwApp(){ }                                  // put start up code here

BOOL CFwApp::InitInstance(){
    CSingleDocTemplate* pDocTemplate;
    pDocTemplate = new CSingleDocTemplate(          // set up a standard document
            IDR_MAINFRAME,
            RUNTIME_CLASS(CFwDoc),                  // use this document class
            RUNTIME_CLASS(CMainFrame),              // main frame window class
            RUNTIME_CLASS(CFwView));                // use this frame class
    AddDocTemplate(pDocTemplate);
    CCommandLineInfo cmdInfo;                        // get the command line
    ParseCommandLine(cmdInfo);                       // parse the command line
    m_pMainWnd->ShowWindow(SW_SHOW);                 // show the main window
    m_pMainWnd->UpdateWindow();                      // draw the client area
        return TRUE;
}
```

Listing E.2. The application class functions.

```
// our aplication's document class
class CFwDoc : public CDocument{
protected:
     CFwDoc();
// allows an instance to be created dynamically
     DECLARE_DYNCREATE(CFwDoc)
public:
     //{{AFX_VIRTUAL(CFwDoc)
// these functions are overridden from base class
public:
     virtual BOOL OnNewDocument();     // standard startup
     virtual void Serialize(CArchive& ar);
     //}}AFX_VIRTUAL
public:
     virtual ~CFwDoc();
protected:                                // Generated message map functions
     //{{AFX_MSG(CFwDoc)
        // NOTE - the ClassWizard will add and remove member functions here.
        //        DO NOT EDIT what you see in these blocks of generated code!
     //}}AFX_MSG
     DECLARE_MESSAGE_MAP()               // declare that this class will handle some
                                         // messages
public:
     afx_msg void OnFileCommand1(); // the handler for a custom menu command
};
```

Listing E.3. The document class.

```
#include "stdafx.h"
#include "Fw.h"                          // applicatoin header
#include "FwDoc.h"                       // this class' header

IMPLEMENT_DYNCREATE(CFwDoc, CDocument) // the document is created dynamically
// the document class handles some commands
BEGIN_MESSAGE_MAP(CFwDoc, CDocument)
     //{{AFX_MSG_MAP(CFwDoc)
     // NOTE - the ClassWizard will add and remove mapping macros here.
     //    DO NOT EDIT what you see in these blocks of generated code!
     //}}AFX_MSG_MAP
// match menu command with functions
     ON_COMMAND(ID_FILE_COMMAND1, OnFileCommand1)
END_MESSAGE_MAP()
```

Listing E.4. The document class functions.

```
CFwDoc::CFwDoc(){}
CFwDoc::~CFwDoc(){}

BOOL CFwDoc::OnNewDocument(){        // called in response to "New" on File menu
        if (!CDocument::OnNewDocument())return FALSE;
    ..// put any code here to be executed when new document is created
        return TRUE;
}
void CFwDoc::Serialize(CArchive& ar){    // This function is used to read
if (ar.IsStoring()){ .. }                // and write documents to/from disk.
else { ... }
}
//////////////// other message handers appear here in the code  //////////////////
// this function matches the MESSAGE_MAP
void CFwDoc::OnFileCommand1(){
  ..// put code to implement the command here
}
```

Listing E.4. (continued).

```
class CFwView : public CView{           // the application's view class
protected:
    CFwView();
    DECLARE_DYNCREATE(CFwView)          // class may be created dynamically
public:
    CFwDoc* GetDocument();              // a functions that will get a document
                                        // pointer

    //{{AFX_VIRTUAL(CFwView)
public:
    virtual void OnDraw(CDC* pDC);  // overridden to draw this view
    virtual BOOL PreCreateWindow(CREATESTRUCT& cs);
    //}}AFX_VIRTUAL
    virtual ~CFwView();
protected:
    //{{AFX_MSG(CFwView)              // Generated message map functions
    // NOTE - the ClassWizard will add and remove member functions here.
    //    DO NOT EDIT what you see in these blocks of generated code !
    //}}AFX_MSG
    DECLARE_MESSAGE_MAP()               // this class also handles some messages
public:
    afx_msg void OnFileCommand2();      // a handler for a command on the menu
};
```

Listing E.5. The view class.

```
// This is the function that gets a pointer to the document. m_PDocument
// is a member variable in the base class.
inline CFwDoc* CFwView::GetDocument() { return (CFwDoc*)m_pDocument; }
```

Listing E.5. (continued).

```
#include "stdafx.h"                    // standard headers
#include "Fw.h"                        // application header
#include "FwDoc.h"
#include "FwView.h"                    // this class' header

// creates an object of this class dynamically
IMPLEMENT_DYNCREATE(CFwView, CView)

BEGIN_MESSAGE_MAP(CFwView, CView)      // here is a message map for this client
    //{{AFX_MSG_MAP(CFwView)
    // NOTE - the ClassWizard will add and remove mapping macros here.
    //    DO NOT EDIT what you see in these blocks of generated code!
    //}}AFX_MSG_MAP
    ON_COMMAND(ID_FILE_COMMAND2, OnFileCommand2) // handle this message
END_MESSAGE_MAP()

CFwView::CFwView(){}
CFwView::~CFwView(){}

BOOL CFwView::PreCreateWindow(CREATESTRUCT& cs){
 ..//  put an application specific code here
 ..//  For example in an OpenGL application the OpenGL start-up code
 ..//  would go here.
 return CView::PreCreateWindow(cs);
}

// To draw a view of the document one needs access
void CFwView::OnDraw(CDC* pDC){
// to the document get a pointer to the document.
  CFwDoc* pDoc = GetDocument();
 ..// Draw the desired view of the document in this function
}

void CFwView::OnFileCommand2(){  // CFwView message handler for menu command
 ..// put code here
}
```

Listing E.6. The view class functions.

Bibliography

[1] S. Stanfield. *Visual C++ 6 How-To.* Indianapolis, IN: Sams, 1997.

[2] J. Swanke. *Visual C++ MFC Programming by Example.* New York: McGraw Hill, 1998.

F   The COM in Practice

In this appendix, we provide a very brief review of the component object model (COM) used by Microsoft's multimedia and graphics program development environments.

F.1   Using COM

As we said in Chapter 12, COM is architecture-neutral. It is a binary standard with a predefined layout for the interfaces a COM object exposes to the outside world; the objects themselves never appear externally. Most COM CoClasses and their interfaces are implemented in C++. The CoClass is just a class (because the COM object itself is never exposed, it can be described in a language-specific way) and a C++ abstract class is perfect for the job. Interfaces are tables of pointers to functions implemented by the COM object. The table represents the interface, and the functions to which it points are the methods of that interface.

The binary standard for an interface table is exactly analogous to C++ *virtual function tables* (known as *vtables*), and therefore if you are building a COM object, this is the format in which you would define your interfaces. Figure F.1 illustrates the relationship between the interface pointer, the *vtable* pointers and the interface's methods.

From the point of view of using a COM interface in a program, provided one uses C++ it couldn't be simpler. Since a C++ pointer to an instance of an object is always passed to the class's methods as an implicit parameter, (the *this* parameter), when any of its methods are called, one can use the pointer to the interface in the same way one would use any pointer to a function to call that function. In C, the *this* argument and the indirection through the *vtable* must be included explicitly.

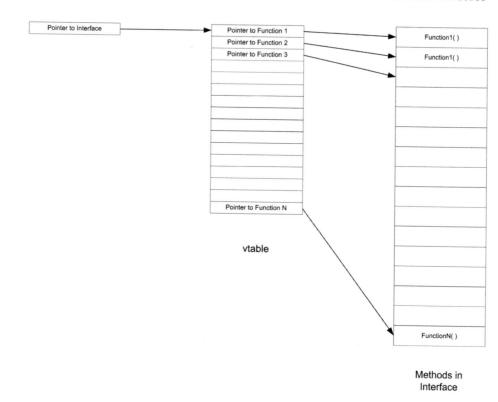

Figure F.1. COM object layout. The interface pointer leads to a table of pointers to the methods of the interface. This is the *vtable*.

So for example, suppose you want to use DirectDraw[1] to retrieve an interface to one of its supported components, for example an interface to Direct3D. Then in C++, it would be coded in the following way:

```
.
LPDIRECTDRAW lpDD;
LPDIRECT3D   lpD3D;
ddres = DirectDrawCreate(NULL, &lpDD, NULL);
if (FAILED(ddres)) ...
// Now query for the Direct3D COM object interface
ddres = lpDD->QueryInterface(IID_IDirect3D,  &lpD3D);
if (FAILED(ddres)) ...
```

---

[1]DirectDraw is one of the base components of DirectX that rarely gets a mention these days. It provides direct access (via a C pointer for example) to the memory address space used by the display screen, and thus it bypasses the normal Windows GDI drawing functions.

To do the same job in C, the following code is required:

```
.
LPDIRECTDRAW lpDD;
LPDIRECT3D   lpD3D;
ddres = DirectDrawCreate(NULL, &lpDD, NULL);
if (FAILED(ddres)) ...
/* Now query for the Direct3D COM object interface        */
/* Note the EXPLICIT use of the object as the first parameter */
/* the use of the "lpVtbl" member of the lpDD structure      */
ddres = lpDD->lpVtbl->QueryInterface(lpDD, IID_IDirect3D,  &lpD3D);
if (FAILED(ddres)) ...
```

Note the explicit use of the `lpVtbl` pointer and the repetition of the `lpDD` pointer (which is the implied `this` pointer in C++). If one were writing any COM program in C, this tedious process would have to be done every time a COM interface method were used—or at least a macro would have to be written to achieve the same thing.

F.2   COM Recipes

With the basics of COM fresh in our minds, this short section will explain a few basic *recipes* that over the next few chapters are going to be deployed to get the applications initialized. Taken together with the frameworks from Appendices D and E, they will help us get our example programs ready for business without having to repeatedly go over what is going on every time we meet COM initialization. Therefore, what we need here are a few lines of explanation of how COM is set up and configured for using Direct3D, DirectInput and DirectShow.

- With the introduction of DirectX 9, getting a 3D rendering application started is a lot simpler than it used to be in the earlier versions. The behavior and operation of Direct3D is controlled by two important interfaces: the *Direct3D System object* interface and the *Direct3D device* interface. It is necessary to get a pointer to the interface exposed by the Direct3D system object and use it to create an object to manage the display. The display manager object exposes an interface (IDirect3DDevice) and its methods control the whole rendering process. Note in the code that the `Direct3DCreate(..)` function is called in a conventional (functional) manner. This is *not* the usual way to get a COM object interface, but it helps to eliminate quite a bit of code from every Direct3D application program.

```
#include <d3dx9.h>  // header file for all things Direct3D
.
//  IDirect3D9 interface pointer
LPDIRECT3D9              g_pD3D              = NULL;
//  IDirect3DDevice9 interface pointer .
LPDIRECT3DDEVICE9        g_pd3dDevice        = NULL;
// create a COM object that exposes the IDirect3D9 interface
// and return pointer to it g_pD3D = Direct3DCreate9( D3D_SDK_VERSION
);
// Set up the structure used to describe the D3DDevice
D3DPRESENT_PARAMETERS d3dpp;
ZeroMemory( &d3dpp, sizeof(d3dpp) );
d3dpp.Windowed = TRUE;
// other structure members set things like a Z buffer
// for hidden surface drawing.
// Create the D3DDevice
g_pD3D->CreateDevice( D3DADAPTER_DEFAULT, D3DDEVTYPE_HAL, hWnd,
                   D3DCREATE_SOFTWARE_VERTEXPROCESSING,
                   &d3dpp, &g_pd3dDevice ) ) );
```

Before the application exits, we must ensure that the interfaces are released so that the COM system can free the memory associated with the Direct3D9 objects if no one else is using them.

```
if( g_pd3dDevice != NULL )g_pd3dDevice->Release();
if( g_pD3D != NULL )g_pD3D->Release();
```

- The DirectShow system is not commonly used as is Direct3D, and therefore there is not such a rich collection of helper functions available. Getting a DirectShow application started is a process much closer to doing native COM programming. As we saw in Chapter 15, DirectShow is based around the *FilterGraph*, and the first thing an application has to do is construct a COM object that will build the FilterGraph for us. Since we don't have a COM object from which to QueryInterface() for the GraphBuilder interface, the conventional COM CoClass instancing function CoCreateInstance() is used instead. (Something similar is going on deep inside the function Direct3DCreate9(), too.)

```
.
// pointer to the interface
IGraphBuilder * pGB;
.
// initialize the COM mechanism
CoInitializeEx(NULL, COINIT_APARTMENTTHREADED);
// Create an instance of the CoClass of the
CoCreateInstance(CLSID_FilterGraph, NULL,
// DirectShow FilterGraph object and return
                 CLSCTX_INPROC_SERVER,
// <- here) a pointer to the GraphBuilder
                 IID_IGraphBuilder, (void **)&pGB));
// interface.
.
.
// clean up
// release the graph builder interface
pGB->Release();
// close the COM mechanism
CoUninitialize();
.
```

You may remember we discussed using ATL (the active template library) to help us avoid the need to remember to release interfaces when we are finished with them. When using the ATL, our DirectShow initialization code would be written as:

```
CoInitializeEx(NULL, COINIT_APARTMENTTHREADED);
CComPtr<IGraphBuilder>          pGB;          // GraphBuilder
pGB.CoCreateInstance(CLSID_FilterGraph, NULL, CLSCTX_INPROC);
.
CoUninitialize();
```

There's no need to remember to release the pGB interface; the ATL handles that for us.

- The DirectInput interfaces did not change in the revision of DirectX from version 8 to 9, so the example programs will refer to DirectInput8.

```
LPDIRECTINPUT8        g_pDI                  = NULL;
LPDIRECTINPUTDEVICE8 g_pJoystick1            = NULL;

hr = DirectInput8Create(g_hInst,DIRECTINPUT_VERSION,
     IID_IDirectInput8,(VOID**)&g_pDI,NULL);
```

If you want to see a full-blown implementation of a COM object and its interfaces in a useful setting, go to the DirectShow SDK and the sample-video processing filter called *EZRGB4*. It is located in `<SDK Location>\ Samples\C++\DirectShow\Filters\EZRGB24`. This filter is a COM object that exposes a special interface called `IPEffect` which allows the parameters of the effect to be changed through the `get_IPEffect(..)` and `put_IPEffect(..)` methods.

Index